AF618802

Operator Theory
Advances and Applications
Vol. 70

Editor
I. Gohberg

Mathematical Results in Quantum Mechanics

International Conference in Blossin (Germany), May 17–21, 1993

Edited by

M. Demuth
P. Exner
H. Neidhardt
V. Zagrebnov

Springer Basel AG

Editors

M. Demuth
Technische Universität Clausthal
Institut für Mathematik
Erzstrasse 1
D-38678 Clausthal-Zellerfeld
Germany

P. Exner
Laboratory of Theoretical Physics
Joint Institute for Nuclear Research
Head Post Office Box 79
Moscow
Russia

H. Neidhardt
Fachbereich Mathematik MA 7-2
Technische Universität Berlin
Strasse des 17. Juni 136
D-10623 Berlin
Germany

V. Zagrebnov
Université d'Aix-Marseille II et
Centre de Physique Théorique
CNRS - Luminy - Case 907
F-13288 Marseille Cedex 9
France

A CIP catalogue record for this book is available from the Library of Congress, Washington D.C., USA

Deutsche Bibliothek Cataloging-in-Publication Data
Mathematical results in quantum mechanics : international conference in Blossin (Germany), May 17 – 21, 1993 / ed. by M. Demuth ... – Basel ; Boston ; Berlin : Birkhäuser, 1994
(Operator theory ; Vol. 70)
ISBN 978-3-7643-5025-3 ISBN 978-3-0348-8545-4 (eBook)
DOI 10.1007/978-3-0348-8545-4
NE: Demuth, Michael [Hrsg.]

Originally published by Birkhäuser Verlag in 1994
Printed on acid-free paper produced from chlorine-free pulp
Cover design: Heinz Hiltbrunner, Basel

ISBN 978-3-7643-5025-3

Contents

3 Stochastic spectral analysis

4 Many-body problems and statistical physics

5 Chaos

6 Operator theory and its application

Preface

The last decades have demonstrated that quantum mechanics is an inexhaustible source of inspiration for contemporary mathematical physics. Of course, it seems to be hardly surprising if one casts a glance toward the history of the subject; recall the pioneering works of von Neumann, Weyl, Kato and their followers which pushed forward some of the classical mathematical disciplines: functional analysis, differential equations, group theory, etc. On the other hand, the evident powerful feedback changed the face of the "naive" quantum physics. It created a contemporary quantum mechanics, the mathematical problems of which now constitute the backbone of mathematical physics. The mathematical and physical aspects of these problems cannot be separated, even if one may not share the opinion of Hilbert who rigorously denied differences between pure and applied mathematics, and the fruitful oscilllation between the two creates a powerful stimulus for development of mathematical physics.

The International Conference on Mathematical Results in Quantum Mechanics, held in Blossin (near Berlin), May 17–21, 1993, was the fifth in the series of meetings started in Dubna (in the former USSR) in 1987, which were dedicated to mathematical problems of quantum mechanics. A primary motivation of any meeting is certainly to facilitate an exchange of ideas, but there also other goals. The first meeting and those that followed (Dubna, 1988; Dubna, 1989; Liblice (in the Czech Republic), 1990) were aimed, in particular, at paving ways to East-West contacts. The most recent conference in Blossin was organized after a three year period during which the old barriers were removed completely. There are, however, other challenges which have nothing to do with the vagaries of politics and geography: in a period of high specialization in scientific thought scientists with a different bent of mind should be gathered under the same "roof" emphasizing similarities, convergences and analogies between ideas they are advocating in their fields of research. For us this "roof" was the Mathematical Results in Quantum Mechanics conference.

The proceedings start with lectures devoted to the traditional core of the Quantum Mechanics — "Schrödinger and Dirac Operators". They touch on spectral problems, asymptotic behaviour of the resolvents, singular potentials and other topics. Following naturally from these subjects is the section "Generalized Schrödinger Operators" wherein lectures from less traditional fields have been collected: the quantum Hall effect, contact perturbations, predissociation, etc. In the section "Stochastic Spectral Analysis" readers will find results strongly motivated by Quantum Mechanics: decay of eigenfunctions, Dirichlet operators and semigroups and others. The section "Many-Body-Problems and Statistical Physics" contains lectures related to the problems of quantum statistical mechanics, including the spectra of reduced density matrices, macroscopic quantum fluctuations, ground-states of the quantum spin chains and spectrum of the spin-boson model etc. Lectures on delicate problems of the quantum evolution irregularities are collected in the section "Chaos". The last section, "Operator Theory and Its Applications", was reserved for lectures motivated by different types of mathematical observations with roots in Quantum Mechanics such as trace formulas for obstacle problems, self-adjoint extensions and singular perturbations, adiabatic reduction theory or p-adic quantum theory.

We hope that the broad areas covered by these proceedings may give readers an impression of the contemporary situation at the intersection of quantum mechanics and mathematical physics at least from the point of view of a not inconsiderable part of the community working in this field.

We would like to thank the Deutsche Forschungsgemeinschaft, Sonderforschungsbereich 288 and the Max-Planck-Gesellschaft for their financial support which made the conference possible. We want to stress, in particular, that it was this support that allowed the scientists of the former USSR to participate.

November 1993, The Editors

Chapter 1

Schrödinger and Dirac operators

Operator Theory:
Advances and Applications, Vol. 70

Discrete spectrum of the periodic Schrödinger operator for non-negative perturbations *

M.Sh.Birman[†]

1. Let us consider the periodic operator

$$A = -\mathrm{div}(g(x)\mathrm{grad}) + p(x) \tag{1}$$

acting in $L_2(\mathbf{R}^d)$. Assume that $g, p \in L_\infty(\mathbf{R}^d)$ are real functions obeying

$$g(x+n) = g(x), \qquad p(x+n) = p(x), \qquad n \in \mathbf{Z}^d$$

and that the matrix function g is positive definite. If the matrix g is constant, then the operator A with the lattice periods $\mathbf{Z}^d$ is equivalent to the Schrödinger operator with arbitrary lattice periods. Further, together with A the perturbed operator

$$A(\alpha) = A + \alpha V, \qquad V(x) \geq 0, \qquad \alpha > 0, \tag{2}$$

is considered where $V \in L_{\infty,loc}$ behaves like

$$V(x) \sim |x|^{-2\sigma} f(\vartheta), \qquad \vartheta = \frac{x}{|x|}, \qquad \sigma > 0 \tag{3}$$

as $|x| \longrightarrow \infty$. The exact definition of the operators (1),(2) is given by quadratic forms. Let the interval (λ_-, λ_+) be a gap of the spectrum of the operator A and let λ be a fixed number in the closed gap, i.e. $\lambda_- \leq \lambda \leq \lambda_+$. By $N(\alpha, \lambda)$ we denote the number of eigenvalues of the operator $A(t)$ which go through the point λ if the coupling constant t increases from 0 to α.

We are interested in the asymptotic behaviour of $N(\alpha, \lambda)$ as $\alpha \longrightarrow \infty$. Similar questions were earlier considered in the remarkable paper [2]. In contrast with [2] we allow the value $\lambda = \lambda_-$ which needs a new technique. Further notes and literature hints can be found in Section 6.

Notation: $Q \subset \mathbf{R}^d$ is the unit cube; $\Omega_* = [-\pi, \pi]^d$ is the closed cube in $\mathbf{R}^d$; Ω is the plane torus which is obtained by identifying the opposite boundaries of Ω_*. Further we set $W = \sqrt{V}$.

---

*Translated by the editors

[†]Supported by the Russian Foundation for Fundamental Researches (Grant 93-011-1697)

2. Let us introduce the operator

$$T(\lambda) = -W(A - \lambda I)^{-1}W, \qquad \lambda_- \le \lambda \le \lambda_+. \tag{4}$$

For $\lambda = \lambda_\pm$ it is assumed that the operator (4) exists as the limit of $T(\lambda)$ in the operator norm for $\lambda \longrightarrow \lambda_\pm$. By $n_+(\cdot, T(\lambda))$ we denote the spectral distribution function with respect of the positive spectrum of the operator $T(\lambda)$. Then

$$N(\alpha, \lambda) = n_+(s, T(\lambda)), \qquad \alpha s = 1, \qquad \lambda_- \le \lambda \le \lambda_+. \tag{5}$$

Let us describe the spectral representation of the resolvent $(A - \lambda I)^{-1}$. In $L_2(\mathcal{Q})$ we consider the family of operators $A(k)$, $k \in \Omega$ which is given by the expression (1) and the boundary condition which guarantees the periodic continuation of the function $u(x)\exp(-ikx)$ for $u \in \mathrm{Dom}(A(k))$. Let $E_j(k)$ be the sequence of eigenvalues of the operator $A(k)$ and let $\psi_j(k, x)$ be the corresponding orthonormal eigenfunctions. Further, let us introduce the operators

$$(\Psi_j u)(k) = (2\pi)^{-d/2} \int \overline{\psi_j(k, x)} u(x) dx$$

which maps $L_2(\mathbf{R}^d)$ onto $L_2(\Omega)$. The operators $P_j = \Psi_j^* \Psi_j$ are pairwise orthogonal projections in $L_2(\mathbf{R}^d)$ obeying $\sum_j P_j = I$. By $[h]$ we denote the multiplication operator which is generated by the function $h : \Omega \longrightarrow \mathbf{C}$. Then

$$T(\lambda) = \sum_j W \Psi_j^* [(\lambda - E_j)^{-1}] \Psi_j W, \qquad \lambda_- \le \lambda \le \lambda_+. \tag{6}$$

Let the number r be determined by the condition

$$\lambda_- = \max_{k \in \Omega} E_r(k), \qquad \lambda_+ = \min_{k \in \Omega} E_{r+1}(k).$$

In the sum (6) the first r summands are positive operators while the other ones are negative. Thus

$$T(\lambda) \le \sum_{j \le r} W \Psi_j^* [(\lambda - E_j)^{-1}] \Psi_j W =: T_r(\lambda)$$

and hence

$$n_+(s, T(\lambda)) \le n(s, T_r(\lambda)), \qquad \lambda \le \lambda \le \lambda_+. \tag{7}$$

3. From (5),(7) we see that the investigation of the asymptotics of the spectrum of the positive operator $T_r(\lambda)$ leads to upper asymtotic estimates for $N(\alpha, \lambda)$. It is useful instead of $T(\lambda)$ to consider a more general class of operators. Let $h_j : \Omega \longrightarrow \mathbf{R}_+$, $j = 1, \dots, r$. We set

$$H = \sum_{j \le r} W \Psi_j^* [h_j] \Psi W. \tag{8}$$

Important for us is the following theorem.

Theorem 1 *Let $2\sigma < d$, $f \in L_q(\mathbf{S}^{d-1})$, $q = d/2\sigma (> 1)$ in (3) and let*

$$h_j \in L_q(\Omega), \qquad j = 1, \ldots, r, \tag{9}$$

in (8). Then

$$\begin{aligned}\lim_{s\to 0} s^q n(s,H) &= \\ M(h_1,\ldots,h_r;f) &:= (2\pi)^{-d} d^{-1} \left(\sum_{j\le r} \int_\Omega h_j^q(k)dk \right) \int_{\mathbf{S}_{d-1}} f^q(\vartheta) dS(\vartheta).\end{aligned} \tag{10}$$

The proof idea of Theorem 1 consists in replacing the operators Ψ_j in (8) by the Fourier transformation Φ. Doing so it can be shown that the leading term of the spectral asymptotics remains unchanged. Moreover it is possible to carry out some other transformations of the operator under investigation. In order to legitimate these transformations the generalized Cwikel estimate (see [4],[5]) is used essentially. By these considerations the problem transforms into the investigation of the spectrum of an orthogonal sum of operators of type

$$[h_j^{1/2}]\Phi V \Phi^* [h_j^{1/2}], \qquad j = 1, \ldots, r \tag{11}$$

which act in $L_2(\Omega_*)$. The operators (11) are Ψ-differential operators of negative order (-2σ). For them the spectral asymptotics can be obtained directly from the general results of the paper [3]. As can be seen from (10) the asymptotic contributions of the different spectral zones of the operator A sum up.

Remark 1 For Ψ-differential operators of type (11) the role of the variables x and k is "non-traditional": The order of the operator is determined by the asymptotics (3) while the functions $h_j(k)$ can be chosen sufficiently arbitrary and behaves like "potentials".

Remark 2 The condition $q > 1$ in Theorem 1 has a technical reason and can be replaced by the condition $q > 0$. However for $q \le 1$ it is necessary to use other estimates which need a certain smoothness (dependent on q) of the data of the problem.

4. From Theorem 1 and from (5),(7) we find the following corollary.

Corollary:

a) For $\lambda_- < \lambda \le \lambda_+$, $2\sigma < d$ the upper asymptotic estimate

$$\limsup_{\alpha\to\infty} \alpha^{-q} N(\alpha,\lambda) \le M(h_1,\ldots,h_r;f) \tag{12}$$

holds where

$$h_j(k) = (\lambda - E_j(k))^{-1}, \qquad j = 1, \ldots, r. \tag{13}$$

b) If the condition (9) is satisfied for the function (13), then the estimate (12) is valid for $\lambda = \lambda_-$ *too.*

Remark 3 With respect to conditions of point a) the functions (13) are bounded. The same takes place in the case $\lambda = \lambda_-$ for those functions of (13) for which $\max E_j(k) < \lambda_-$. However, the condition (9) is essential for those h_j for which

$$\max_{k\in\Omega} E_j(k) = \lambda_-. \tag{14}$$

Assume in particular that (14) is satisfied only for $j = r$ and, moreover, that max in (14) is obtained only in a finite number of points and that these maximal points non-degenerated. Then obviously the condition (9) is valid if and only if $\sigma > 1$. It is not hard to see that under the conditions of point b) it must be satisfied always $\sigma > 1$.

Remark 4 Let us make some remarks concerning the validity of lower asymptotic estimate of type

$$\liminf_{\alpha\to\infty} \alpha^{-q} N(\alpha, \lambda) \geq M(h_1, \ldots, h_r; f) \tag{15}$$

under the conditions of the corollary. Till now the author can prove it only assuming that the functions E_j, ψ_j are sufficiently smooth on the torus Ω. However, it is hard to verify these conditions in the general case, although, for $r = 1$ this is satisfied automatically. Therefore we prefer to use the results of the paper [2] in order to obtain (15) for $\lambda = \lambda_-$.

5. In this subsection we believe that the matrix g in (1) is constant (case of the Schrödinger operator) and that the function f of (3) is continuous. Under these conditions and for $\lambda \in (\lambda_-, \lambda_+)$ the asymptotics was found in [2], i.e., both estimates (12), (15) are verified where the functions $h_1, \ldots, h_r$ are given by (13). Since $N(\alpha, \lambda)$ monotonuously dependents on λ the estimate (15) can be extended to $\lambda = \lambda_-$ for functions obeying

$$h_j = (\lambda_- - E_j)^{-1} \in L_q(\Omega), \qquad j = 1, \ldots, r. \tag{16}$$

Summing up we get the following theorem.

Theorem 2 *Assume that the matrix* g *is constant in (1). Let* $1 < \sigma < d/2$ *in (3) and let* f *be a continuous function. Then under the condition (16) the asymptotics*

$$\lim_{\alpha\to\infty} \alpha^{-q} N(\alpha, \lambda_-) = M(h_1, \ldots, h_r; f) \tag{17}$$

are valid.

Remark 5 The condition $2\sigma < d$ (i.e. $q > 1$) is not used in the asymptotic theorem of [2]. Therefore, taking into account Remark 2 the asymptotics (17) can be proved for $q \leq 1$.

6. Comments. In [2] it were considered only *inner* points of the gap. This fact is very important for the technique which is used there. In [2] the Schrödinger operator must not be necessarily periodic. It is only assumed that the state density exists for it in terms of which then the asymptotics are written down. In the periodic case the asymptotic coefficient can be transformed into the form (10). Let us mentioned the recent paper [1] where the operator of type (1) is perturbed by a non-negative differential operator of second order.

For $d = 1$ essential asymptotic results were obtained in [6] for $N(\alpha, \lambda)$ in both cases $\alpha > 0$ and $\alpha < 0$. The Schrödinger operator is assumed to be periodic. For λ the values λ_- and λ_+ are allowed. Hoewever, in [6] it is essentially used that the problem is an one-dimensional one.

References

[1] S.Alama, M.Avellaneda, P.A.Deift, R.Hempel: On the existence of eigenvalues of a divergence form operator $A + \lambda B$ in a gap of $\sigma(A)$. To appear in *Asymptotic Analysis.*

[2] S.Alama, P.A.Deift, R.Hempel: Eigenvalue branches of the Schrödinger operator $H - \lambda W$ in a gap of $\sigma(H)$. *Commun. Math. Phys.* **121**, 291-321 (1989).

[3] M.Sh.Birman, M.Z.Solomyak: Asymptotics of the spectrum of pseudo-differential operators with anisotropic homogenuous symbols. *Vestnik Leningrad State Unversity,* I. **N 13**, 1977, 13-21; II. **N 13**, 1979, 5-10 (in Russian).

[4] M.Sh.Birman, M.Z.Solomyak: Schrödinger operator. Estimate for number of bounded states as a function-theoretical problem. *Amer. Math. Transl. (2)* **vol. 150**, 1992, 1-54.

[5] M.Sh.Birman, G.E.Karadzhov, M.Z.Solomyak: Boundedness conditions and spectrum estimates for the operators $b(X)a(D)$ and their analogs. *AMS, Adv. Sov. Math.*, **vol. 7**, 1991, 85-106.

[6] A.V.Sobolev: Weyl asymptotics for the disrecte spectrum of the perturbed Hill operator. *AMS, Adv. Sov. Math.*, **vol. 7**, 1991, 159-178.

Author's address:
M.Sh.Birman
St.-Petersburg Unversity
Department of Physics
198904 St.-Petersburg
Russia

Operator Theory:
Advances and Applications, Vol. 70

The discrete spectrum in a gap of the continuous one for compact supported perturbations

M.Sh. Birman T. Weidl

1. We start from the traditional problem on the negative spectrum of the Schrödinger operator in $\mathbf{R}^d, d \geq 3$. Let $A = -\Delta$,

$$A(\alpha) = -\Delta - \alpha V, \quad V(x) \geq 0, \ \alpha > 0, \tag{1}$$

and $\lambda \leq 0$. By $N_A(\alpha, \lambda)$ we denote the number of eigenvalues of the operator (1) on the left-hand side of the point λ. Then for potentials $V \in L_{d/2}(\mathbf{R}^d)$ we have the well-known asymptotics

$$N_A(\alpha, \lambda) \sim (2\pi)^{-d} \omega_d \alpha^{d/2} \int V^{d/2} dx, \quad \alpha \to \infty, \tag{2}$$

with ω_d the volume of the unit ball in $\mathbf{R}^d$. We call potentials $V \in L_{d/2}(\mathbf{R}^d)$ *"regular"* perturbations of the operator A (cf. [BS1]). The asymptotic (2) do not depend on $\lambda \leq 0$, it's character is determined by the behavior of the symbol $|\xi|^2 - \alpha V(x)$ for large $|\xi|$ only. In [BS1] the asymptotics of $N_A(\alpha, \lambda)$ are discussed precisely for potentials violating the assumption $V \in L_{d/2}(\mathbf{R}^d)$ because of a slow decrease as $|x| \to \infty$ (*"non-regular perturbations"*). There typically $N_A(\alpha, \lambda) = o(N(\alpha, 0)), \lambda < 0$, is found; the main asymptotical term of $N(0, \lambda)$ for $\alpha \to \infty$ is given by the symbol of $A(\alpha)$ for small $|\xi|$ (threshold effect). So, for instance, for $V \in L_\infty, V(x) \sim |x|^{-2}(\ln|x|)^{-1/q}$, $2q > d$, $|x| \to \infty$, we have $N(\alpha, 0) \sim c(d)\alpha^q$, $N(\alpha, \lambda) = O(\alpha^{d/2} \ln \alpha)$, and the latter estimate can be refined. Here we discuss the inverse case, when $V \notin L_{d/2}(\mathbf{R}^d)$ because of local singularities. In detail we assume

$$V \in L_1(\mathbf{R}^d), \quad supp\, V \subset K_R := \{x : |x| < R\}, \quad V \geq 0. \tag{3}$$

We call potentials of the form (3) *"quasi-regular"*.

2. For $V \notin L_{d/2}(\mathbf{R}^d)$ the number of eigenvalues $N_A(\alpha, \lambda)$ can show non-powerlike asymptotics. Our second aim is to show that the technical tools developed in [W1], [W2] allow us to consider non-powerlike asymptotics, too. We call a function $f : \mathbf{N} \to \mathbf{R}_+$ a *normal estimation function (NEF)*, if $f \uparrow \infty$ for $n \to \infty$ and if the function f^κ is subadditive for some $\kappa > 0$. We introduce the functionals

$$\Delta_f(\lambda, A) := \lim_{\alpha\to\infty} \sup \alpha^{-1} f(N_A(\alpha; \lambda)), \tag{4}$$

$$\delta_f(\lambda, A) := \lim_{\alpha\to\infty} \inf \alpha^{-1} f(N_A(\alpha; \lambda)). \tag{5}$$

Theorem 1. *Let (3) be fullfilled and assume, that for some $\lambda \leq 0$ and for a NEF f $\Delta_f(\lambda, A) < \infty$ holds. Then for every $\mu \leq 0$ we have the equalities*

$$\Delta_f(\lambda, A) = \Delta_f(\mu, A), \quad \delta_f(\lambda, A) = \delta_f(\mu, A). \tag{6}$$

In particular (6) is fullfilled for $\lambda < 0, \mu = 0$; this explains why we call potentials V satisfying assumption (3) *"quasi-regular"*. We remark, that under assumptions of theorem 1 the functionals (4), (5) are determined by the behavior of the symbol of $A(\alpha)$ for large $|\xi|$ only.

3. Further we consider the operator

$$H = -\Delta + p(x), \quad p \in L_\infty(\mathbf{R}^d), \ d \geq 3, \tag{7}$$

as unperturbed. The spectrum $\sigma(H)$ may be interrupted by gaps. Let $\Lambda = (\lambda_-, \lambda_+)$ be such a gap. For a large class of potentials V decreasing to zero sufficiently fast for $|x| \to \infty$ the spectrum of the perturbed operator

$$H(\alpha) = H - \alpha V, \ \alpha > 0, \ V(x) \geq 0,$$

in the gap Λ is discrete. For λ, $\lambda_- \leq \lambda \leq \lambda_+$, we introduce $N_H(\alpha, \lambda)$ - the number of eigenvalues of $H(t)$ which passed the point λ for coupling constant t increasing from 0 to α, (for operator $A(\alpha)$ and $\lambda \leq 0$ the function $N_A(\alpha, \lambda)$ coincides with the function N_A from subsection 1). In [B1] an abstract theorem was presented, which gives the equality of the asymptotical functionals Δ_f, δ_f for $A, \mu < 0$ and $H, \lambda \in \Lambda$, *in the case of powerlike estimation functions f.* We state here an analogue of this theorem for arbitrary NEF and apply it to the Schrödinger operator. Next we prepare some material required in the corresponding formulations.

4. Let $\mathcal{H}$ be a Hilbert space, $T \in \mathcal{S}_\infty(\mathcal{H})$ (i.e. T is a compact operator on $\mathcal{H}$); and let $\{s_k(T)\}_{k \in \mathbf{N}}$ denote the sequence of singular numbers of the operator T. For some NEF f we introduce the operator classes

$$\Sigma_f = \{T \in \mathcal{S}_\infty : |T|_f := \sup_{n \in \mathbf{N}} s_n(T) f(n) < \infty\}.$$

The class Σ_f is a complete, non-separable space with respect to the quasi-norm $|\cdot|_f$. We denote by Σ_f^0 the subspace of operators $T \in \Sigma_f$, for which $s_n(T) f(n) \to 0$. The set of finite rank operators is dense in Σ_f^0. For $T \in \Sigma_f$ we define the functionals

$$\Delta_f(T) := \lim_{n \to \infty} \sup s_n(T) f(n), \qquad \delta_f(T) := \lim_{n \to \infty} \inf s_n(T) f(n).$$

For $T = T^*$ analogous functionals $\Delta_f^{(\pm)}, \delta_f^{(\pm)}$ can be introduced by the sequences $\{\lambda_n^{(\pm)}(T)\}$, e.g. the sequences of positive eigenvalues of the operator $\pm T$. All six functionals $\Delta_f, \Delta_f^{(\pm)}, \delta_f, \delta_f^{(\pm)}$ are continuous on Σ_f. In fact they are well defined and continuous on the factor space Σ_f / Σ_f^0, too. The class Σ_f is a two-sided ideal in the space of bounded operators on $\mathcal{H}$. The material of this subsection was developed in [W1].For similar powerlike ideals see [BS2].

5. Let $A = A^* > 0, P = P^*$ be operators in $\mathcal{H}$; $\mathbf{a}, \mathbf{p}$ are the corresponding bilinear forms. We assume that

$$|\mathbf{p}[u,v]|^2 \leq \mathbf{p}_*[u,u]\mathbf{p}_*[v,v], \tag{8}$$

where $\mathbf{p}_*$ is a nonnegative form such that

$$\mathbf{p}_*[u,u] \leq \varepsilon \mathbf{a}[u,u] + c(\varepsilon)\|u\|^2, \quad \forall \varepsilon > 0. \tag{9}$$

Put $H = A+P$ (in form sense). Let, moreover, W be a closed operator, $Dom\ W \supset Dom\ A^{1/2}$ and $W(A+I)^{-1/2} \in \mathcal{S}_\infty$. We set $\mathbf{v}[u,u] = \|Wu\|^2$ and consider operators $H(\alpha) = H - \alpha V$; the perturbation V is given by the form $\mathbf{v}$. Further $\rho(\cdot)$ denotes the resolvent set of an operator.

Put

$$X_\lambda(A) = W(A-\lambda I)^{-1}W^*, \quad \lambda \in \rho(A) \cap \mathbf{R},$$

and the operator $X_\mu(H)$, $\mu \in \rho(H) \cap \mathbf{R}$ is defined analogously.

Theorem 2. *Let assumptions (8), (9) be fullfilled and let*

$$W(A+I)^{-1/2} \in \Sigma_f, \tag{10}$$

$$W(A+I)^{-1} \in \Sigma_f^0, \tag{11}$$

for some NEF f. Then for $\lambda = \overline{\lambda} \in \rho(A)$, $\mu = \overline{\mu} \in \rho(H)$ we have

$$X_\lambda(A) - X_\mu(H) \in \Sigma_{f2}^0.$$

For $T = T^* \in \mathcal{S}_\infty$ we use the notation $n_+(s,T) = card\{k : \lambda_k^{(+)}(T) > s\}, s > 0$. We recall the well-known relation

$$N_H(\alpha,\mu) = n_+(\alpha^{-1}, X_\mu(H)), \quad \mu = \overline{\mu} \in \rho(H),$$

and the analogous equality for A. Then from theorem 2 we claim the following "stability theorem".

Theorem 3. *Under the conditions of theorem 2 for $g = f^2$ the following identities for the fucntionals (4),(5) hold*

$$\Delta_g(\lambda, A) = \Delta_g(\mu, H), \quad \delta_g(\lambda, A) = \delta_g(\mu, H). \tag{12}$$

For powerlike functions f theorems 2,3 were proven in [B1].

6. Next we apply theorem 3 to operators (1), (??) in case of quasi-regular potentials. It remains true that

Proposition 4. *Put $A = -\Delta$ and for a potential V suppose the condition (3) holds. Let assumption (10) be fullfilled for some NEF f. From this follows (11).*

Together with theorem 3 this proposition leads to

Theorem 5. *Put $A = -\Delta$ and let H be the operator (7). Let $\Delta_g(\lambda, A)$ be finite-valued for some NEF g and for some $\lambda < 0$. Then for all $\mu = \overline{\mu} \in \rho(H)$ equalities (12) hold.*

We remark, that theorem 5 can be sharpened and remains true for weaker conditions on V too. More significantly for *periodic* p relations (12) can also be shown for $\mu = \lambda_\pm$; for regular perturbations this is discussed in detail in [B2].

Theorem 3 can be used not only in regular or quasi-regular situations. We give an example of a non-regular perturbation. Put $V(x) = (1 + |x|^2)^{-1}$. The asymptotical behavior of $N_A(\alpha, \lambda), \lambda < 0$, is known (and doesn't depend on $\lambda < 0$). Theorem 3 allows us to carry this asymptotic into a gap of the operator (7). In fact, for $\mu \in \Lambda$ we have

$$N_H(\alpha, \mu) \sim c(d)\alpha^{d/2} \log \alpha, \quad c(d) = 2d^{-1}(2\pi)^{-d}\omega_d^2.$$

We can describe a sufficiently large class of non-quasi-regular potentials, to which theorem 3 can be applied, but do not discuss it here in detail.

References

[B1] M.Sh.Birman *"Discrete Spectrum in the Gaps of a Continuous One for Perturbations with large Coupling Constant"*, Adv. Sov. Math. v.7, AMS 1991

[B2] M.Sh.Birman *"On the discrete spectrum in gaps of a perturbed periodic operator of the second order"* (Russian), Func. Anal. t. 25 vyp. 2, 1991

[BS1] M.Sh.Birman, M.Z.Solomyak *"Estimates for the Number of Negative Eigenvalues of the Schrödinger-Operator and Its Generalizations"*, Adv. Sov. Math. v.7, AMS 1991

[BS2] M.Sh.Birman, M.Z.Solomyak *"Spectral Theory of Self-Adjoint Operators in Hilbert Space"*, D. Reidel Publishing Company, 1986

[W1] T.Weidl *"General operator ideals of the weak type"* (Russian), Algebra i Analiz t.4 vyp. 3, 1992

[W2] T.Weidl *"Estimates for operators of the type $b(x)a(D)$ in non- powerlike ideals"*, Mittag-Leffler-Report 4, 1992/93, to be published in Algebra i Analiz

M.Sh. Birman
Department of Physics
St.Petersburg State University
Ulyanovskaya 1, Stary Peterhof
St. Petersburg 198904
Russia

T. Weidl
Max-Planck-Arbeitsgruppe
"Partielle Differentialgleichungen"
Universit"at Potsdam
Am Neuen Palais 10
O-1571 Potsdam, Germany

Operator Theory:
Advances and Applications, Vol. 70

Schrödinger Operators with Strong Local Magnetic Perturbations: Existence of Eigenvalues in Gaps of the Essential Spectrum

Rainer Hempel* and Jörg Laitenberger

1 Description of Our Main Result

In the present note, we show that perturbations by strong magnetic fields of compact support may produce eigenvalues inside a spectral gap of a (periodic) Schrödinger operator. Here we will discuss the following situation:

In the Hilbertspace $\mathcal{H} = L_2(\mathbf{R}^\nu)$, we consider the Schrödinger operator $H = -\Delta + V$, with a fixed potential $V : \mathbf{R}^\nu \to \mathbf{R}$, V bounded and $V \geq 1$, where H is defined as the unique self-adjoint extension of $(-\Delta + V)\,|C_c^\infty(\mathbf{R}^\nu)$. We shall make the basic assumption that H has a (non-trivial) gap in its essential spectrum; more precisely, we assume that there exist $b > a > \inf \sigma_{\text{ess}}(H)$ with $[a, b] \cap \sigma(H) = \emptyset$. In view of the applications in solid state physics, one may think of H as a periodic Schrödinger operator.

Suppose now that we are given a vector potential $\mathbf{a} = (a_j)_{j=1,\dots,\nu}$, $a_j \in C^1(\mathbf{R}^\nu)$ real-valued and of compact support. Introducing also a coupling $\lambda \in \mathbf{R}$, we define the associated magnetic Schrödinger operator as

$$H(\lambda\mathbf{a}) := (i\nabla - \lambda\mathbf{a})^2 + V(x) = \sum_{j=1}^{\nu} (i\partial_j - \lambda a_j)^2 + V(x).$$

Again, there are classical results assuring that $H(\lambda\mathbf{a})$ is essentially self-adjoint on $C_c^\infty(\mathbf{R}^\nu)$. The domains of $H = H(\mathbf{0})$ and $H(\lambda\mathbf{a})$ coincide with the Sobolev space $\mathcal{H}^2$. (For a general account of magnetic Schrödinger operators, see, e. g., [AHS], [CFrKS], [S].)

The operators $H(\lambda\mathbf{a})$, with $\nu = 2$, provide a simple mathematical model for a thin layer or wafer of solid matter which is locally penetrated by a magnetic field.

For $\mathbf{a} \in C^1$, $\mathbf{a}$ of compact support, it is easy to see that the magnetic terms $-i\lambda(2\mathbf{a} \cdot \nabla + \operatorname{div} \mathbf{a}) + \lambda^2|a|^2$ are a relatively compact perturbation of H; as a

[1]On leave from Math. Inst. der Univ. München, Theresienstr. 39, D-8000 München 2
Address after September 1, 1993: Dep. of Maths., Univ. of Alabama, Birmingham, AL 35294

consequence, the essential spectrum does not change as we switch on the magnetic field, i. e., we have

$$\sigma_{\text{ess}}(H(\lambda\mathbf{a})) = \sigma_{\text{ess}}(H), \quad \lambda \in \mathbf{R}.$$

However, as we let λ increase from 0 to ∞, discrete eigenvalues may move into the gap. As in our previous work ([DH], [ADH], [AADH], [H1]), we now fix some "control point" $E \in (a,b)$ and ask whether there exist coupling constants λ with the property that $E \in \sigma(H(\lambda\mathbf{a}))$. In other words, it is our aim to produce lower bounds for the eigenvalue counting function

$$\mathcal{N}(\lambda; \mathbf{a}, E) := \sum_{0<\mu<\lambda} \dim \ker(H(\mu\mathbf{a}) - E).$$

In the present brief note, we concentrate on the paradigmatic case where we have a *constant field* inside a cube $Q_R = \{x \in \mathbf{R}^\nu; |x_j| < R, j = 1, \ldots, \nu\}$ while the field is zero outside the cube Q_{R+1}, for some $R \geq 1$. To construct such a situation, we choose, for any $R \geq 1$, a cut-off function $\varphi_R \in C_c^\infty(\mathbf{R}^\nu)$ satisfying

$$\varphi_R(x) = 1, \quad x \in Q_R, \qquad \varphi_R(x) = 0, \quad x \notin Q_{R+1},$$

and we fix a (real-valued) vector potential $\mathbf{b}$ on $\mathbf{R}^\nu$ which generates a (non-zero) constant magnetic field.

Since our mechanism will basically rely on the "repulsive" effect of the magnetic perturbation $\lambda\varphi_R\mathbf{b}$ on the cube Q_R, we need a (rather weak) condition on the density of states of H below E: for any self-adjoint operator A, let $P_{(-\infty,E)}(A)$ denote the associated spectral family, so that $\dim P_{(-\infty,E)}(A)$ is just the number of eigenvalues of A below E, if A has compact resolvent. Letting H_R denote the operator $-\Delta + V$, acting in $L_2(Q_R)$ with Dirichlet boundary conditions, we shall assume that there exist constants $c_0 > 0$ and $R_0 \geq 1$ such that

$$\dim P_{(-\infty,E)}(H_R) \geq c_0 R^\nu, \qquad R \geq R_0. \tag{$*$}$$

This condition is satisfied for periodic potentials V, for example. Now our main result on the eigenvalues of $H(\lambda\varphi_R\mathbf{b})$ can be stated as follows:

Theorem 1.1. *Let H, E, $\mathbf{b}$ and φ_R be as above; in particular, let us suppose that $(*)$ holds. Then, given any $k \in \mathbf{N}$, there exist $R \geq 1$ and $\lambda > 0$ such that*

$$\mathcal{N}(\lambda; \varphi_R\mathbf{b}, E) \geq k.$$

A more complete presentation of these and related results will be given in a forthcoming paper. In the subsequent remarks, we put the result of Theorem 1.1 in perspective:

1. There exist extensive studies ([DH], [ADH], [H1], [GS], [B]) dealing with the situation where a Schrödinger operator $H = -\Delta + V$ with spectral gap is perturbed by a multiplication operator W, W a real-valued function of compact

support or with some decay at infinity. While the case where W does not change sign allows the application of the Birman-Schwinger-principle, the general case $W = W_+ - W_-$ requires a thorough analysis of the competition between the "attractive" potential well versus the "repulsive" barrier created by W_+ and W_-, respectively, in $H - \lambda W$, for $\lambda > 0$. Note that in Theorem 1.1, above, the magnetic terms have **no** definite sign, so we are a priori in deep water.

2. In a context related to control theory,[2] [AADH] have obtained the existence of eigenvalues in gaps in situations where the unperturbed operator A is given by $A = -\sum \partial_j a_{ij}(x)\partial_i$, a second order elliptic divergence form operator with a spectral gap (examples are constructed in [H2]), and the perturbation $B = -\sum \partial_j b_{ij}(x)\partial_i$, with $(b_{ij}) \geq 0$ decays at ∞. Again, one asks for eigenvalues of $A + \lambda B$, for $\lambda > 0$. In this problem, one looses relative compactness of the perturbation but, at least, the perturbation does not change sign.

In the present work, we are now dealing with perturbations given by a *first order* differential expression where we have relative compactness but no monotonicity and, in general, no Dirichlet-Neumann-bracketing. (Note that Birman and Raikov [BR] have studied eigenvalues in gaps under the influence of magnetic fields; in their work, however, the eigenvalues are still "produced" by a potential W while the whole system is "bathed" in a (constant) magnetic field.)

3. Due to the particular structure of the magnetic terms, the coupling λ enters in a non-linear way. This seems to rule out any use of Birman-Schwinger-type arguments.

2 Sketch of the Proof

Let us first sketch the basic strategy of our proof: again, we use a sequence of approximating problems on large cubes Q_n with suitable operators $\tilde{H}_n$ and $\tilde{H}(\lambda\varphi_R\mathbf{a})$, constructed as in [ADH], [H1]: if H_n denotes the operator $-\Delta + V$ acting in $L_2(Q_n)$, with Dirichlet boundary conditions, one knows that the eigenfunctions of H_n associated with eigenvalues in the interval $[a', b']$ (where $a < a' < E < b' < b$) decay exponentially away from the boundary ∂Q_n. Since, on the other hand, $\dim P_{[a',b']}(H_n) \leq cn^\nu$, it can be shown that the operators

$$\tilde{H}_n := H_n + \psi_n P_{[a',b']}(H_n)\psi_n,$$

with $\psi_n \in C^\infty(Q_n)$, $\psi_n(x) = 1$, for $x \in Q_n \setminus Q_{3n/4}$, and $\psi_n(x) = 0$, for $x \in Q_{n/2}$, enjoy the following two properties: 1. There exist $\alpha < E < \beta$ such that, for n large enough, $\sigma(\tilde{H}_n) \cap [\alpha, \beta] = \emptyset$. 2. The "non-local" part of $\tilde{H}_n$ vanishes on $Q_{n/2}$.

We now compare the number of eigenvalues below E for the operators $\tilde{H}_n = \tilde{H}_n(\mathbf{0})$ and

$$\tilde{H}_n(\lambda\,\varphi_R\mathbf{b}) := H_n(\lambda\,\varphi_R\mathbf{b}) + \psi_n\,P_{[a',b']}(H_n)\,\psi_n\,,$$

[2]Here one asks to which extent localized impurities in periodic microstructures will affect the controllability of the wave equation.

using Neumann decoupling on the boundaries of Q_R and of Q_{R+1}:

Proposition 2.1. *Let* **a** *and* φ_R *be as in Section 1. Then, for any given* $k \in \mathbf{N}$, *we can find* $R_k \geq 1$, $\Lambda_k > 0$ *and* $n_k \in \mathbf{N}$ *such that*

$$\dim P_{(-\infty,E)}\left(\tilde{H}_n(\Lambda_k \varphi_{R_k}\mathbf{b})\right) \leq \dim P_{(-\infty,E)}\left(\tilde{H}_n\right) - k, \quad n \geq n_k.$$

From Proposition 2.1, it is easy to conclude by Kato-Rellich perturbation theory that at least k eigenvalues of the family $(\tilde{H}_n(\lambda\varphi_{R_k}\mathbf{b}); 0 < \lambda \leq \Lambda_k)$ must cross the level E. Hence, there exist couplings $0 < \lambda_j^{(n)} \leq \Lambda_k$, $j = 1, \ldots, k$, such that E is an eigenvalue of $\tilde{H}_n(\lambda_j^{(n)} \varphi_{R_k}\mathbf{b})$, and it follows via a convergence argument of [DH], [H1] that there exist at least k coupling constants $\lambda_1, \ldots, \lambda_k \in (0, \Lambda_k]$ such that E is an eigenvalue of $H(\lambda_j \varphi_{R_k}\mathbf{b})$, $j = 1, \ldots, k$, which proves our theorem. We will now try to give an idea of how to obtain Proposition 2.1.

Proof of Proposition 2.1 We decouple the region Q_n into the three pieces Q_R, $Q_{R+1} \setminus Q_R$, and $Q_n \setminus Q_{R+1}$, by adding in a Neumann boundary condition on ∂Q_R and on ∂Q_{R+1}; this will increase the number of eigenvalues of the operators $\tilde{H}_n(\lambda\varphi_R\mathbf{b})$ below E. We introduce the notation $\tilde{H}_{R+1,n;ND}$, $H_{R,R+1;N}(\lambda\varphi_R\mathbf{b})$, $H_{R;N}(\lambda\mathbf{b})$, to denote, respectively, the parts of the decoupled operator acting in $L_2(Q_n \setminus Q_{R+1})$, $L_2(Q_{R+1} \setminus Q_R)$, and in $L_2(Q_R)$, with Dirichlet boundary conditions on ∂Q_n and Neumann b. c. on ∂Q_{R+1} and on ∂Q_R. Note that there is no magnetic part on $Q_n \setminus Q_{R+1}$, while there is no contribution from $\psi_n P_{[a',b']}\psi_n$ living on Q_{R+1}, provided $n \geq 2(R+1)$; furthermore, $\varphi_R | Q_R = 1$. As a consequence, we obtain

$$\begin{aligned}\dim P_{(-\infty,E)}\left(\tilde{H}_n(\lambda\varphi_R\mathbf{b})\right) \quad \leq \quad & \dim P_{(-\infty,E)}\left(H_{R;N}(\lambda\mathbf{b})\right) \\ & + \dim P_{(-\infty,E)}\left(H_{R,R+1;N}(\lambda\varphi_R\mathbf{b})\right) \\ & + \dim P_{(-\infty,E)}\left(\tilde{H}_{R+1,n;ND}\right).\end{aligned}$$

The last term is not contaminated by any magnetic terms and we may employ the estimate of [H1]

$$\dim P_{(-\infty,E)}\left(\tilde{H}_{R+1,n;ND}\right) \leq \dim P_{(-\infty,E)}\left(\tilde{H}_n\right) - \dim P_{(-\infty,E)}\left(H_{R+1}\right) + CR^{\nu-1},$$

with a constant C which is independent of R and n, as long as $R \geq 1$ and $n \geq 4R$. By the assumption made in Theorem 1.1, $\dim P_{(-\infty,E)}\left(H_{R+1}\right) \geq c_0 R^\nu$.

In order to estimate the contribution coming from the dangerous "transition region" $Q_{R+1} \setminus Q_R$, we employ Lemma 2.2, given below, which is based on a trace class estimate of Simon [S]. As may be expected, we find a "surface term", i. e., there exists a constant C' such that

$$\dim P_{(-\infty,E)}\left(H_{R,R+1;N}(\lambda\varphi_R\mathbf{b})\right) \leq C' R^{\nu-1},$$

where the constant C' is independent of λ and the special choice of the cutoffs φ_R.

For k given, we now first choose R_k large enough to ensure that $c_0 R^\nu - (C + C') R^{\nu-1} \geq k$, so that, by the above estimates,

$$\dim P_{(-\infty,E)}\left(\tilde{H}_n(\lambda\varphi_{R_k}\mathbf{b})\right) \leq \dim P_{(-\infty,E)}\left(\tilde{H}_n\right) - k + \dim P_{(-\infty,E)}\left(H_{R;N}(\lambda\mathbf{b})\right).$$

By adapting the estimates of [AHS] (cf. also [CdV]) to the case of Neumann boundary conditions, it follows that the infimum of the spectrum of $H_{R;N}(\lambda\mathbf{b})$ goes to infinity, as $\lambda \to \infty$, and the result follows. ∎

Lemma 2.2. *Let $Q \subset \mathbf{R}^\nu$ denote a cube of sidelength 1, centered at the origin, and let $H_{Q;N}(\mathbf{a}) = (i\nabla - \mathbf{a})^2$ acting in $L_2(Q)$ with Neumann boundary conditions. Then, given $E > 0$, there exists a constant C_E such that*

$$\dim P_{(-\infty,E)}\left(H_{Q;N}(\mathbf{a})\right) \leq C_E\,,$$

independently of the real-valued vector potential $\mathbf{a} \in C^1(\mathbf{R}^\nu)^\nu$.

Proof. We shall only discuss how the corresponding estimate is obtained in the case of Dirichlet boundary conditions; the Neumann estimate follows from the Dirichlet case by extending the Neumann eigenfunctions to a larger cube, applying a cut-off procedure and using min-max arguments.

Define a sequence of auxiliary potentials $v_k(x)$, $k \in \mathbf{N}$, by setting $v_k(x) := 0$, for $x \in Q$, and $v_k(x) := k(1 + |x|^2)$, for $x \notin Q$. Then, by a classical estimate due to Simon (see [S]), we have ("tr" denoting the trace)

$$\operatorname{tr}\left(\mathrm{e}^{-t(H(\mathbf{a})+v_k)}\right) \leq \operatorname{tr}\left(\mathrm{e}^{-t(H(\mathbf{0})+v_k)}\right), \quad t > 0,$$

for any real C^1-vector potential $\mathbf{a}$; it is clear that the RHS is finite for all $k \in \mathbf{N}$ and for all $t > 0$. By standard convergence arguments ([W]), $H(\mathbf{a}) + v_k$ converges in norm resolvent sense to $H_{Q;D}(\mathbf{a})$, the operator $(i\nabla - \mathbf{a})^2$ on Q with Dirichlet boundary conditions. We now put $t = 1$ and conclude that

$$\operatorname{tr}\left(\mathrm{e}^{-H_{Q;D}(\mathbf{a})}\right) \leq \operatorname{tr}\left(\mathrm{e}^{-H_{Q;D}(\mathbf{0})}\right),$$

independently of $\mathbf{a}$. As each eigenvalue of $H_{Q;D}(\mathbf{a})$ below E will contribute at least e^{-E} to the trace on the LHS, our claim follows. ∎

Acknowledgements. R. Hempel would like to thank T. Hoffmann-Ostenhof for the kind invitation to the Erwin Schrödinger Institute in Vienna and I. Herbst for enlightening conversations on magnetic fields.

References

[AADH] S. Alama, M. Avellaneda, P. A. Deift and R. Hempel, *On the existence of eigenvalues of a divergence form operator $A + \lambda B$ in a gap of A.* Asymptotic Anal., to appear.

[ADH] S. Alama, P. A. Deift and R. Hempel, *Eigenvalue branches of the Schrödinger operator $H - \lambda W$ in a gap of $\sigma(H)$.* Commun. Math. Phys. **121** (1989), 291—321.

[AHS] J. Avron, I. Herbst and B. Simon, *Schrödinger operators with singular magnetic fields,* I. *General Interactions.* Duke Math. J. **45** (1978), 847—883.

[B] M. Sh. Birman, *Discrete spectrum in the gaps of the continuous one for perturbations with large coupling constant.* Advances in Soviet Mathematics **7**, pp. 57—74. Amer. Math. Soc., Providence, 1991.

[BR] M. Birman and G. D. Raikov, *Discrete spectrum in the gaps for perturbations of the magnetic Schrödinger operator.* Advances in Soviet Mathematics **7**, p.p. 75 - 84. Amer. Math. Soc., Providence, 1991.

[CdV] Y. Colin de Verdiere, *L'asymptotique de Weyl pour les bouteilles magnétiques.* Commun. Math. Phys. **105** (1986), 327—325.

[CFrKS] H. L. Cycon et al., *Schrödinger operators.* Springer, New York 1987.

[DH] P. A. Deift and R. Hempel, *On the existence of eigenvalues of the Schrödinger operator $H - \lambda W$ in a gap of $\sigma(H)$.* Commun. Math. Phys. **103** (1986), 461—490.

[GS] F. Gesztesy and B. Simon, *On a theorem of Deift and Hempel.* Commun. Math. Phys. **116** (1988), 503—505.

[H1] R. Hempel, *Eigenvalues in gaps and decoupling by Neumann boundary conditions.* J. Math. Anal. Appl. **169** (1992), 229—259.

[H2] —, *Second order perturbations of divergence type operators with a spectral gap.* Operator Theory: Advances and Applications, Vol. **47**. Birkhäuser, Basel 1992.

[S] B. Simon, *Functional Integration and Quantum Physics.* Academic Press, New York 1979.

[W] J. Weidmann, *Stetige Abhängigkeit der Eigenwerte und Eigenfunktionen elliptischer Differentialoperatoren vom Gebiet.* Math. Scand. **54** (1984), 51—69.

Rainer Hempel, Erwin Schrödinger International Institute for Mathematical Physics, Pasteurgasse 4/7, A-1090 Vienna, Austria
Jörg Laitenberger, Math. Inst. der Univ. München, Theresienstr. 39, D-8000 Mü nchen 2, Germany

Operator Theory:
Advances and Applications, Vol. 70

Regularity of the nodal sets of solutions to Schrödinger equations

M. Hoffmann-Ostenhof, T. Hoffmann-Ostenhof
and N. Nadirashvili

1 Introduction

Let Ω be an open set in $\mathbb{R}^n$ and let $V : \Omega \to \mathbb{R}$ with $V \in L^1_{\mathrm{loc}}(\Omega)$. We consider real valued solutions $u \neq 0$ which satisfy

$$\Delta u = Vu \text{ in } \Omega \tag{1.1}$$

in the distributional sense.

In a recent paper two of us [HO2] investigated the local behaviour of such solutions under rather mild assumptions on the potential V, namely we assumed that $V \in K^{n,\delta}(\Omega)$ for some $\delta > 0$, see e.g. [AS, S], where the class $K^{n,\delta}$ is defined by requiring that

$$\lim_{\varepsilon \downarrow 0} \sup_{x \in \mathbb{R}^n} \int_{|x-y|<\varepsilon} \chi_\Omega \frac{|V(y)|}{|x-y|^{n-2+\delta}} dy = 0 \tag{1.2}$$

Here χ_Ω denotes the characteristic function of Ω.

One of our main results was

Theorem 1.1. *Suppose $u \neq 0$ is a real valued solution of (1.1). Let $x_0 \in \Omega$ then either there is a harmonic homogenous polynomial $P_M \neq 0$ of degree M such that*

$$u(x) = P_M(x - x_0) + \Phi(x) \tag{1.3}$$

with

$$\Phi(x) = O(|x - x_0|^{M+min(1,\delta')}) \quad \forall \delta' < \delta \quad \text{for } x \to x_0 \tag{1.4}$$

or u vanishes at x_0 faster than polynomially, that is

$$\overline{\lim_{x \to x_0}} |x - x_0|^{-\alpha} |u(x)| = 0$$

for every $\alpha > 0$.

It is known [AS, S] that $V \in K^{n,\delta}(\Omega)$, $\delta < 1$ implies $u \in C^{0,\delta}(\Omega)$ where $C^{0,\delta}$ denotes Hölder continuity. Suppose u has a zero of first order at x_0 so that $u = P_1(x - x_0) + \Phi$ in a neighbourhood of x_0 according to Theorem 1.1. (1.3) and (1.4) implies that for $\delta' < \delta \leq 1$

$$\lim_{x \to x_0} \frac{|u(x) - P_1(x - x_0)|}{|x - x_0|^{1+\delta'}} = 0$$

so that u is at x_0 'smoother' than at points for which $u \neq 0$. So the question arises whether the zero sets of solutions of Schrödinger equations are in fact smoother than the corresponding solutions.

Let us illustrate this with an explicit example. According to the theorem of Cauchy and Kowalewski there is a small disk

$$B_\rho = \{(x, y) \in \mathbb{R}^2 : x^2 + y^2 < \rho^2\}$$

such that

$$\Delta v = v \text{ in } B_\rho$$

with

$$v = x - y + \frac{1}{6}x^3 - \frac{1}{2}x^2 y + \text{ higherorderterms}$$

and with $v(0, y) = -y$, $\frac{\partial v}{\partial x}(0, y) = 1$. Now let in B_ρ, u be defined by

$$u = \begin{cases} x - y & \text{for } x \leq 0 \\ v & \text{for } x > 0 \end{cases}$$

then $\Delta u = Vu$ with

$$V = V(x, y) = \begin{cases} 1 & x > 0 \\ 0 & x \leq 0 \end{cases}$$

and a simple calculation shows that the nodal line of u is given by

$$\begin{array}{lll} y = x & \text{for} & x \leq 0 \\ y = x - \frac{1}{3}x^3 + O(x^5) & \text{for} & x > 0. \end{array}$$

Hence $y(x)$ has a second derivative and one sided third derivatives. But u itself already has a jump in the second derivative for every (x, y) with $x = 0$ and $|y| \neq 0$.

Results on this additional regularity of nodal sets together with some proofs will be presented in this announcement, the full paper will appear elsewhere.

2 Regularity of nodal sets

Without loss of generality we consider (1.1) in $B_{R_0} = \{x \in \mathbb{R}^n : |x| < R_0\}$ and we assume $V \in K^{n,\delta}(B_{R_0})$. Let

$$N_u = \{x \in B_{R_0} : u(x) = 0\} \tag{2.1}$$

and let $N_u^{(1)} = \{x \in N_u : u \text{ vanishes of first order at } x\}$ so that for each $x_0 \in N_u^{(1)}$ there is a $P_1^{(x_0)}(x - x_0) \neq 0$ with

$$u(x) = P_1^{(x_0)}(x - x_0) + \Phi(x) \tag{2.2}$$

for $x \to x_0$ according to Theorem 1.1.

Theorem 2.1. *Pick $x_0 \in N_u^{(1)}$ and assume $0 < \delta \leq 1$. Then for each $\delta' < \delta$ and for sufficiently small $\varepsilon > 0$, $N_u^{(1)} \cap B_\varepsilon(x_0)$ is a $C^{1,\delta'}$-hypersurface.*

Remarks.
(i) By a $C^{1,\delta'}$ hypersurface we mean that $N_u^{(1)} \cap B_\varepsilon(x_0)$ can be represented as the graph of a $C^{1,\delta'}$- function.

(ii) Theorem 2.1 is sharp in the sense that $\delta' > \delta$ is not possible. We do not know whether $\delta' = \delta$ might be allowed.

(iii) We shall later discuss the case of smoother potentials, say $V \in K^{n,\delta}(\Omega)$ for $\delta \in (1,2)$ or $V \in C^{k,\alpha}(\Omega)$. $C^{k,\alpha}$ denotes the usual Hölder spaces.

Sketch of the proof.
We first state a Lemma which is a sharpening of Theorem 1.1.

Lemma 2.1. *Let $x_0 \in N_u \cap B_{R_1}(\mathbf{0})$ with $R_1 = R_0/2$ and suppose $\sup_{x \in B_{R_0}} |u| = C_1$ then for every $\delta' < \delta$, there is a C_2 such that for $|x| < R_0$*

$$|u(x) - P_1^{(x_0)}(x - x_0)| \leq C_2 |x - x_0|^{1+\delta'}, \tag{2.3}$$

where

$$C_2 = C_2(V, C_1, \delta - \delta', n, R_0)$$

Remark. The V-dependence can be made explicit via a suitable norm of V. The important fact is that C_2 does not depend on x_0. If x_0 happens to be a higher order zero of u then $P_1^{(x_0)}(x - x_0) \equiv 0$. The proof of Lemma 2.1 relies heavily on the techniques which have been developed in [HO2] in order to prove Theorem 1.1. Some additional technical complications arise, causing however no entirely new problems. Naturally a complete proof is, as already the proof of Theorem 1.1 somewhat involved.

Since for $x_0, x_1 \in N_u$

$$(\nabla u)(x_i) = (\nabla P^{(x_i)})(x_i), \quad i = 0, 1 \tag{2.4}$$

it can be shown via Lemma 2.1

Lemma 2.2. *Let $x_0, x_1 \in N_u \cap B_{R_2}$, $R_2 = R_0/4$ then for $\delta' < \delta$ there is a constant C_3 such that*

$$|(\nabla u)(x_1) - (\nabla u)(x_2)| \leq C_3 |x_0 - x_1|^{\delta'}. \tag{2.5}$$

with $C_3 = C_3(C_2)$, C_2 the constant given according to (2.3).

For later purposes we prove the following more general statement.

Lemma 2.2′ *Suppose $P_M^{(0)}$ and $P_M^{(1)}$ are polynominals of degree M with*

$$|P_M^{(0)}(x) - P_M^{(1)}(x - x_1)| \leq c|x_1|^{M+\delta'}$$

for $|x| \leq 2|x_1|$. Then for every M-th partial derivative there exists a constant $C(M,n)$ such that

$$|\partial_M(P_M^{(0)}(x) - P_M^{(1)}(x - x_1))| \leq cC(M,n)|x_1|^{\delta'}$$

Proof. Let $Q_M(x) = P_M^{(0)}(x) - P_M^{(1)}(x-x_1)$. In the one-dimensional case $|Q_M(x)| \leq c|x_1|^{M+\delta'}$ for $|x| \leq 2|x_1|$ implies via a classical inequality of Chebyshev that

$$|\frac{d^M}{dx^M}Q_M(x)| \leq cC_M|x_1|^{\delta'}$$

To obtain the corresponding estimate for the n-dimensional case we consider directional derivatives of M-th order. There the one-dimensional estimate obviously holds. The partial derivatives can then be estimated by linear combinations of the directional derivatives [BO].

We shall now proceed in the following way: we assume that $x_0 \in N_u^{(1)}$ and that

$$(\nabla u)(x_0) = (\lambda, 0, \dots, 0) \equiv \lambda e_1, \quad \lambda \neq 0. \tag{2.6}$$

We shall first show that in a neighbourhood of x_0 the nodal set $u(x) = 0$ can be represented as the graph of a uniquely determined continuous function

$$\varphi : B_\gamma(\pi x_0) \subset \mathbb{R}^{n-1} \to \mathbb{R}$$

such that

$$u(\varphi(y), y) = 0 \quad \forall y \in B_\gamma(\pi x_0). \tag{2.7}$$

Thereby $\pi x := (x_2, \dots, x_n)$, for every $x = (x_1, x_2, \dots, x_n)$ and $B_\gamma(\pi x_0) = \{y \in \mathbb{R}^{n-1} : |\pi x_0 - y| < \gamma\}$ with $\gamma > 0$ small enough.

Proposition 2.1. *For sufficiently small $\rho > 0$*

$$\pi : N_u \cap B_\rho(x_0) \to \mathbb{R}^{n-1}$$

is injective.

Proof. We start with some rather obvious observations and definitions. Since $u(x_0) = 0, \quad x_0 \in \partial G$ where $G = \{x \in B_{R_2} : u(x) > 0\}$. Let $\tilde{x} \in \partial G$. We say $\varepsilon_{\tilde{x}}$ is an $n-1$ dimensional affine hyperplane to ∂G at $\tilde{x}$ if for every sequence of points $x_i \in \partial G$ with $x_i \to \tilde{x}$

$$\text{dist}(x_i, \varepsilon_{\tilde{x}}) = o(|x_i - \tilde{x}|) \text{ for } i \to \infty.$$

Let $y \in \partial G$ and $d_y(x) := (x - y, (\nabla u)(y))$ then $\varepsilon_{x_0} = \{x \in \mathbb{R}^n : d_{x_0}(x) = 0\}$. Now define for $x \in \partial G \cap N_u^{(1)}$

$$(\tilde{\nabla} u)(x) = \frac{(\nabla u)(x)}{|(\nabla u)(x)|}$$

then by Lemma 2.2 it is straight forward to show that for small ρ'

$$((\tilde{\nabla} u)(x_0), (\tilde{\nabla} u)(x)) \geq \frac{1}{2} \tag{2.8}$$

for $x \in N_u \cap \overline{B}_{\rho'}(x_0)$ with $\overline{B}_{\rho'}(x_0) = \{x \in \mathbb{R}^n : |x - x_0| \leq \rho'\}$. Set $\rho = \rho'$ in Proposition 2.1.

Now suppose that Proposition 2.1 is wrong. Then there exist $\bar{x}, \underline{x} \in B_{\rho'}(x_0) \cap N_u, \bar{x} \neq \underline{x}$ such that $\pi(\bar{x}) = \pi(\underline{x}) := \tilde{x} = (0, \tilde{x}_2, \tilde{x}_3, \ldots, \tilde{x}_n)$. Let $E = \pi^{-1}(\tilde{x}) \cap \overline{B}_\rho(x_0) \cap N_u$. E is a closed set with cardinality ≥ 2. If E has an accumulation point y then $\pi^{-1}(\tilde{x}) \subset \varepsilon_y$, but $\varepsilon_y = \{x \in \mathbb{R}^n : (x - y, (\nabla u)(y)) = 0\}$ and this implies $((\tilde{\nabla} u)(x_0), (\tilde{\nabla} u)(y)) = 0$ contradicting (2.8). So there are $z_1, z_2 \in E$ where z_1, z_2 only differ in the first coordinate such that

$$g(z_1, z_2) = \{x \in \mathbb{R}^n : x = tz_1 + (1-t)z_2, t \in (0,1)\}$$

satisfies $g(z_1, z_2) \cap N_u = \emptyset$ implying $u(x) \neq 0 \quad \forall x \in g(z_1, z_2)$. But

$$((\nabla u)(x_0), (\nabla u)(z_j)) = \frac{\partial u}{\partial x_1}(x_0) \frac{\partial u}{\partial x_1}(z_j)$$

and

$$\text{sgn } \frac{\partial u}{\partial x_1}(z_1) \neq \text{sgn } \frac{\partial u}{\partial x_1}(z_2), \quad j = 1, 2,$$

hence sgn $((\tilde{\nabla} u)(x_0), (\tilde{\nabla} u)(z_1)) \neq$ sgn $((\tilde{\nabla} u)(x_0), (\tilde{\nabla} u)(z_2))$ again contradicting (2.8). This proves the proposition.

Now we have to show that $\forall y \in B_\gamma(\pi x_0)$, γ small enough, there is a $t \in \mathbb{R}$ such that $(t, y) \in B_\rho(x_0) \cap N_u$. But this is an immediate consequence of the continuity of u : Denote $x_0 = (x_{01}, \ldots, x_{0n})$, let $y \in B_\gamma$ and $x_\pm := (x_{0,1} \pm \varepsilon, y)$ such that $x_\pm \in B_\rho(x_0)$. Since $\nabla u(x_0) = \lambda e_1$ and $u(x_0) = 0$, sgn $u(x_{0,1} + \varepsilon, x_{0,2} \ldots x_{0,n}) \neq$ sgn $u(x_{0,1} - \varepsilon, x_{0,2} \ldots x_{0n})$ for ε small enough. Since u is continuous we also have $\forall y \in B_\gamma$, for γ sufficiently small, sgn $u(x_{0,1} + \varepsilon, y) \neq$ sgn $u(x_{0,1} - \varepsilon, y)$. Hence by the intermediate value theorem there is a t with $|t - x_{0,1}| < \varepsilon$ such that $u(t, y) = 0$. This implies (2.7).

Further via [H, Satz 170.1] we conclude that $\varphi \in C^{1,\delta'}(B_\gamma)$ so that $N_u^{(1)}$ is indeed locally the graph of a $C^{1,\delta'}$ function.

We give now a brief discussion of the higher regularity of nodal sets if V is assumed to be more regular.

Theorem 2.2. *(i) Suppose (1.1) holds with $V \in K^{n,\delta}(\Omega)$ with $\delta \in (1,2)$. Then $N_u^{(1)}$ is locally a $C^{2,\delta'}$-hypersurface for $\delta' < \delta - 1$.*

(ii) Suppose (1.1) holds with $V \in C^{k,\alpha}(\Omega)$ with $\alpha \in (0,1)$ then $N_u^{(1)}$ is locally a $C^{k+3,\alpha}$-hypersurface.

The idea of the proof is basically the same as for the proof of Theorem 2.1. However we have to replace Theorem 1.1 by a more detailed result.

Theorem 2.3. *(i) Suppose (1.1) holds and $V \in K^{n,\delta}(\Omega), \delta \in (1,2)$. Let $x_0 \in \Omega$ then either there exist two harmonic homogenous polynomials $P_M \neq 0, P_{M+1}$ of degree $M, M+1$ respectively such that*

$$u(x) = P_M(x - x_0) + P_{M+1}(x - x_0) + \Phi(x)$$

with

$$\Phi(x) = O(|x - x_0|^{M+\delta'}) \qquad \text{for } x \to x_0, \qquad \forall \delta' < \delta$$

or u vanishes at x_0 faster than polynomially.

(ii) Suppose (1.1) holds and

$$V \in C^{k,\alpha}(\Omega), \quad \alpha \in (0,1).$$

Let $x_0 \in \Omega$, then there is a polynomial of degree $M + k + 2$ such that $p(x) = P_M(x) + P_{M+1}(x) + p_1$ with P_M, P_{M+1} again harmonic and homogenous of degree $M, M+1$ respectively and $P_M \neq 0$, and $p_1(x)$ vanishes at least of order $M+2$ at zero. We have then

$$u(x) = p(x - x_0) + \Phi(x)$$

with

$$\Phi(x) = O(|x - x_0|)^{M+k+2+\alpha} \qquad \text{for} \qquad x \to x_0$$

Remarks. (a) Under the conditions of part (ii) of this theorem strong unique continuation is well known. Also (ii) is related to the classical Schauder estimates, see e.g. [GT].

(b) The proof of Theorem 2.3 again uses the techniques of [HO2] but some iterations are necessary.

(c) For the coulombic case a more detailed version of Theorem 1.1 was recently shown in [HO2S1] and [HO2S2]. An investigation of the regularity of the nodal sets for this important case should be possible along the present lines.

Starting from Theorem 2.3 the proof of Theorem 2.2 follows the same ideas as the proof of Theorem 2.1. So one first proves a suitable analog of Lemma 2.1, uses Lemma 2.2' and proceeds essentially as we did above in order to prove Theorem 2.1.

References

[AS] M. Aizenman, B. Simon, *Brownian motion and Harnack inequality for Schrödinger operators*, Commun Pure Applied Mathematics **35** (1982), 209–273.

[BO] J. Boman, *Differentiability of a function and of its composition with functions of one variable*, Math. Scand. **20** (1967), 249–286.

[GT] D.Gilbarg, N.S.Trudinger, *Elliptic partial differential equations*, Springer, Berlin, 1983.

[HO2] M. Hoffmann-Ostenhof, T. Hoffmann-Ostenhof, *Local properties of solutions of Schrödinger equations*, Comm. PDE **17** (1992), 491–522.

[HO2S1] M.Hoffmann-Ostenhof, T.Hoffmann-Ostenhof and H.Stremnitzer, *Electronic wavefunctions near coalescence points*, Phys. Rev. Letters **68** (1992), 3857–3860.

[HO2S2] M. Hoffmann-Ostenhof, T. Hoffmann-Ostenhof and H. Stremnitzer, *Local properties of Coulombic wavefunctions*, submitted for publication.

[H] H. Heuser, *Lehrbuch der Analysis Teil 2*, B. G. Teubner, Stuttgart, 1988.

[S] B. Simon, *Scrödinger semigroups*, Bull. Am. Math. Soc. **7** (1982), 447–526.

M. Hoffmann-Ostenhof: Institut für Mathematik, Universität Wien, Strudlhofgasse 4, A-1090 Wien, Austria

T. Hoffmann-Ostenhof: Institut für Theoretische Chemie, Universität Wien, Währinger Strasse 17, A-1090 Wien, Austria and International Erwin Schrödinger Institute for Mathematical Physics, Pasteurgasse 6/7, A-1090 Wien, Austria
E-mail address: hoho@itc.univie.ac.at

N. Nadirashvili: International Erwin Schrödinger Institute for Mathematical Physics, Pasteurgasse 6/7, A-1090 Wien, Austria and Institute of Earth Physics, Moscow, Russia

Operator Theory:
Advances and Applications, Vol. 70

Results in the spectral theory of Schrödinger operators with wide potential barriers

Günter Stolz

Introduction

Motivated by the beautiful results but also open problems in the spectral theory of Schrödinger operators with random or almost periodic potentials, in recent years there was rising interest in spectral properties of deterministic potentials with some kind of irregular asymptotics near infinity. Here we consider (rapid enough) decay of a potential to zero or (asymptotic) periodicity as regular asymptotics, typically leading to isolated eigenvalues and absolute continuity of the rest of the spectrum.

That potentials consisting of an infinite number of suitably placed barriers can give rise to completely different types of spectra was first found by Pearson [9], who showed that sparsely distributed barriers in dimension $d = 1$ lead to purely singular continuous spectrum. Simon and Spencer [10] showed the absence of absolutely continuous spectrum for (i) high barriers in one-dimensional Schrödinger operators and their discrete counterparts (see also [6] and [13]), (ii) wide barriers and low energy in the discrete case for arbitrary d. Kirsch, Molchanov and Pastur [7], [8] have shown that in some of these situations one actually gets dense pure point spectrum. Another interesting result was found in [5], where it is shown for $d > 1$ that operators of the type $-\Delta + \cos|x|$ have alternating intervals of absolutely continuous spectrum and dense pure point spectrum.

In the following we present some contributions to the study of Schrödinger operators with wide potential barriers. We start with a result on absolute continuity for slowly oscillating perturbations of periodic potentials in $d = 1$ (Section 1). This generalizes results of Behncke [1], who did not consider a periodic background.

In Section 2 we give a result on the absence of absolutely continuous spectrum for $1d$-Schrödinger operators, which can be applied to (i) potentials with wide barriers, and (ii) perturbations of periodic potentials. The results of Sections 1 and 2 are illustrated in Section 3 by studying the examples $-d^2/dx^2 + \cos x^\alpha$ and $-d^2/dx^2 + \cos x + \lambda \cos x^\alpha$ with $\alpha \in (0,1)$ and $\lambda > 0$.

In Section 4 we give a generalization of the result of Section 2 to arbitrary dimension d. This final result is joint work with P. Stollmann. A closely related investigation was carried out by Combes and Hislop in [2].

1. Pure absolute continuity in $d = 1$

We start with a number of definitions in order to make precise what we mean by a slowly oscillating potential. Let

$$\ell_+^\infty(L^1) := \left\{ f : \sup_{n \geq 0} \int_n^{n+1} |f| \, dx < \infty \right\}$$

and for $1 \leq p < \infty$

$$\ell_+^p(L^1) := \left\{ f : \sum_{n=0}^{\infty} \left(\int_n^{n+1} |f| \, dx \right)^p < \infty \right\}.$$

For $\omega > 0$ let the difference operator Δ_ω be defined by $(\Delta_\omega f)(t) = f(t+\omega) - f(t)$ and for $k = 1, 2, \dots$ let the class $\mathcal{D}_\omega^k$ be defined by

$$f \in \mathcal{D}_\omega^k \quad :\Longleftrightarrow \quad \Delta_\omega^j f \in \ell_+^{k/j}(L^1), \; j = 0, \dots, k.$$

Finally we introduce the class $\mathcal{D}_\omega := \bigcup_{k \geq 1} \mathcal{D}_\omega^k$ of *slowly oscillating functions* (with respect to ω).

Examples: (i) If $V \in L^1(0, \infty)$ or V is smooth with $V' \in L^1(0, \omega)$, then $V \in \mathcal{D}_\omega$ for all ω.

(ii) If V is bounded, $V' \in L^p(0, \infty)$ for some $p < \infty$ and $V^{(i)} \in L^1(0, \infty)$ for some $i \geq 2$, then $V \in \mathcal{D}_\omega$ for all ω. This includes the particular examples $V(x) = \cos(x^\alpha)$, where $\alpha \in (0, 1)$, explaining the notion *slowly oscillating.*

(iii) Let $V \in \mathcal{D}_\omega$ and q bounded and ω-periodic, then $qV \in \mathcal{D}_\omega$. This shows that $\mathcal{D}_\omega$ depends on ω (in particular: $V = 1$).

Theorem 1 *Let $H = -d^2/dx^2 + V_0 + V$ be defined as a self-adjoint operator in $L^2(\mathbf{R})$, where $V_0, V \in L^1_{loc}$ and real valued, V_0 is ω-periodic, $V \in \mathcal{D}_\omega$, $V = V_1 + V_2$, where V_1 is bounded near $+\infty$, and $\lim_{x \to \infty} \int_x^{x+1} |V_2(t)| \, dt = 0$. Define $\underline{V_1} = \liminf_{x \to \infty} V_1(x)$ and $\overline{V_1} = \limsup_{x \to \infty} V_1(x)$.*

If (α, β) is a stability interval of $-d^2/dx^2 + V_0$, then H is purely absolutely continuous in $(\alpha + \overline{V_1}, \beta + \underline{V_1})$.

Remarks: (i) Note that no condition on the potential is needed near $-\infty$ (despite L^1_{loc} and real). This includes situations, where the self-adjoint realization of $-d^2/dx^2 + V_0 + V$ is not unique (*limit circle case* at $-\infty$). In this case the Theorem holds for every self-adjoint realization. It is also true for self-adjoint realizations in $L^2(0, \infty)$ with arbitrary boundary condition at 0.

(ii) A similar result holds for discrete Schrödinger operators (Jacobi matrices) h of the form $(hu)(n) = u(n-1) + u(n+1) + (V_0 u)(n) + (Vu)(n)$ in $\ell^2(\mathbf{Z})$.

The general method underlying the *proof* of Theorem 1 is the *method of Subordinacy* introduced by Gilbert and Pearson in [4]. By this method the proof of the Theorem can be reduced to showing that all solutions (in o.d.e. sense) of $Hu = \lambda u$ are bounded near $+\infty$ for all $\lambda \in (\alpha + \overline{V_1}, \beta + \underline{V_1})$. Boundedness of solutions of $Hu = \lambda u$ follows from the boundedness of $\left\|\prod_{n=1}^{N} T_n\right\|$ in N. where the T_n are the *transfer matrices* defined by

$$\begin{pmatrix} u((n+1)\omega) \\ u'((n+1)\omega) \end{pmatrix} = T_n \begin{pmatrix} u(n\omega) \\ u'(n\omega) \end{pmatrix}.$$

The result now follows by subjecting the products $\prod_{n=1}^{N} T_n$ to an iterative diagonalization procedure. Details of this can be found in [14] and [15]. The illustrative special case $V \in \mathcal{D}_{\omega}^{1}$ was already treated in [12].

2. Absence of absolute continuity in $d = 1$

Our next result can be used to show that many Schrödinger operators have intervals of essential spectrum, which do not contain any absolutely continuous spectrum.

We consider operators $H_0 = -d^2/dx^2 + V_0$ and $H = -d^2/dx^2 + V$ in $L^2(\mathbf{R})$, where we assume $V_0 \in L^1_{loc}$ to be bounded from below and $V \in L^1_{loc}$ such that $-d^2/dx^2 + V$ is limit point at $\pm\infty$. In both cases this gives unique self-adjoint operators.

Theorem 2 *Let $I_n \subset (0, \infty)$ and $I_{-n} \subset (-\infty, 0)$ be intervals for $n = 1, 2, \ldots,$ such that $|I_n| \to \infty$ as $|n| \to \infty$ and $V(x) = V_0(x)$ for $x \in \bigcup_n I_n$.*

Then $\sigma_{ac}(H) \subset \sigma_{ess}(H_0)$.

The *proof* of this result, which can be found in [15], uses the same principal ideas then the proof in [10] for a similar result in the discrete case: decoupling by Dirichlet boundary conditions, trace estimates for resolvent differences and trace class methods from scattering theory. In obvious form the Theorem also holds for operators in $L^2(0, \infty)$.

Applications of this result are as follows:

(i) Let $V(x) \geq c$ for $x \in \bigcup I_n$ with I_n as above. Then $\sigma_{ac}(H) \cap (-\infty, c) = \emptyset$. This can be seen by choosing $V_0(x) = \max\{V(x), c\}$.

(ii) Let V_0 be periodic with (γ, δ) a gap in $\sigma(H_0)$. If $\int_n^{n+1} |V'(t)|\, dt \to 0$ as $|n| \to \infty$, then $\underline{V} := \liminf_{|x|\to\infty} V(x)$ and $\overline{V} := \limsup_{|x|\to\infty} V(x)$ are finite and Theorem 2 can be used to show that $\sigma_{ac}(H) \cap (\gamma + \underline{V}, \delta + \overline{V}) = \emptyset$.

3. A typical example

A typical example where the results of both Theorem 1 and Theorem 2 can be applied is given by $H = -d^2/dx^2 + \cos x + \lambda \cos x^{\alpha}$ with $\alpha \in (0, 1)$, $\lambda > 0$. Theorem 1 says that the absolutely continuous bands of $-d^2/dx^2 + \cos x$ shrink by λ from both sides to give absolutely continuous regions of H. On the other hand the spectrum as a set grows by λ at the ends of every band and Theorem 2 assures that intervals of length 2λ at all the band edges do not contain absolutely continuous spectrum, i.e. are purely singular:

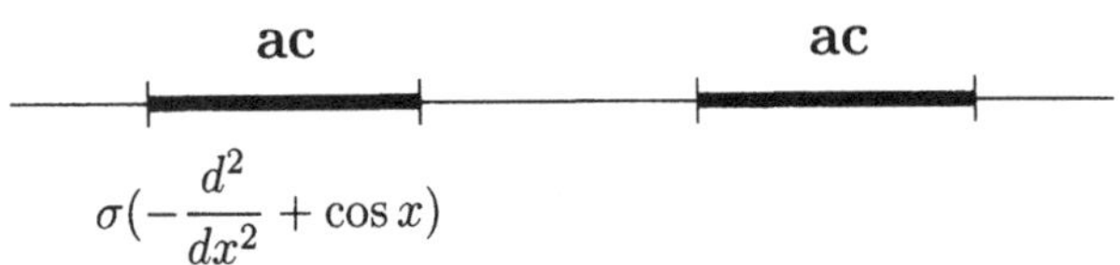

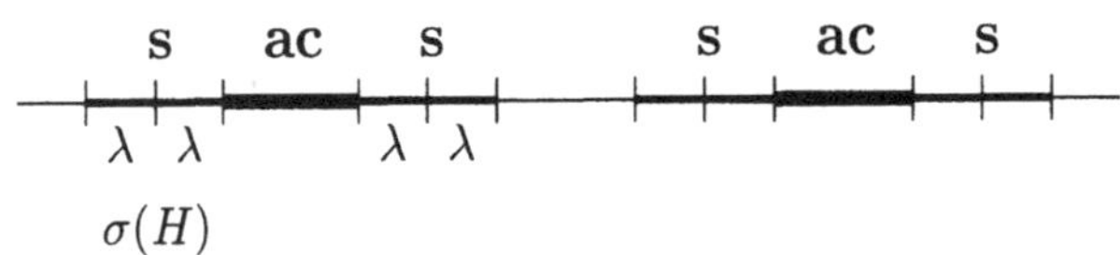

Similarly one gets that $-d^2/dx^2 + \cos x^\alpha$ has purely singular spectrum in $[-1, 1)$ and purely absolutely continuous spectrum in $(1, \infty)$.

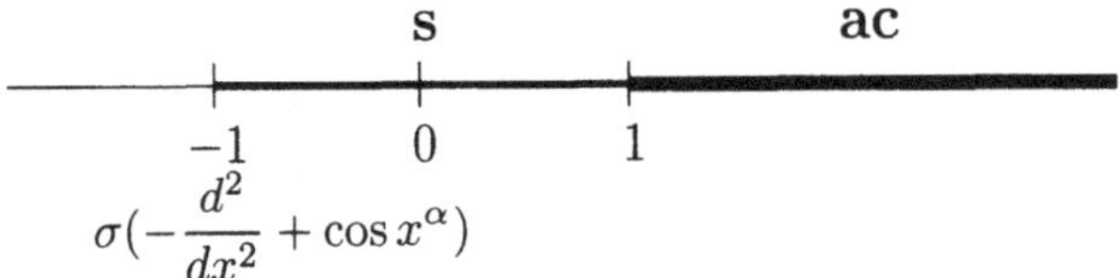

We do not know the nature of the spectrum in the singular intervals, i.e. whether there is singular continuous or point spectrum. There are interesting answers to this question for the half line operator corresponding to the last example:

Let $H_\Theta := -d^2/dx^2 + \cos x^\alpha$ in $L^2(0, \infty)$ with boundary condition $f(0)\cos\Theta - f'(0)\sin\Theta = 0$, $\Theta \in [0, \pi)$. Then for almost every Θ with respect to Lebesgue measure we have that $\sigma(H_\Theta) \cap [-1, 1)$ is dense pure point! See [8]. Moreover, a result announced in [3] shows that $\sigma(H_\Theta) \cap [-1, 1)$ is purely singular continuous for $\Theta \in L$, where $L \subset [0, \pi)$ is a dense G_δ-set, i.e., in particular, locally uncountable!!

4. Absence of absolute continuity in $d > 1$

Let $H_0 = -\Delta + V_0$ and $H = -\Delta + V$ be defined as self-adjoint form sums in $L^2(\mathbf{R}^d)$, where it is assumed that $V_{0,+}$, $V_+ \in L^1_{loc}$ and $V_{0,-}$, $V_- \in K_d$, the d-dimensional Kato class.

A sequence of subsets S_n of $\mathbf{R}^d$, $n = 1, 2, \ldots$, is called a *decomposition* of $\{x \in \mathbf{R}^d : V_0(x) \neq V(x)\}$, if (i) S_n is compact with Lebesgue measure $|S_n| = 0$ for all n, (ii) $\mathbf{R}^d \setminus \bigcup_n S_n = \bigcup_i U_i$ (disjoint union), where the U_i are open subsets of $\mathbf{R}^d$, and (iii) $|U_i| < \infty$ if $U_i \cap \{V_0 \neq V\} \neq \emptyset$.

Finally, we say that some compact $S \subset \mathbf{R}^d$ has *generalized measure* $\sigma(S) > 0$, if there exists $\alpha \leq d$ such that

$$|\{x : r \leq \text{dist}(x, S) \leq r + 1\}| \leq \sigma(S)(r^\alpha + 1)$$

for all $r \geq 0$.

In the particular example of spheres $S = \{|x| = R\}$ one can choose $\sigma(S) = c_d(R^{d-1} + 1)$, i.e. $\sigma(S)$ is a surface measure for large R, which also holds for much more general S.

Theorem 3 *Let (S_n) be a decomposition of $\{V_0 \neq V\}$, $\delta_n :=$ dist$(S_n, \{V_0 \neq V\}) \geq \delta_0 > 0$ and $\sigma_n := \sigma(S_n)$ the generalized area of S_n. If*

$$\sum_n \sigma_n e^{-\varepsilon \delta_n} < \infty \quad \textit{for all } \varepsilon > 0,$$

then $\sigma_{ac}(H) \cap (-\infty, \inf \sigma_{ess}(H_0)) = \emptyset$.

If, in addition, (S_n) is a total *decomposition, i.e. $|U_i| < \infty$ for all i, then $\sigma_{ac}(H) \subset \sigma_{ess}(H_0)$.*

The *proof* of this Theorem uses similar ideas than the proof of Theorem 2, a main step being the successive introduction of Dirichlet boundary conditions on the S_n. The most important change is that differences of the corresponding resolvents are no longer finite dimensional. A way of getting trace estimates directly is to replace resolvents by semigroups, which can be calculated via the Feynman-Kac formula. Details are given in [11]. For the trace estimates of semigroups see also the contribution of P. Stollmann to this volume.

References

[1] H. Behncke: Absolute continuity of Hamiltonians with von Neumann-Wigner potentials II, Manuscripta Math. **71** (1991) 163-181

[2] J.M. Combes, P.D. Hislop: Some transport and spectral properties of disordered media, in: Proceedings from the workshop on Schrödinger operators, Aarhus 1991, Springer Lecture Notes in Physics, ed. E. Balslev

[3] R. del Rio, S. Jitomirskaya, N. Makarov, B. Simon: Singular continuous spectrum is generic, Preprint 1993

[4] D.J. Gilbert, D.B. Pearson: On subordinacy and analysis of the spectrum of one-dimensional Schrödinger operators, J. Math. Anal. Appl. **128** (1987) 30-56

[5] R. Hempel, I. Herbst, A.M. Hinz, H. Kalf: Intervals of dense point spectrum for spherically symmetric Schrödinger operators of the type $-\Delta + \cos|x|$, J. London Math. Soc. **43** (1991) 295-304

[6] P.D. Hislop, S. Nakamura: Stark Hamiltonian with unbounded random potentials, Rev. Math. Phys. **2** (1990) 479-494

[7] W. Kirsch, S.A. Molchanov, L.A. Pastur: One-dimensional Schrödinger operator with unbounded potential: The pure point spectrum, Funct. Anal. Appl. **24** (1990) 176-186

[8] W. Kirsch, S.A. Molchanov, L.A. Pastur: One-dimensional Schrödinger operators with high potential barriers, in: Operator Theory: Advances and Applications, Vol. 57, 163-170, Birkhäuser 1992

[9] D.B. Pearson: Singular continuous measures in scattering theory, Commun. Math. Phys. **60** (1978) 13-36

[10] B. Simon, T. Spencer: Trace class perturbations and the absence of absolutely continuous spectra, Commun. Math. Phys. **125** (1989) 113-125

[11] P. Stollmann, G. Stolz: Singular spectrum for multidimensional Schrödinger operators with potential barriers, Preprint 1993

[12] G. Stolz: On the absolutely continuous spectrum of perturbed periodic Sturm-Liouville operators, J. reine angew. Math. **416** (1991) 1-23

[13] G. Stolz: Note to the paper by P.D. Hislop and S. Nakamura: Stark Hamiltonian with unbounded random potentials, to appear in Rev. Math. Phys. **5** (1993)

[14] G. Stolz: Spectral theory for slowly oscillating potentials, I. Jacobi matrices, Preprint 1992

[15] G. Stolz: Spectral theory for slowly oscillating potentials, II. Schrödinger operators, Preprint 1992

Author's address: Universität Frankfurt, Fachbereich Mathematik,
Postfach 11 19 32, D-60054 Frankfurt am Main.
E-mail: stolz@mathematik.uni-frankfurt.dbp.de

Operator Theory:
Advances and Applications, Vol. 70

Stark ladders and perturbation theory

V. Grecchi, M. Maioli and A. Sacchetti

Abstract

We consider the Bloch problems with a finite number of open gaps and we prove, for any external weak enough electric field, the existence of a finite number of Stark ladders, given by complex translation, decoupled band approximation and regular perturbation theory.

Some of the results reported in this lecture have been announced in [6] and given with more details in [7]. We don't discuss here other well known rigorous results on this field (see for instance [2] and [4]).

Let us consider a one-dimensional Bloch-Stark problem with Hamiltonian:

$$H_F = H_B + Fx, \tag{1}$$

where H_B is the Bloch operator :

$$H_B = -\Delta + V(x), \quad V(x) = V(x+a), \quad F, a > 0,$$

in the space $\mathcal{H} = L^2(\mathbf{R})$, with $\sigma(H_B) = \cup_{n=1}^{\infty} B_n$, where $B_n = [E_n^m, E_n^M]$ is a band and $G_n = (E_n^M, E_{n+1}^m)$ is the n-th gap. Let $k(E) = a^{-1}\arccos(D(E)/2)$, where $D(E)$ is the discriminant of the Bloch operator, be the crystal momentum and $\epsilon_n(k)$ the energy functions defined on the torus $B = \mathbf{R}\backslash b$ where $b = 2\pi/a$.

A naive approach to the problem was followed in [5](where the problem is extended to the disordered case). Let $V(x) = \sum_j u(x - ja)$, where $u(x)$ is both translation and dilation analytic, somewhere negative and vanishing fast enough at infinity, so that the atomic operators

$$H_{A,j} = -\Delta + u(x - ja)$$

have one isolated eigenvalue λ. Thus the atomic Stark operators

$$H_{A,j,F} = H_{A,j} + Fx$$

have one resonance near $\lambda + Faj$. We expect (and prove [5]!) the existence of a ladder of resonances of H_F near $\{\lambda + Faj\}_j$ for F small fixed and the parameter a large.

In order to consider fixed Bloch problems ($a = 2\pi$) for any F small enough we must use other intuitions and other techniques. The other intuition we shall consider is more deep and goes back to the Zener tilted bands picture and to the more precise Buslaev adiabatic approximation. Near $x = x_0$ and for F small enough H_F acts locally as the Bloch operator with effective energy $E_0 = E - Fx_0 = f(x_0)$. Thus if we connect each band to the next one turnig around a branch point [3] we have effective (Zener) barriers in the inverse image of the gaps: $f^{-1}(G_n)$, with effective potential $U(x) = E + \mathbf{Im}[k(f(x))]^2$. Because of this picture we can expect bound states or sharp resonances, but we don't have any suggestion on the position of the energy levels. For this purpose we should consider the Wannier [9] idea of decoupling the one-band spaces at finite field F. The compression of the operator on a one-band space gives a compact resolvent operator with a ladder as its spectrum. The prove of the last statement is easy in the Crystal Momentum Representation (CMR). Let us define the unitary operator U on a function ψ belonging to $\mathcal{S}$:

$$\mathbf{a} = U\psi, \quad \mathbf{a} = \{a_n\}_{n=1}, a_n \in L^2(B),$$
$$a_n(k) = \eta(\psi_n^k, \psi)_{\mathcal{H}}, \quad H_B\psi_n^k = \epsilon_n(k)\psi_n^k, \tag{2}$$

where $\eta = (2\pi)^{-\frac{1}{2}}$ and ψ_n^k are the Bloch vectors orthonormalized on $\mathcal{X} = L^2([0, 2\pi], \frac{dx}{2\pi})$. The transformed operator in CMR is given by:

$$\tilde{H}_F = \mathrm{diag}(\epsilon_n + iFd_k)_n + F\mathbf{X}, \quad (\mathbf{X})_{n,m}(k) = X_{n,m}^k, \tag{3}$$

where $X_{n,m}^k = i < d_k\mathbf{W}_n^k, \mathbf{W}_m^k >_{l^2}$, and $\mathbf{W}_n^k$ is the sequence of Fourier coefficients of $e^{-ikx}\psi_n^k(x)$, $k \in B = \mathbf{R}\backslash 1$.

The Decoupled Band (DB) approximation H_F^{DB} of Wannier has the eigenvalues $\{\lambda_{n,j}\}_{n,j} = \{< \epsilon_n > + 2\pi FjJ\}_{n,j}$ with eigenvectors

$$\psi_{n,j}(k) = \exp \frac{i}{F} \int_0^k (\epsilon_n(h) - \lambda_{n,j})dh$$

in $L^2(B)$. Since the spectrum of the DB operator is generically dense in $\mathbf{R}$, it is not suitable as a strarting point of perturbation theory. As suggested by Avron [1] we restrict ourself to the Bloch operators with a finite number of open gaps, which in some sense are a dense subset, and are translation analytic (we now assume inversion symmetry $x \to -x$). In this case the DB approximation gives a finite number of ladders imbedded on a continuous spectrum coming from the infinite band. A good starting point is obtained by the same models transformed by the complex translation $xJ \to x + \alpha$, with $|\mathbf{Im}(\alpha)| < 2\alpha_0 + \delta$. In order to control the spectrum we define the Extended Crystal Momentum Representation (ECMR) which is directly applied to the translated operator. Let for simplicity to have one finite band and $\psi_{1,\alpha}^k$, $k \in B$, $\psi_{2,\alpha}^p$, $p \in \mathbf{R}$ be the translated Bloch vectors, defined as above. . We define ϵ_2 continuous (and analytic) on $\mathbf{R}$, and we have $e^{-ipx}\psi_2^p(x) = 1 + O(\frac{1}{p})$. Let ψ be a translation analytic vector of class $\mathcal{S}$ as well as its translated vectors, we define :

$$U_\alpha \psi = \{a_{1,\alpha}, a_{2,\alpha}\}$$
$$\text{where} \tag{4}$$
$$a_{1,\alpha}(k) = \eta(\psi^k_{1,\overline{\alpha}}, \psi), \quad \alpha_{2,\alpha}(p) = \eta e^{ip\alpha}(\psi^p_{2,\overline{\alpha}}, \psi)$$

The translated Bloch-Stark operator $H_{F,\alpha}$ in the ECMR becomes

$$\tilde{H}_{F,\alpha} = \text{diag}(\epsilon_1(k) + iFd_k, \epsilon_2(p) + iFd_p + F\alpha) + F\mathbf{X}_\alpha \tag{5}$$

on $L^2(B) \oplus L^2(\mathbf{R})$. The DB approximation gives now the operator

$$\tilde{H}^{DB}_{F,\alpha} = \tilde{H}_{F,\alpha} - F\mathbf{X}_\alpha$$

with discrete spectrum $\{\lambda_{1,j}\}_j$ on the real axis and essential spectrum on all the line $F\alpha + \mathbf{R}$. The coupling operator $\mathbf{X}_\alpha$ and U_α, $(U_\alpha)^{-1}$ are bounded analytic on a stripe $|\mathbf{Im}(\alpha)| < 2\alpha_0$.

We now consider the auxiliary perturbative parameter f, with $|f| \leq F$, and

$$\tilde{H}_{F,\alpha}(f) = \tilde{H}^{DB}_{F,\alpha} + f\mathbf{X}_\alpha \tag{6}$$

so that $\tilde{H}_{F,\alpha}(F) = \tilde{H}_{F,\alpha}$. Let us now fix α and F and drop them from the labels. We have:

Theorem. *Let $\tilde{H}(f)$ be as above, we have:*
i) The essential spectrum does not depend on f because of the relative compactness of $\mathbf{X}$
ii) The ladder is stable for small fixed F and any f, $|f| \leq F$, and the resonance $\lambda_{1,j}(f)$ is analytic in the disk and given by the perturbation formula:

$$\lambda_{1,j}(f) = \frac{\int_\Gamma(\psi_{1,j}, R_{1,1}(f,z)\psi_{1,j})zdz}{\int_\Gamma(\psi_{1,j}, R_{1,1}(f,z)\psi_{1,j})dz} \tag{7}$$

where Γ is a closed path encircling only one point $(\lambda_{1,j})$ of the unperturbed spectrum, and staying at the maximal distance from the unperturbed spectrum. $R_{1,1}(f,z)$ is the resolvent of the operator in (6) at parameter z and compressed on the space of the first band.

The proof is based on explicit expressions of the unpertubed components of the resolvent, integration by parts, and the estimates of $\mathbf{X}$ based on the estimates of the vectors $\mathbf{W}^k_n$, $n = 1, 2$.

A control on the full resolvent shows that there isn't any other sharp resonance.

If we expand the perturbation coefficients in power series of F, taking $f = F$, we get explicit expression of the coefficients of the asymptotic series studied by Nenciu [8]. The second perturbation coefficient gives the width of the resonances in the Fermi Golden Rule approximation which coincides with the adiabatic evaluation of the full width [3] up to a factor $\pi^2/9$ which seems to be universal in a class of semiclassical perturbations [6].

Acknowledgments. This work is partially supported by MURST and GNFM. One of us (V.G.) thanks the organizers of this Conference for the kind invitation.

References

[1] Avron J, (1982) The lifetime of Wannier ladder states. Ann. Phys. **143** 33-53.

[2] Bentosela F, Grecchi V, (1991) Stark-Wannier ladders. Commun. Math. Phys. **142** 169-192.

[3] Buslaev V S, Dmitrieva L A, (1990) A Bloch electron in an external field. Leningrad Math. J. **1** 287-320.

[4] Combes J M, Hislop P D, (1991) Stark ladder resonances for small electric fields. Commun. Math. Phys. **140** 291-320.

[5] Grecchi V, Maioli M, Sacchetti A, (1992) Stark resonances in disordered systems. Commun. Math. Phys. **146** 231-240.

[6] Grecchi V, Maioli M, Sacchetti A, (1993) Wannier ladders and perturbation theory. J. Phys. A: Math. Gen. (letters to the Editor)**26**

[7] Grecchi V, Maioli M, Sacchetti A, (1993) Stark ladders of resonances: Wannier ladders and perturbation theory. To appear on Commun. on Math. Phys.

[8] Nenciu G, Adiabatic theorem and spectral concentration: I Arbitrary order spectral concentration for the Stark effect in atomic physics. Commun. Math. Phys.**82** (1981) 121-135.

[9] Wannier G H, Wave functions and effective Hamiltonian for Bloch electrons in a electric field. Phys. Rev. **117** (1960), 432-439.

V. Grecchi, Università di Bologna, Dipartimento di Matematica, I-40127 Bologna, Italy
M. Maioli, A. Sacchetti, Università di Modena, Dipartimento di Matematica, I-41100 Modena, Italy

Operator Theory:
Advances and Applications, Vol. 70

Singular potentials: Algebraization

Miloslav Znojil

Abstract

Energies of certain spiked oscillators (possessing a singular repulsive core in the origin) are shown obtainable as roots of a set of algebraic equations. These equations represent an adequate generalization of the so called Hill-determinant non-variational prescription applicable, under certain assumptions, in the regular cases.
Keywords. Schrödinger equation, strongly singular potentials, binding energies, algebraic equations

1 Introduction

Quantum systems are often described by the differential Schrödinger equation

$$[-\triangle + V(r)]\,\psi = E\,\psi.$$

It is numerically solvable, usually without any serious troubles. A few exactly solvable models $V(r)$ may also prove useful [1].

Sextic oscillator

$$V(r) = a\,r^2 + b\,r^4 + c\,r^6$$

represents one of the simplest "unsolvable" examples which have attracted a lot of attention in the methodical context. In accord with the proposal of Singh et al [2] and its later modifications and proofs [3], it may admit various continued-fraction re-summations of the perturbation series. Moreover, its spectrum of binding energies may coincide with roots of certain (so call Hill) secular determinants with non-variational origin [4].

All the latter results may be generalized to a broad class of potentials which are regular or at most weakly singular in the origin [5]. Here, in the similar methodical setting, we intend to study the challenging, strongly singular counterpart

$$V(r) = a\,r^2 + b\,r^{-4} + c\,r^{-6}, \quad a = 4\,\mu^2, \quad c = 4\,\nu^2 \tag{1}$$

of the sextic oscillator.

The challenge also comes from the needs of physics: The presence of a singularity at $r \to 0$ is a typical feature of phenomenological potentials $V(r)$ in the nuclear and molecular physics. At the same time, it is very unpleasant computationally. Here, we shall describe and recommend its Hill-determinant-like full algebraization.

2 Bound States and Infinite Series

Let us contemplate our differential Schrödinger equation or rather its ordinary-equation radial projection

$$\left[-\frac{\mathrm{d}^2}{\mathrm{d}r^2} + \frac{l(l+1)}{r^2} + V(r)\right] \psi(r) = E\,\psi(r), \quad l = 0, 1, \dots , \tag{2}$$

in a variational setting. Then, we usually choose an orthonormalized basis (*e.g.*, harmonic oscillators) $\chi_n(r)$ and decompose

$$\psi(r) = \sum_n \chi_n(r)\, h_n.$$

This, in accord with the standard textbooks, converts our equations into the matrix eigenvalue problem

$$\det \langle \chi_m | (H - E) | \chi_n \rangle = 0.$$

This – variational – form of algebraization introduces no problems but the solutions remain purely numerical.

Non-variationally, we may try to search for (or classify) the exactly solvable cases via a systematic non-orthogonal and non-normalized, much simpler choice of $\chi_n(r)$'s. With the weaker pair-of-series Ansätze

$$\psi^{(1,2)}(r) = \sum_n \chi_n(r)\, h_n^{(1,2)} \tag{3}$$

only superpositions

$$\psi(r) = c_1\,\psi^{(1)}(r) + c_2\,\psi^{(2)}(r)$$

may provide the solutions acceptable as physical.

Usually, we intend to succeed in an explicit representation of the coefficients $h_n^{(1,2)}$ (and subsequent use of special functions) in such a context. Indeed, the ambiguity in the c_i's has to be removed via boundary conditions

$$\psi(0) = 0, \qquad \psi(\infty) = 0 \tag{4}$$

in such a case and, up to the exaxctly solvable examples, this would amount again to a purely numerical prescription.

The method of Hill determinants [4] lies somewhere in between the variational and special-function extremes. For regular potentials, it starts from non-orthogonal $\chi_n(r)$'s, *e.g.*,

$$\chi_n(r) = r^{2n+\gamma} \exp\left[-f(r)\right] \tag{5}$$

with a quasi-variational parameter γ and with an auxiliary Riccati-like function $f(r)$. Nevertheless, it avoids the use of c_i's, basically, by making all the χ's individually compatible with the physical boundary condition in the origin, $\chi_n(0) = 0$, and in infinity, $\chi_n(\infty) = 0$.

The method of Hill determinants shares a number of features with the solutions obtainable in terms of special functions [5]. In particular, its transparency and simplicity must be "paid for" by the necessary rigorous proofs of its validity. In some cases, this price is reasonable. Here, we intend to offer, describe and recommend one of the possible extensions of this method to the strongly singular forces [6].

3 Regular Potentials as a Methodical Guide

All the forces regular in the origin enable us to choose and classify $\gamma^{(1)} = l + 1$ as physical and $\gamma^{(2)} = -l$ as unphysical. Then, we may choose $\psi^(r) \equiv \psi^{(1)}(r)$ with

$$\psi^{(1)}(r) = \psi^{(regular)}(r) \sim r^{l+1} + corrections$$

and make the first boundary condition redundant ($c_2 = 0$).

In general, our regular solution is numerically obtainable, *e.g.*, by the Runge-Kutta algorithm. Of course, after we use a polynomial exponent $f(r)$ and succeed in constructing an analytic, Taylor series representation of the wavefunctions,

$$\psi^{(regular)}(r) = \sum_{n=0}^{\infty} \chi_n(r)\, h_n^{(regular)},$$

energies may also become roots of the related "Hill determinants" (cf. [4] for more details). Here, analogous simplifications will be studied for Laurent series and oscillators with a strongly repulsive core.

As a guide, we may just review the separate steps of construction in the regular cases: Routinely, we

(1) convert differential Schrödinger equation (for ψ's) into its difference, discrete conterpart (for h_n's);

(2) find an explicit special-function-like formula for coefficients,

$$h_n \equiv h_n(E) = g(n) \ \det Q^{[n]}(E)$$

where, often, n-dimensional matrices $Q^{[n]}(E)$ have a sparse, Hessenberg structure;

(3) treat the trivial identities $h_{-1} = h_{-2} = h_{-3} = \ldots = 0$ as a counterpart of the original $r \to 0$ boundary condition in the origin;

(4) prove, under certain assumptions, that we may also replace the continuous, $r \to \infty$ physical boundary condition by its discrete analogue

$$h_{N+1} = 0, \quad N \to \infty$$

(5) notice that we may re-write the above discrete asymptotic boundary condition in the Hill-determinant language,

$$\det Q^{[\infty]}(E) = 0.$$

As a consequence, we may also re-express Schrödinger equation in a matrix, quasi-variational form

$$Q^{[\infty]}(E)\; h^{(physical)} = 0.$$

It may prove suitable for perturbative purposes [7].

4 Singular Interactions as a Challenge

Our example (1) remains fairly simple in the continuous coordinate r. In particular, its independent solutions may exhibit the contrasted threshold behaviour

$$\psi^{(regular)}(r) \sim \exp[-\nu r^{-2} + O(\ln r)]$$

$$\psi^{(irregular)}(r) \sim r^{(b-6\nu)/8\nu} \exp[\,\nu r^{-2} + O(r^2)]$$

or, alternatively, asymptotic behaviour

$$\psi^{(Jost)}(r) \sim \exp[-\mu r^2 + O(\ln r)]$$

$$\psi^{(non-Jost)}(r) \sim r^{-(E+2\mu)/8\mu} \exp[\,\mu r^2 + O(1/r^2)].$$

Thus, any Laurent power series $\psi^{(1,2)}(r) = \psi^{(\gamma_1,\gamma_2)}(r) = \sum_{-\infty}^{\infty} \dots$ (3) with the Floquet parameter $\gamma = \gamma_{1,2}$ and exponent $f(r) = \mu r^2 + \nu r^{-2}$ (5) may routinely be decomposed near the origin,

$$\psi^{(1)}(r) = \alpha_1\, \psi^{(irregular)}(r) + \beta_1\, \psi^{(regular)}(r)$$

$$\psi^{(2)}(r) = \alpha_2\, \psi^{(irregular)}(r) + \beta_2\, \psi^{(regular)}(r)$$

as well as in infinity,

$$\psi^{(1)}(r) = \gamma_1\, \psi^{(non-Jost)}(r) + \delta_1\, \psi^{(Jost)}(r)$$

$$\psi^{(2)}(r) = \gamma_2\, \psi^{(non-Jost)}(r) + \delta_2\, \psi^{(Jost)}(r).$$

Now, we may re-formulate the analytic physical boundary conditions (4) in the form $c_1\,\alpha_1 + c_2\,\alpha_2 = 0$ and $c_1\,\gamma_1 + c_2\,\gamma_2 = 0$, *i.e.*,

$$\det \begin{pmatrix} \psi^{(1)}(0) & \psi^{(2)}(0) \\ \psi^{(1)}(\infty) & \psi^{(2)}(\infty) \end{pmatrix} = 0. \tag{6}$$

Numerical tests confirm reasonable precision of results (energies) obtained from this condition [6].

4.1 The Discrete Boundary Conditions

Unexpectedly, apparent analogues

$$h_{\infty}^{(\gamma_1,\,\gamma_2)} = 0, \qquad h_{-\infty}^{(\gamma_1,\,\gamma_2)} = 0$$

of eq. (4) do not reflect any physics at all: These conditions are trivial, responsible just for the convergence of our power-series Ansatz. An explanation may be found in close analogy with Mathieu functions. Thus, "secular equation"

$$\det Q^{[-\infty,\,\infty]}(\gamma,\,E) = 0$$

with the doubly infinite matrix

$$Q^{[-M,\,N]}(\gamma,\,E) \to Q^{[-\infty,\,\infty]}(\gamma,\,E), \qquad M, N \to \infty$$

determines just the pair of Floquet parameters $\gamma = \gamma_1(E),\, \gamma_2(E)$.

The analogy with the regular case seems to fail completely: The pair of the doubly infinite matrix equations

$$Q^{[-\infty,\,\infty]}(\gamma_j,\,E) \times \begin{pmatrix} \dots \\ h_n^{(j)} \\ \dots \end{pmatrix} = 0, \qquad j = 1,\,2$$

leaves the problem of boundary conditions unresolved.

4.2 Main Result: The Determination of Energies

The suitable $n \geq 0$ and $n < 0$ changes of variables

$$h_n = \frac{(2\mu)^n \Gamma(n+A)}{\Gamma(n+B^{(+)})\Gamma(n+B^{(-)})}\, p_n$$

$$A = \gamma/2 - E/8\mu + 1/4, \quad B^{(\pm)} = \gamma/2 \pm D + 3/4, \quad 4d^2 = (l+1/2)^2 + 8\mu\nu$$

and

$$h_{-m} = \frac{(2\nu)^m \Gamma(m+S)}{\Gamma(m+T^{(+)})\Gamma(m+T^{(-)})}\, q_m$$

$$S = -\gamma/2 + b/8\nu + 3/4, \qquad T^{(\pm)} = -\gamma/2 \pm D + 5/4,$$

respectively, convert our difference Schrödinger equation or recurrences

$$Q_{n,\,n-1}\, h_{n-1} + Q_{n,\,n}\, h_n + Q_{n,\,n+1}\, h_{n+1} = 0$$

$$Q_{n,\,n-1} = 2\mu(4n + 2\gamma - 3) - E, \qquad Q_{n,\,n+1} = b - 2\nu(4n + 2\gamma + 1)$$

$$Q_{n,\,n} = 8\mu\nu + l(l+1) - (2n+\gamma)(2n+\gamma-1)$$

into equations exactly solvable in the $|n| \gg 1$ asymptotic region. Thus, we get

$$p_n = p_\infty \exp[4\mu\nu/n + O(1/n^2)], \quad n \gg 1$$

or

$$q_m = q_\infty \exp[4\mu\nu/m + O(1/m^2)], \quad m \gg 1$$

and our main

Theorem. The fully algebraized secular equation reads

$$\det \begin{pmatrix} (2\mu)^{-\gamma_1/2} p_\infty(\gamma_1) & (2\mu)^{-\gamma_2/2} p_\infty(\gamma_2) \\ (2\nu)^{\gamma_1/2} q_\infty(\gamma_1) & (2\nu)^{\gamma_2/2} q_\infty(\gamma_2) \end{pmatrix} = 0$$

and fixes the energies.

Proof. We shall show that

$$(2\mu r^2)^{1/4+E/8\mu+\gamma/2} \sum h_n(\gamma) r^{2n} = p_\infty(\gamma) \exp[2\mu r^2 + O(1/r^2)], \quad r \gg 1$$

$$(r^2/2\nu)^{3/4-b/8\nu+\gamma/2} \sum h_n(\gamma) r^{2n} = q_\infty(\gamma) \exp[2\nu r^{-2} + O(r^2)], \quad r \ll 1$$

For brevity, we take just $r \gg 1$ – the $r \ll 1$ case is fully analogous.

In the first step, we insert our power-series wavefunctions in the original boundary conditions and omit an exponentially small error, $\sum_{-\infty} \to \sum_{n_0}$, $n_0 \gg 1$. Then, due to the positivity of all its terms, we may replace the new, simply infinite series by an integral,

$$\sum_{n=n_0}^{\infty} h_n(\gamma) r^{2n+\gamma} \approx \int_{n_0}^{\infty} \exp g(n)\, \mathrm{d}\, n.$$

In the second step, a technical core of the proof is found in a second-order saddle-point method. Indeed, as long as a move of n_0 to $-\infty$ only introduces a small error, we approximate

$$g(n) \approx g(N) + (n-N)\, g'(N) + \tfrac{1}{2}(n-N)^2\, g''(N)$$

and evaluate the integral.

In detail, the latter step uses the fact that the integrand

$$g(N) = (2N+\gamma)\ln r - N \ln N + N(\ln 2\mu + 1) -$$

$$-(\tfrac{1}{2}\gamma + \frac{3}{4} + \frac{E}{8\mu}) \ln N + \ln p_\infty(\gamma) - \ln 2\pi + O(1/N)$$

is peaked around such N that $g'(N) = 0$. Hence, we implicitly define $N = N(r)$ and/or $r = r(N)$,

$$N(r) \exp[(\tfrac{1}{2}\gamma + \frac{3}{4} + \frac{E}{8\mu})/N(r)] = 2\mu r^2 + O(1/r^2)$$

which – after insertions – gives the final result. **QED.**

We may summarize that in the strongly singular cases $(r^2 V(r) \gg 1$ for $r \ll 1)$, Hill determinants are related to mere convergence of the Laurent series and define just the Floquet exponents $\gamma = \gamma(E)$. Our full algebraization changes the representation, $\psi^{(1,2)}(r) \to h_n^{(1,2)}$, transforms continuous coordinates $r \in (0, \infty)$ into their discrete analogue $n \in \mathbb{Z}$, and generates the discretized Schrödinger equation for $h_n^{(1,2)}$'s which is asymptotically solvable. This leads to the transparent, intuitively acceptable and rigorously understood boundary conditions which degenerate back to the old HD method for regular potentials.

References

[1] R. G. Newton, Scattering Theory of Waves and Particles (Springer, New York, 1982)

[2] V. Singh, S. N. Biswas and K. Datta, Phys. Rev. D 18 (1978) 1901.

[3] M. Znojil, Phys. Rev. D 26 (1982) 3750 and D 34 (1986) 1224 and Phys. Lett. A 169 (1992) 415.

[4] A. Hautot, Phys. Lett. A 38 (1972) 305 and Phys. Rev. D 33 (1986) 437.

[5] M. Znojil, Lett. Math. Phys. 5 (1981) 405 and J. Math. Phys. 29 (1988) 1433.

[6] M. Znojil, J. Math. Phys. 30 (1989) 23 and 31 (1990) 108

[7] Phys. Lett. 150 (1990) 67.

Operator Theory:
Advances and Applications, Vol. 70

Asymptotic Behavior of the Resolvent of the Dirac Operator

Chris Pladdy, Yoshimi Saitō and Tomio Umeda

1 Introduction

We consider the Dirac operator

$$H = -i \sum_{j=1}^{3} \alpha_j \frac{\partial}{\partial x_j} + \beta + Q(x), \tag{1.1}$$

which appears in the relativistic quantum mechanics. For the detailed definition of the Dirac operator (1.1) see §2. It is well-known that the liming absorption principle holds for the Dirac operator (1.1) and, as a result, that the extended resolvents $R_{\pm}(\lambda)$ exist for any real value λ with $|\lambda| > 1$. The limiting absorption principle has a close connection with the spectral and scattering theory for the Dirac operator.

Our aim here is to investigate the asymptotic behavior of $R_{\pm}(\lambda)$ as $|\lambda| \to \infty$. Our results indicate that the extended resolvents of the Dirac operator decay much more slowly, in a certain sense, than those of the Schrödinger operator. (Compare Theorems 2.3 and 2.4 with Theorem 3.1.)

We are introducing the notation which will be used in the note. For $x \in \mathbf{R}^3$, $|x|$ denotes the Euclidean norm of x and

$$\langle x \rangle = \sqrt{1 + |x|^2}. \tag{1.2}$$

For $s \in \mathbf{R}$, we define the weighted Hilbert spaces $L_{2,s}(\mathbf{R}^3)$ and $H^1_s(\mathbf{R}^3)$ by

$$L_{2,s}(\mathbf{R}^3) = \{f \ / \ \langle x \rangle^s f \in L_2(\mathbf{R}^3)\}, \tag{1.3}$$

and

$$H^1_s(\mathbf{R}^3) = \{f \ / \ \langle x \rangle^s \partial_x^\alpha f \in L_2(\mathbf{R}^3), |\alpha| \le 1\}, \tag{1.4}$$

where $\alpha = (\alpha_1, \alpha_2, \alpha_3)$ is a multi-index, $|\alpha| = \alpha_1 + \alpha_2 + \alpha_3$, and

$$\partial_x^\alpha = \frac{\partial^{|\alpha|}}{\partial x_1^{\alpha_1} \partial x_2^{\alpha_2} \partial x_3^{\alpha_3}}. \tag{1.5}$$

The inner products and norms in $L_{2,s}(\mathbf{R}^3)$ and $H^1_s(\mathbf{R}^3)$ are given by

$$\begin{cases} (f,g)_s = \displaystyle\int_{\mathbf{R}^3} \langle x\rangle^{2s} f(x)\overline{g(x)}\,dx, \\ \|f\|_s = \left[(f,f)_s\right]^{1/2}, \end{cases} \tag{1.6}$$

and

$$\begin{cases} (f,g)_{1,s} = \displaystyle\int_{\mathbf{R}^3} \langle x\rangle^{2s} \left[\nabla f(x)\cdot\overline{\nabla g(x)} + f(x)\overline{g(x)}\right]dx, \\ \|f\|_{1,s} = \left[(f,f)_{1,s}\right]^{1/2}, \end{cases} \tag{1.7}$$

respectively. The spaces $\mathcal{L}_{2,s}$ and $\mathcal{H}^1_s$ are defined by

$$\begin{cases} \mathcal{L}_{2,s} = \left[L_{2,s}(\mathbf{R}^3)\right]^4, \\ \mathcal{H}^1_s = \left[H^1_s(\mathbf{R}^3)\right]^4, \end{cases} \tag{1.8}$$

i.e., $\mathcal{L}_{2,s}$ and $\mathcal{H}^1_s$ are direct sums of the Hilbert spaces $L_{2,s}(\mathbf{R}^3)$ and $H^1_s(\mathbf{R}^3)$, respectively. The inner products and norms in $\mathcal{L}_{2,s}$ and $\mathcal{H}^1_s$ are also denoted by $(\ ,\ )_s$, $\|\ \|_s$ and $(\ ,\ )_{1,s}$, $\|\ \|_{1,s}$, respectively. When $s=0$, we simply write

$$\begin{cases} \mathcal{L}_2 = \mathcal{L}_{2,0}, \\ \mathcal{H}^1 = \mathcal{H}^1_0. \end{cases} \tag{1.9}$$

For a pair of Hilbert spaces X and Y, $\mathbf{B}(X,Y)$ denotes the Banach space of all bounded linear operators from X to Y, equipped with the operator norm

$$\|T\| = \sup_{x\in X\setminus\{0\}} \|Tx\|_Y/\|x\|_X, \tag{1.10}$$

where $\|\ \|_X$ and $\|\ \|_Y$ are the norms in X and Y.

This is an expository note. The detailed proofs and discussions will be found in a paper to be published elsewhere.

2 Main results

We first consider the free Dirac operator

$$H_0 = -i\sum_{j=1}^{3} \alpha_j \frac{\partial}{\partial x_j} + \beta, \tag{2.1}$$

where $i=\sqrt{-1}$ and $x=(x_1,x_2,x_3)\in\mathbf{R}^3$. Here α_j,β are 4×4 Hermitian matrices satisfying the anticommutation relations

$$\alpha_j\alpha_k + \alpha_k\alpha_j = 2\delta_{jk}I, \qquad (j,k=1,2,3,4) \tag{2.2}$$

with the convention $\alpha_4 = \beta$, δ_{jk} being Kronecker's delta and I being the 4×4 identity matrix. It is known that H_0 restricted on $[C_0^\infty(\mathbf{R}^3)]^4$ is essentially selfadjoint in $\mathcal{L}_2$ and its selfadjoint extension, which will be denoted by H_0 again, has the domain $\mathcal{H}^1$.

We make the following assumption on the potential.

Assumption 2.1.
(i) $Q(x) = (q_{jk}(x))$ is a 4×4 Hermitian matrix-valued C^1 function on $\mathbf{R}^3$;
(ii) There exist positive constants ϵ and K such that

$$\langle x \rangle^{1+\epsilon} |q_{jk}(x)| + \sum_{\ell=1}^{3} \left|\frac{\partial q_{jk}}{\partial x_\ell}\right| \le K \tag{2.3}$$

for $j, k = 1, 2, 3, 4$.

Under Assumption 2.1 the multiplication operator $Q = Q(x)\times$ is a bounded selfadjoint operator in $\mathcal{L}_2$. Hence, by the Kato-Rellich theorem (Kato[3], p.287), H restricted on $[C_0^\infty(\mathbf{R}^3)]^4$ is also essentially selfadjoint in $\mathcal{L}_2$ and its selfadjoint extension, which will be denoted by H again, has the same domain $\mathcal{H}^1$ as H_0. We write

$$R_0(z) = (H_0 - z)^{-1}, \tag{2.4}$$

and

$$R(z) = (H - z)^{-1}. \tag{2.5}$$

As we mentioned in the Introduction, the limiting absorption principle holds for the Dirac operator H.

Theorem 2.2 *(Yamada [7]).*
Suppose that Assumption 2.1 is satisfied and let $s > 1/2$.
Then for $\lambda \in (-\infty, -1) \cup (1, \infty)$, there exist the extended resolvents $R_\pm(\lambda) \in \mathbf{B}(\mathcal{L}_{2,s}, \mathcal{H}^1_{-s})$ such that

$$\text{s-}\lim_{\eta \downarrow 0} R(\lambda \pm i\eta) = R_\pm(\lambda) \qquad \text{in } \mathcal{H}^1_{-s}. \tag{2.6}$$

Moreover, for $f \in \mathcal{L}_{2,s}$, $R_\pm(\lambda)f$ is an $\mathcal{L}_{2,-s}$-valued, continuous function on the set $(-\infty, -1) \cup (1, \infty)$. We now state the main theorems, which are concerned with the asymptotic behavior of the extended resolvents $R_{0\pm}(\lambda)$ of the free Dirac operator.

Theorem 2.3. *Let $s > 1/2$. Then*

$$\|R_{0\pm}(\lambda)\| = O(1) \qquad (|\lambda| \to \infty), \tag{2.7}$$

where $\|R_{0\pm}(\lambda)\|$ denotes the operator norm of $R_{0\pm}(\lambda)$ in $\mathbf{B}(\mathcal{L}_{2,s}, \mathcal{L}_{2,-s})$.

As we shall see in Remark 2.5, $\|R_{0\pm}(\lambda)\|$ cannot be small no matter how $|\lambda|$ is large. In this sense the estimate in Theorem 2.3 is best possible. However, $R_{0\pm}(\lambda)$ can become small as $|\lambda| \to \infty$ in a weaker sense than in Theorem 2.3.

Theorem 2.4. *Let* $s > 1/2$.
Then $R_{0\pm}(\lambda)$ *converge strongly to* 0 *as* $|\lambda| \to \infty$, *i.e.,*

$$\lim_{|\lambda|\to\infty} R_{0\pm}(\lambda)f = 0 \qquad \text{in } \mathcal{L}_{2,-s} \tag{2.8}$$

for any $f \in \mathcal{L}_{2,s}$.

Remark 2.5. Yamada [8] proved the following:
Let $s > 1/2$. *Then there exists a sequence* $\{f_n\}$ *in* $\mathcal{L}_{2,s}$ *such that*

$$\sup_n \|f_n\|_s < \infty, \tag{2.9}$$

and

$$\lim_{n\to\infty} (R_{0+}(n)f_n, f_n)_0 \neq 0. \tag{2.10}$$

Since

$$|(R_{0+}(n)f_n, f_n)_0| \leq \|R_{0+}(n)\| \, \|f_n\|_s{}^2, \tag{2.11}$$

Yamada's example implies that $\|R_{0+}(\lambda)\|$ cannot converge to 0 as $\lambda \to \infty$.

Based on Theorems 2.3 and 2.4, the Dirac operator with a small coupling constant can be handled; one can use the Neumann series expansion. Let

$$H_\tau = -i\sum_{j=1}^{3} \alpha_j \frac{\partial}{\partial x_j} + \beta + \tau Q(x), \tag{2.12}$$

where τ is a real number. The extended resolvents of H_τ will be denoted by $R_{\tau\pm}(\lambda)$. Then we have the following

Theorem 2.6.
Suppose that $Q(x)$ *satisfies Assumption 2.1 and* $1/2 < s < (1+\epsilon)/2$. *Let* $R_{\tau\pm}(\lambda)$ *be the extended resolvents of* H_τ. *Then for sufficiently small* τ

(i) The operator norm of $R_{\tau\pm}(\lambda)$ *in* $\mathbf{B}(\mathcal{L}_{2,s}, \mathcal{L}_{2,-s})$ *is bounded as* $|\lambda| \to \infty$;

(ii) $R_{\tau\pm}(\lambda)$ *converge strongly to* 0 *as* $|\lambda| \to \infty$.

3 A Known result for Schrödinger operator

The limiting absorption principle for Schrödinger operators has been extensively studied. We will use a result due to Saitō [5, 6]. For this reason, we make a review. Let T denote the selfadjoint operator which is defined to be the closure of $-\Delta + V(x)$ restricted on $C_0^\infty(\mathbf{R}^n)$, where $V(x)$ is a real-valued function satisfying

$$|V(x)| \leq C\langle x\rangle^{-1-\epsilon} \tag{3.1}$$

for $C > 0$ and $\epsilon > 0$. Let $\tilde{R}(z) = (T - z)^{-1}$. Then it is well-known that the limiting absorption principle holds for T, that is, for any $\lambda > 0$, there correspond the extended resolvents $\tilde{R}_\pm(\lambda)$ in $\mathbf{B}(L_{2,s}(\mathbf{R}^n), L_{2,-s}(\mathbf{R}^n))$ such that

$$\text{s-}\lim_{\eta\downarrow 0} \tilde{R}(\lambda \pm i\eta)f = \tilde{R}_\pm(\lambda)f \qquad \text{in } L_{2,-s} \tag{3.2}$$

for any f in $L_{2,s}(\mathbf{R}^n)$. Furthermore, it is known that $\tilde{R}_\pm(\lambda)f$ are $L_{2,-s}(\mathbf{R}^n)$-valued continuous functions in $(0, \infty)$. (Saitō [5], Ikebe-Saitō [2] and Agmon [1].) As for asymptotic behaviors of $\tilde{R}_\pm(\lambda)$, we have

Theorem 3.1 *(Saitō [5, 6]).*
Let $\|\tilde{R}_\pm(\lambda)\|$ be the operator norm of $\tilde{R}_\pm(\lambda)$ in $\mathbf{B}(L_{2,s}(\mathbf{R}^n), L_{2,-s}(\mathbf{R}^n))$. Then

$$\|\tilde{R}_\pm(\lambda)\| = O(\lambda^{-1/2}) \qquad (\lambda \to \infty). \tag{3.3}$$

More precisely, Saitō proved

Theorem 3.2 *(Saitō [5, 6]).*
Let $s > 1/2$. Then for any $a > 0$ there exists a positive constant $C > 0$ such that

$$\|\tilde{R}(\kappa^2)\| \leq C/|\kappa| \tag{3.4}$$

for all κ with $|\mathrm{Re}\,\kappa| > a$ and $\mathrm{Im}\,\kappa > 0$, where $\|\tilde{R}(\kappa^2)\|$ is the operator norm of $\tilde{R}(\kappa^2)$ in $\mathbf{B}(L_{2,s}(\mathbf{R}^n), L_{2,-s}(\mathbf{R}^n))$.

4 Pseudo-differential operators

The proof of Theorem 2.3 is based on the resolvent estimate for the Schr"odinger operator (Theorem 3.2) as well as the theory of pseudo-differential operators. So we need to introduce a class of symbols of pseudo-differential operators which are suitable to our purpose.

Definition 4.1.
A C^∞ function $p(x,\xi)$ on $\mathbf{R}^3 \times \mathbf{R}^3$ is said to be in the class $S_{0,0}^0$ if for any pair α and β of multi-indecies there exists a constant $C_{\alpha\beta} \geq 0$ such that

$$\left|\left(\frac{\partial}{\partial\xi}\right)^\alpha \left(\frac{\partial}{\partial x}\right)^\beta p(x,\xi)\right| \leq C_{\alpha\beta} \tag{4.1}$$

for all $x, \xi \in \mathbf{R}^3$.

Remark 4.2.
The class $S^0_{0,0}$ is a Fréchet space equipped with the semi-norms

$$|p|^{(0)}_\ell = \max_{|\alpha|,|\beta|\le\ell} \sup_{x,\xi} \left\{ \left| \left(\frac{\partial}{\partial\xi}\right)^\alpha \left(\frac{\partial}{\partial x}\right)^\beta p(x,\xi) \right| \right\} \qquad (\ell = 0,1,2,\cdots). \tag{4.2}$$

A pseudo-differential operator $p(x, D)$ with symbol $p(x,\xi)$ is defined by

$$p(x,D)f(x) = (2\pi)^{-3} \int_{\mathbf{R}^3} e^{ix\cdot\xi}\, p(x,\xi)\hat{f}(\xi)\, d\xi \tag{4.3}$$

for $f \in \mathcal{S}(\mathbf{R}^3)$, the space of all rapidly decreasing functions on $\mathbf{R}^3$. Here $\hat{f}(\xi)$ denotes the Fourier transform of f. In connection with the limiting absorption principle, pseudo-differential operators which are bounded in $L_{2,s}$ are important.

Lemma 4.3.
Let $p(x,\xi) \in S^0_{0,0}$. Then for any $s > 0$ there exist a positive constant $C\,(= C_s)$ and a positive integer $\ell\,(= \ell_s)$ such that

$$\|p(x,D)f\|_s \le C|p|^{(0)}_\ell \|f\|_s \qquad (f \in \mathcal{S}(\mathbf{R}^3)). \tag{4.4}$$

We shall omit the proof of Lemma 4.3, which is based on the Calderón-Vaillancourt theorem and some techniques in the theory of pseudo-differential operators. Now we need to extend Lemma 4.3 to a system of pseudo-differential operators. Let

$$P(x,\xi) = \big(p_{jk}(x,\xi)\big)_{1\le j,k\le 4} \tag{4.5}$$

be a 4×4 matrix-valued symbol. Then we define

$$P(x,D) = \big(p_{jk}(x,D)\big)_{1\le j,k\le 4} \tag{4.6}$$

by

$$P(x,D)f(x) = (2\pi)^{-3} \int_{\mathbf{R}^3} e^{ix\cdot\xi}\, P(x,\xi)\hat{f}(\xi)\, d\xi \tag{4.7}$$

for $f \in [\mathcal{S}(\mathbf{R}^3)]^4$. Here an explanation must be needed. For a $\mathbf{C}^4$-valued function $f(x) = (f_1(x), f_2(x), f_3(x), f_4(x))$ on $\mathbf{R}^3$, the Fourier transform $\hat{f}(\xi) = \mathcal{F}f(\xi)$ is defined by $\hat{f}(\xi) = (\hat{f}_1(\xi), \hat{f}_2(\xi), \hat{f}_3(\xi), \hat{f}_4(\xi))$. If $p_{jk}(x,\xi) \in S^0_{0,0}$, $1 \le j,k \le 4$, we define

$$|P|^{(0)}_\ell = \Big\{ \sum_{j,k=1}^{4} (|p_{jk}|^{(0)}_\ell)^2 \Big\}^{1/2} \tag{4.8}$$

for $\ell = 0,1,2,\ldots$, where $|p_{jk}|^{(0)}_\ell$ are the semi-norms introduced in Remark 4.2. We then have a natural extension of Lemma 4.3.

Lemma 4.4.
Let $p_{jk}(x,\xi) \in S^0_{0,0}$ for j, $k = 1,2,3,4$. Then for any $s > 0$ there exist a positive constant $C\,(= C_s)$ and a positive integer $\ell\,(= \ell_s)$ such that

$$\|P(x,D)f\|_s \leq C\,|P|^{(0)}_\ell\, \|f\|_s \tag{4.9}$$

for $f \in [\mathcal{S}(\mathbf{R}^3)]^4$.

5 Outline of the proof of Theorem 2.3

In view of Theorem 2.2, we see that Theorem 5.1 below implies Theorem 2.3. Note that $\|R_\pm(\lambda)\|$ are bounded on any compact interval in $(-\infty,-1)\cup(1,\infty)$. Before giving the outline of the proof of Theorem 5.1, we make a few remarks on the free Dirac operator H_0.

For $f \in [\mathcal{S}(\mathbf{R}^3)]^4$,

$$H_0 f = \mathcal{F}^{-1}\hat{L}_0(\xi)\mathcal{F}f, \tag{5.1}$$

where

$$\hat{L}_0(\xi) = \sum_{j=1}^{3} \alpha_j\xi_j \,+\, \beta. \tag{5.2}$$

It is easy to see that

$$\left(\hat{L}_0(\xi)\right)^2 \,=\, \langle\xi\rangle^2\, I. \tag{5.3}$$

Using (5.3), we get

$$R_0(z)f \,=\, \mathcal{F}^{-1}\Big[\frac{\hat{L}_0(\xi)+z}{\langle\xi\rangle^2 - z^2}\Big]\mathcal{F}f \tag{5.4}$$

for $f \in [\mathcal{S}(\mathbf{R}^3)]^4$ and $z \in \mathbf{C}\backslash\mathbf{R}$.

Theorem 5.1.
Suppose that $s > 1/2$. Then

$$\sup\,\big\{\,\|R_0(\lambda \pm i\eta)\| \,/\, 2 \leq |\lambda|,\, 0<\eta<1\big\} \,<\infty, \tag{5.5}$$

where $\|R_0(\lambda \pm i\eta)\|$ denotes the operator norm of $R_0(\lambda \pm i\eta)$ in $\mathbf{B}(\mathcal{L}_{2,s},\mathcal{L}_{2,-s})$.

Outline of the proof.
Set

$$J = \big\{\,z \in \mathbf{C}\,/\,2 \leq |\mathrm{Re}\,z|,\, 0<|\mathrm{Im}\,z|<1\,\big\}. \tag{5.6}$$

Choose $\rho \in C_0^\infty(\mathbf{R})$ so that

$$\rho(t) = \begin{cases} 1, & \text{if } |t| < 1/2\,; \\ 0, & \text{if } |t| > 1 \end{cases} \tag{5.7}$$

For each $z \in J$, we define a cutoff function $\gamma_z(\xi)$ on $\mathbf{R}^3$ by

$$\gamma_z(\xi) = \begin{cases} \rho(\langle\xi\rangle - \operatorname{Re} z), & \text{if } \operatorname{Re} z \geq 2; \\ \rho(\langle\xi\rangle + \operatorname{Re} z), & \text{if } \operatorname{Re} z \leq -2. \end{cases} \tag{5.8}$$

Using (5.4) and $\gamma_z(\xi)$, we decompose the resolvent of H_0 into three parts:

$$R_0(z) = (-\Delta + 1 - z^2)^{-1} A_z + B_z + z(-\Delta + 1 - z^2)^{-1} \tag{5.9}$$

where

$$\begin{aligned} A_z &= \mathcal{F}^{-1}\Big[\gamma_z(\xi)\hat{L}_0(\xi)\Big]\mathcal{F}, \\ B_z &= \mathcal{F}^{-1}\Big[\frac{1-\gamma_z(\xi)}{\langle\xi\rangle^2 - z^2}\hat{L}_0(\xi)\Big]\mathcal{F}. \end{aligned} \tag{5.10}$$

Applying Lemma 4.4 to A_z, we get

$$\|A_z f\|_s \leq C_1|z|\|f\|_s, \tag{5.11}$$

where C_1 is independent of $z \in J$. Combining (5.11) with Theorem 3.2, we see that

$$\|(-\Delta + 1 - z^2)^{-1} A_z f\|_{-s} \leq C_2\|f\|_s, \tag{5.12}$$

where C_2 is independent of $z \in J$. It is easy to see that there exists a constant $C_3 > 0$ such that

$$\Big|\frac{1-\gamma_z(\xi)}{\langle\xi\rangle^2 - z^2}\hat{L}_0(\xi)\Big| \leq C_3 \qquad (\xi \in \mathbf{R}^3) \tag{5.13}$$

for all $z \in J$. Using (5.13), we have

$$\|B_z f\|_{-s} \leq C_3\|f\|_s \tag{5.14}$$

for all $z \in J$. It follows from Theorem 3.2 that

$$\|z(-\Delta + 1 - z^2)^{-1} f\|_{-s} \leq C_4\|f\|_s, \tag{5.15}$$

where C_4 is independent of $z \in J$. Combining (5.15), (5.14) and (5.12), we get the desired conclusion.

6 Outline of the proof of Theorem 2.4

We use the following two lemmas.

Lemma 6.1. *Define*

$$\mathcal{X}_0 = \{\, f \in [\mathcal{S}(\mathbf{R}^3)]^4 \,/\, \mathcal{F}f \in [C_0^\infty(\mathbf{R}^3)]^4 \,\}. \tag{6.1}$$

Then $\mathcal{X}_0$ is dense in $\mathcal{L}_{2,s}$ for any $s \in \mathbf{R}$.

Lemma 6.2. *For $z \in \mathbf{C}$, put*

$$R(\xi; z) = \frac{\hat{L}_0(\xi) + z}{\langle\xi\rangle^2 - z^2}. \tag{6.2}$$

Then for any $K > 1$ and any multi-index α there exists a constant $C_{K\alpha} > 0$ such that

$$\left|\left(\frac{\partial}{\partial\xi}\right)^\alpha R(\xi; z)\right| \leq C_{K\alpha}/|z| \tag{6.3}$$

for all ξ and z satisfying $\langle\xi\rangle \leq K$ and $|z| \geq 2K$.

Outline of the proof of Theorem 2.4.
In view of Theorem 2.2, it is sufficient to show that

$$\lim_{|\lambda|\to\infty} R_{0\pm}(\lambda)f = 0 \qquad \text{in } \mathcal{L}_{2,-s} \tag{6.4}$$

for any $f \in \mathcal{X}_0$. Let $f \in \mathcal{X}_0$. Choose $K > 1$ so that

$$\text{supp}[\mathcal{F}f] \subset \{\xi \in \mathbf{R}^3 \mid \langle\xi\rangle \leq K\}. \tag{6.5}$$

Then we see that for $\lambda \in \mathbf{R}$ with $|\lambda| \geq 2K$

$$R_{0\pm}(\lambda)f = \mathcal{F}^{-1}[R(\xi; \lambda)]\mathcal{F}f, \tag{6.6}$$

and that $R_{0\pm}(\lambda)f \in [\mathcal{S}(\mathbf{R}^3)]^4$. Moreover, for any non-negative integer ℓ, there exists a constant $C_\ell > 0$ such that

$$|R_{0\pm}(\lambda)f|_{\ell,\mathcal{S}} \leq C_\ell/|\lambda| \tag{6.7}$$

for any $\lambda \in \mathbf{R}$ with $|\lambda| \geq 2K$, where $|\cdot|_{\ell,\mathcal{S}}$ are the semi-norms in $[\mathcal{S}(\mathbf{R}^3)]^4$ induced naturally from $\mathcal{S}(\mathbf{R}^3)$. Then (6.7) implies (6.4). This completes the proof.

References

[1] S. Agmon, Spectral properties of Schrödinger operators and scattering theory, Ann. Scoula Norm. Sup. Pisa (**4**)2 (1975), 151-218.

[2] T. Ikebe and Y. Saitō, Limiting absorption method and absolute continuity for the Schrödinger operators, J. Math. Kyoto Univ. **7** (1972), 513-542.

[3] T. Kato, *Perturbation Theory for Linear Operators, 2nd ed.*, Springer-Verlag, 1976.

[4] Y. Saitō, The principle of limiting absorption for second-order differential operators with operator-valued coefficients, Publ. Res. Inst. Math. Sci. Kyoto Univ. **7** (1972), 581-619.

[5] Y. Saitō, The principle of limiting absorption for the non-selfadjoint Schrödinger operator in $\mathbf{R}^N$ $(N \neq 2)$, Publ. Res. Inst. Math. Sci. Kyoto Univ. **9** (1974), 397-428.

[6] Y. Saitō, The principle of limiting absorption for the non-selfadjoint Schrödinger operator in $\mathbf{R}^2$, Osaka J. Math. **11** (1974), 295-306.

[7] O. Yamada, On the principle of limiting absorption for the Dirac operators, Publ. Res. Inst. Math. Sci. Kyoto Univ. **8** (1972/73), 557-577.

[8] O. Yamada, private communication, 1992.

Chris Pladdy, Yoshimi Saitō, Department of Mathematics, University of Alabama at Birmingham, Birmingham, Alabama 35294, USA
Tomio Umeda, Department of Mathematics, Himeji Institute of Technology, Himeji 671–22, Japan

Operator Theory:
Advances and Applications, Vol. 70

Discrete spectrum of the perturbed Dirac operator

M. Sh. Birman and A. Laptev

1. Let $\gamma = (\gamma_1, \gamma_2, \gamma_3)$ and γ_0 be (4×4) Dirac matrices; let ∞ be the unit matrix. The Dirac matrices satisfy the equations

$$\gamma_j \gamma_k + \gamma_k \gamma_j = 2\delta_{jk}\infty, \quad j, k = 0, 1, 2, 3.$$

Let us consider the "free" Dirac operator in $L_2(\mathbb{R}^3; \mathbb{C}^4)$

$$\mathcal{D}_0 = \gamma \cdot \boldsymbol{\mathcal{D}} + \gamma_0, \quad \boldsymbol{\mathcal{D}} = -i\nabla,$$

and its pertubation by an operator of multiplication by the electric potential $V(x)$

$$\mathcal{D}(\alpha) = \mathcal{D}_0 - \alpha V \infty, \quad \alpha > 0,$$

$$V \in L_3(\mathbb{R}^3), \quad V(x) \geq 0. \tag{1}$$

The spectrum of the operator $\mathcal{D}_0$ is continuous and covers the complement of the interval (gap) $(-1, 1)$. The continuous spectrum of the operator $\mathcal{D}(\alpha)$ coincides with the continuous spectrum of $\mathcal{D}_0$. Besides, the operator $\mathcal{D}(\alpha)$ has discrete spectrum in the gap. The eigenvalues of $\mathcal{D}(\alpha)$ are monotonically moving to the left, when α is increasing.

For a fixed λ, $|\lambda| \leq 1$, we denote by $N(\alpha, \lambda)$ the number of eigenvalues of the operator $\mathcal{D}(t)$ passing the point λ when the coupling constant t is increasing from 0 to the value $t = \alpha$. We study the asymptotic behaviour of $N(\alpha, \lambda)$ when $\alpha \to \infty$. In the case $\lambda = \pm 1$ we need some additional assumptions on V. In particular, these assumptions guarantee $N(\alpha, 1)$ to be finite for all $\alpha > 0$.

The starting point of our paper is the following result of [3].

Theorem 1 ([3]) *Let us assume that, as well as (1), we have*

$$V \in L_{3/2}(\mathbb{R}^3). \tag{2}$$

Then the following asymptotic formula holds[1]

$$\lim_{\alpha \to \infty} \alpha^{-3} N(\alpha, \pm 1) = \frac{1}{3\pi^2} \int V^3 \, dx. \tag{3}$$

---

[1]The coefficient in [3] is incorrect and this mistake is repeated in [5].

In this paper we claim the following statements in addition to the result of Klaus.

a) The asymptotic formula (3) holds for $N(\alpha,\lambda)$, $|\lambda|<1$, whenever we only have condition (1).

b) The asymptotic formula (3) survives under some weaker (compare with (2)) additional restrictions on V.

c) There are potentials satisfying (1), such that $N(\alpha,1)$ has an asymptotics of the order α^q, $q>3$.

d) There are potentials, satisfying (1), such that $N(\alpha,1)\sim c\alpha^3$, but the coefficient $c>J$, where

$$J:=\frac{1}{3\pi^2}\int V^3\,dx.$$

2. Let us clarify the previous statements. We begin with the necessary notations. Before stating the precise results we notice the following: let us consider the (compact) operator

$$X(\lambda)=W(\mathcal{D}_0-\lambda I)^{-1}W,\quad W=V^{1/2},\quad |\lambda|\le 1. \tag{4}$$

Let $n_+(.,X(\lambda))$ be the counting function of the positive spectrum of the operator (4). We use the standard relation

$$N(\alpha,\lambda)=n_+(s,X(\lambda)),\quad \alpha s=1,\quad |\lambda|\le 1.$$

The operator (4) decomposes (see [K]) into the sum

$$X(\lambda)=Y(\lambda)+Z(\lambda),\quad |\lambda|\le 1,$$

$$Y(\lambda)=W(\gamma\cdot\boldsymbol{\mathcal{D}}(-\Delta+\sigma I)^{-1})W,\quad \sigma=1-\lambda^2, \tag{5}$$

$$Z(\lambda)=W(\gamma_0+\lambda\infty)(-\Delta+\sigma I)^{-1}W,\quad \sigma=1-\lambda^2. \tag{6}$$

In particular, for $\lambda=\pm1$

$$Y(\pm1)=W(\gamma\cdot\boldsymbol{\mathcal{D}}|\boldsymbol{\mathcal{D}}|^{-2})W, \tag{7}$$

$$Z(\pm1)=W(\gamma_0\pm\infty)|\boldsymbol{\mathcal{D}}|^{-2}W. \tag{8}$$

The operators (5), (6) are pseudodifferential operators of order (-1) and (-2) respectively. The symbols of the operators (7), (8) are homogeneous. Besides, $\pm Z(\pm1)\ge 0$.

The general results of the paper [1], about the asymptotics for the spectrum of pseudodifferential operators of negative order, imply that under condition (1) the following (quasiclassical) formula holds

$$\lim_{s\to 0} s^3 n_+(s,Y(\lambda))=J,\quad |\lambda|\le 1.$$

The following study is reduced to the analysis of the contribution of the operator $Z(\lambda)$ into the asymptotic formula of the spectrum of the operator $X(\lambda)$. If $\lambda = \pm 1$ then there are different cases depending on some additional (see (1)) restrictions on V.

In what follows we denote by $L_{p,q}(\mathbb{R}^d)$, $0 < p < \infty$, $0 < q \leq \infty$ the functional Lorentz classes (see for example [6]). Recall that $L_{p,p} = L_p$. By Φ we denote the Fourier operator.

3. We begin with the case of dominating contribution of the operator $Y(\lambda)$

Theorem 2. *Let the condition (1) be fulfilled. Then*

$$\lim_{\alpha \to \infty} \alpha^{-3} N(\alpha, \lambda) = J, \quad |\lambda| < 1.$$

Theorem 3. *Let us assume that together with (1) we have*

$$V \in L_{3/2,3}(\mathbb{R}^3). \tag{9}$$

Then the asymptotic formula (3) holds.

The condition (9) obviously is less restrictive then (2). Futhermore, let $\varphi \in L_{3/2,\infty}(\mathbb{R}^3)$, $\varphi(x) > 0$; in particular we can take $\varphi(x) = |x|^{-2}$. Then the asymptotic formula (3) holds if, together with (1), the following condition is fulfilled for some q [2]

$$\int V^q \varphi^{\frac{3}{2}-q}\, dx < \infty, \quad 3 \leq 2q \leq 6. \tag{10}$$

4. Let us now consider the case when the second term dominates in the sum

$$X(\pm 1) = Y(\pm 1) + Z(\pm 1). \tag{11}$$

Let us introduce the Schrödinger operator with the potential V

$$H = -\Delta - \alpha V, \quad \alpha > 0, \tag{12}$$

and denote by $N_H(\alpha)$ the number of the negative eigenvalues of the operator H. The asymptotic behaviour of $N_H(\alpha)$ was studied in detail in [2].

Theorem 4. *Let the condition (1) be satisfied and* $N_H(\alpha) = O(\alpha^q)$, $q > 3$. *Then*

$$\begin{aligned} \lim_{\alpha\to\infty} \sup \alpha^{-q} N(\alpha, 1) &= 2^{q+1} \lim_{\alpha\to\infty} \sup \alpha^{-q} N_H(\alpha), \\ \lim_{\alpha\to\infty} \inf \alpha^{-q} N(\alpha, 1) &= 2^{q+1} \lim_{\alpha\to\infty} \inf \alpha^{-q} N_H(\alpha, 0), \\ \lim_{\alpha\to\infty} \alpha^{-q} N(\alpha, -1) &= 0. \end{aligned} \tag{13}$$

[2]Regarding conditions of the type (10) see [2]. It follows from the estimates obtained there that conditions (10) imply $n_+(s, Z(\pm 1)) = o(s^{-3})$, $s \to 0$. However, it is possible to show that any of conditions (10) follows (9). This was varified by T.Weidl.

We introduce more special conditions providing the asymptotic formula of the type

$$N(\alpha, 1) \sim c\,\alpha^q, \quad \alpha \to \infty, \quad q > 3. \tag{14}$$

Let us consider the potential

$$V_\tau(x) = \begin{cases} |x|^{-2}(\ln|x|)^{-\tau}\Psi(\theta), & |x| > 2, \\ 0, & |x| \le 2, \end{cases} \tag{15}$$

where $\theta = x/|x|$, $\tau > 0$ and $\Psi \in L_q(\mathbb{S}^2)$, $q\tau = 1$, $2q > 3$.

The asymptotics of $N_H(\alpha)$ for the operator (12) with the potential $V = V_\tau$ was obtained in [4] (the case $\Psi = 1$ earlier was obtained in [2]). On the basis of this result and Theorem 4 we can establish (14) for $V = V_\tau$. Indeed, let us consider in $L_2(\mathbb{S}^2)$ the operator with a parameter $s > 0$

$$-\Delta_\theta - s^{-\tau}\Psi(\theta), \tag{16}$$

where Δ_θ is the Laplace operator on the unit sphere. Let $\{\nu_l^{(\tau)}(s)\}$ be the sequence of the eigenvalues of the operator (16) and

$$M_\tau = \frac{1}{\pi}\sum_l \int_0^\infty \left(\nu_l^{(\tau)}(s) + \frac{1}{4}\right)_-^{1/2} ds.$$

Theorem 5. *Let $V \in L_{3,\mathrm{loc}}$ and*

$$V = V_\tau(1 + o(1)), \quad |x| \to \infty, \tag{17}$$

where V_τ is the potential (15), $\tau^{-1} = q > 3$. Then (13) is fulfilled and the asymptotic formula (14) holds with $c = 2^{q+1}M_\tau$.

5. The most interesting case is

$$N_H(\alpha) = O(\alpha^3), \quad \alpha \to \infty. \tag{18}$$

Under condition (18) both terms on the right hand side of (11) give the same contribution α^3 into the asymptotic formula for $N(\alpha, 1)$. For a sufficiently large class of potentials we succeed in proving that these contributions lead to the summation of the respective asymptotic coefficients. This is caused by the fact that the main contribution of the operator $Y(\pm 1)$ into the spectrum under the condition (1) is given by large momenta, but the contribution of the operator $Z(\pm 1)$ (for V considered below) is given by small momenta. Let us impose a supplementary condition on V; it is convenient to express this condition in terms of the Fourier transform $\hat{W} = \Phi W$, $W = V^{1/2}$:

$$\int_{|\xi|>\varepsilon} |\hat{W}(\xi)|^2\, d\xi < \infty, \quad \varepsilon > 0. \tag{3.12}$$

Theorem 6. *Let conditions (1), (18) and (19) (for some $\varepsilon > 0$) be fulfilled. Then*

$$\begin{aligned} \lim_{\alpha\to\infty} \sup \alpha^{-3} N(\alpha, 1) &= J + 16 \lim_{\alpha\to\infty} \sup \alpha^{-3} N_H(\alpha), \\ \lim_{\alpha\to\infty} \inf \alpha^{-3} N(\alpha, 1) &= J + 16 \lim_{\alpha\to\infty} \inf \alpha^{-3} N_H(\alpha), \\ \lim_{\alpha\to\infty} \alpha^{-3} N(\alpha, -1) &= J. \end{aligned} \tag{20}$$

From Theorem 6 it is not difficult to deduce the following analogy of Theorem 5.

Theorem 7. *Let $V \in L_{3,\mathrm{loc}}$ and (17) be fulfilled with $\tau = 1/3$. Then (20) is satisfied and the following asymptotic formula holds*

$$\lim_{\alpha\to\infty} \alpha^{-3} N(\alpha, 1) = J + 16 M_{1/3}.$$

References

[1] M.Sh.Birman, M.Z.Solomyak, *Spectral asymptotics of pseudodifferential operators with anisotropic homogeheous symbols.* I, Vestn. Leningr. Univ., Mat. Mekn. Astron. **13** (1977), 13–21; II, Vestn. Leningr. Univ., Mat. Mekn. Astron. **13** (1979), 5–10. (Russian)

[2] M.Sh.Birman, M.Z.Solomyak, *Estimates for the number of negative eigenvalues of the Schrödinger operator and its generalizations*, AMS. Adv. Sov. Math. **7** (1991), 1–55.

[3] M.Klaus *On the point spectrum of Dirac operators*, Helvetica Physica Acta **53** (1980), 453–462.

[4] A. Laptev, *Asymptotics of the negative discrete spectrum of a class of Schrödinger operators with large coupling constant*, Proc. of AMS.

[5] B. Thaller, *The Dirac equation*, Texts and Monographs in Physics, Springer-Verlag, 1992.

[6] H. Triebel, *Introduction Theory. Function spaces. Differential operators*, VEB Deutscher Verlag der Wissenschaften, Berlin, 1978.

St.-Petersburg University, Dept. of Physics, Ulyanov str. 1, 198904 St.-Petersburg, Russia

Department of Mathematics, Linkoping University, S-581 83 Linkoping, Sweden
E-mail address: arlapmath.liu.se

Chapter 2

Generalized Schrödinger operators

Operator Theory:
Advances and Applications, Vol. 70

The Spectrum of Schrödinger Operators in $L_p(\mathbf{R}^d)$ and in $C_0(\mathbf{R}^d)$ *

Rainer Hempel and Jürgen Voigt

Introduction

The aim of this paper is to present results on the independence of the spectrum of Schrödinger operators in different spaces. We treat Schrödinger operators of a very general kind, namely $-\frac{1}{2}\Delta$ perturbed by certain measures μ.

In Section 1 we recall what measures can be used and we review results stating the p-independence of the spectrum of the realizations of $-\frac{1}{2}\Delta + \mu$ in $L_p(\mathbf{R}^d)$, $1 \leq p \leq \infty$.

In Section 2 we show that the realizations of $-\frac{1}{2}\Delta + \mu$ in spaces of continuous functions, e.g., the bounded uniformly continuous functions or the continuous functions vanishing at infinity, again have the same spectrum, for suitable μ. In fact, this is derived in a much more general context, utilizing the semigroup dual of a Banach space with respect to a strongly continuous semigroup.

In Section 3 it is shown thal Shnol's method of constructing singular sequences can also be employed in a proof of the inclusions $\sigma(H_{2,V}) \subset \sigma(H_{p,V})$ and $\sigma(H_{2,V}) \subset \sigma(H_{C_0,V})$, for suitable potentials V. This establishes the connection between the spectrum in L_p and C_0 and the existence of polynomially bounded generalized eigenfunctions.

1. Review of L_p-results.

In order to state the results we have to recall some notations. Let

$$M_0 := \{\mu : \mathcal{B} \to [0,\infty];\ \mu\ \sigma\text{-additive},\ \mu(B) = 0 \text{ for all sets } B \in \mathcal{B} \text{ with capacity zero}\},$$

where $\mathcal{B}$ denotes the σ-algebra of Borel subsets of $\mathbf{R}^d$.

For the definition of the extended Kato class $\hat{S}_K \subset M_0$ of measures and of the constant $c(\mu)$ defined for $\mu \in \hat{S}_K$ we refer to [StV]. We recall that for $\mu_+ \in M_0$, $\mu_- \in \hat{S}_K$ with $c(\mu_-) < 1$ a closed form in $L_2(\mathbf{R}^d)$ is defined by

$$(\mathbf{h} - \mu_- + \mu_+)[u,v] := \frac{1}{2}\int \nabla u \cdot \overline{\nabla v} dx - \int \tilde{u}\, \overline{\tilde{v}}\, d\mu_- + \int \tilde{u}\, \overline{\tilde{v}}\, d\mu_+,$$

*Presented at the meeting by J. Voigt

with domain

$$D(\mathbf{h}-\mu_- + \mu_+) = \{u \in W_2^1(\mathbf{R}^d); \int |u^{\tilde{}}|^2 \, d\mu_+ < \infty\}$$

($u^{\tilde{}}$ denoting a quasi-continuous version of u). The closure of $D(\mathbf{h}-\mu_- + \mu_+)$ in $L_2(\mathbf{R}^d)$ is of the form $L_2(Y)$, for a suitable set $Y \in \mathcal{B}$. The operator $H_\mu = H_{\mu_+ - \mu_-}$ is the self-adjoint operator in $L_2(Y)$ associated with $\mathbf{h}-\mu_- + \mu_+$. It is shown in [StV; Corollary 4.2] that the semi-group $\left(e^{-tH_\mu};\, t \geq 0\right)$ on $L_2(Y)$ acts also as a strongly continuous semigroup $U_{p,\mu}(.)$ on $L_p(Y)$, for all $p \in [1,\infty)$; the generators of these semigroups will be denoted by $-H_{p,\mu}$. Also, $H_{\infty,\mu} := H_{1,\mu}^*$. The corresponding unperturbed operators (for $\mu = 0$) will be denoted by H_p.

1.1. Theorem. *With the notations introduced so far, we have*

$$\sigma(H_{p,\mu}) = \sigma(H_{2,\mu})$$

for all $p \in [1,\infty]$.

We are going to give an outline of the proof of this result. In order to do so we first collect several facts which are needed in the proof.

1.2. Remark. (a) Let $\varepsilon > 0$. There exist constants C, ω such that

$$\|e^{\xi\cdot x}\, e^{-tH_{p,\mu}}\, e^{-\xi\cdot x}\|_{p,q} \leq C\, t^{-\gamma} e^{\omega t}$$

for all $t > 0$, $1 \leq p \leq q \leq \infty$, $\xi \in \mathbf{R}^d$ with $|\xi| \leq \varepsilon$, where $\gamma = \frac{d}{2}(\frac{1}{p} - \frac{1}{q})$. (Here $\|\cdot\|_{p,q})$ denotes the norm in $L(L_p, L_q)$.)

The proof of this fact consists in two steps. In both of these steps it is essential that there exists $a > 1$ such that $a\mu$ is also in the class considered above (in particular, $c(a\mu) < 1$).

(i) One shows the inequality for $\xi = 0$, using Stein interpolation; cf. [StV; Theorem 5.1 (b)].

(ii) From the fact that the desired statement is true for the unperturbed heat semigroup ($\mu = 0$) one concludes it for the perturbed semigroup, again using Stein interpolation; cf. [ScV; Remark 3.4 (b), (c)].

(b) Let $\epsilon > 0, \omega$ be as in (a). Then there exists C such that

$$\left\|e^{\xi\cdot x}(H_\mu - w)^{-1} e^{-\xi\cdot x}\right\|_{p,q} \leq C\left(\frac{1}{1-\gamma} + \frac{1}{-w-\omega}\right)$$

for all $w \in \mathbf{R}$ with $w < -\omega$, $p \leq q$ with $\gamma = \frac{d}{2}(\frac{1}{p} - \frac{1}{q}) < 1$, $|\xi| \leq \epsilon$. Further, $(-\infty, -\omega) \subset \rho(H_{p,\mu})$ for all $p \in [1,\infty]$, and

$$(H_{p,\mu} - w)^{-1} = (H_\mu - w)^{-1}$$

on $L_p(Y) \cap L_2(Y)$, for $w < -\omega$.

The proof consists in integrating the inequality in (a) after multiplying by e^{wt}; cf. [HV; Proposition 3.7], [ScV; Remark 3.4 (d)].

1.3. Lemma. ([ScV; Corollary 3.3]) *Let $1 \leq p \leq q \leq \infty$, $0 < \epsilon' < \epsilon''$. Then there exists $C \geq 0$ such that for each linear operator*

$$A : L_{\infty,c}(\mathbf{R}^d) \to L_{\infty,\mathrm{loc}}(\mathbf{R}^d)$$

($L_{\infty,c}$ denoting L_∞-functions with compact support) satisfying

$$\|e^{\xi\cdot x} A e^{-\xi\cdot x}\|_{p,q} \leq 1 \quad \textit{for all} \quad \xi \in \mathbf{R}^d \quad \text{with} \quad |\xi| \leq \epsilon''$$

one has

$$\left\|e^{\xi\cdot x} A e^{-\xi\cdot x}\right\|_{r,r} \leq C$$

for $p \leq r \leq q$, $|\xi| \leq \epsilon'$.

The inclusion $\rho(H_{p,\mu}) \subset \rho(H_{2,\mu})$ in Theorem 1.1 is obtained as in [HV; section 2], using Remark 1.2 (a) for $\xi = 0$.

Sketch of the proof of the inclusion $\rho(H_{2,\mu}) \subset \rho(H_{p,\mu})$ (compare [ScV]).

It is sufficient to prove the assertion for all $p \in [1, 2]$. According to Remark 1.2 (b) we find w $(< -\omega)$, C such that

$$\|e^{\xi\cdot x}(H_\mu - w)^{-1} e^{-\xi\cdot x}\|_{p,q} \leq C$$

whenever $1 \leq p \leq q \leq 2$, $\frac{d}{2}(\frac{1}{p} - \frac{1}{q}) \leq \frac{1}{2}$, $|\xi| \leq 1$.

Let $K \subset \rho(H_{2,\mu})$ be compact, $\overset{\circ}{K}$ connected, $K = \overline{\overset{\circ}{K}}$, $w \in \overset{\circ}{K}$. Then there exist $\epsilon \in (0, 1]$ and a constant C' such that $K \subset \rho(e^{\xi\cdot x} H_{2,\mu} e^{-\xi\cdot x})$ for $|\xi| \leq \epsilon$, and

$$\begin{aligned} \|e^{\xi\cdot x}(H_{2,\mu} - z)^{-1} e^{-\xi\cdot x}\| &= \|(e^{\xi\cdot x} H_{2,\mu} e^{-\xi\cdot x} - z)^{-1}\| \\ &\leq C' \qquad (|\xi| \leq \epsilon,\ z \in K). \end{aligned}$$

This follows from perturbation theory and analytic continuation. (Note that the equality

$$e^{\xi\cdot x}(H_{2,\mu} - z)^{-1} e^{-\xi\cdot x} = \left(e^{\xi\cdot x} H_{2,\mu} e^{-\xi\cdot x} - z\right)^{-1}$$

on $L_2(Y) \cap L_{2,c}(\mathbf{R}^d)$, whose validity for $z = w$ is obtained by Laplace transform, has to be extended to K by analytic continuation. The absence of this argument in [HV] was pointed out to the authors by W. Arendt.)

Using the resolvent equation

$$(H_{2,\mu} - z)^{-1} = (I + (z - w)(H_{2,\mu} - z)^{-1})(H_{2,\mu} - w)^{-1}$$

together with Lemma 1.3 one concludes the existence of C'' such that

$$\|e^{\xi\cdot x}(H_{2,\mu} - z)^{-1} e^{-\xi\cdot x}\|_{p,q} \leq C''$$

for $z \in K$, $1 \leq p \leq 2$ with $\frac{d}{2}(\frac{1}{p} - \frac{1}{q}) \leq \frac{1}{2}$, $|\xi| \leq \frac{\epsilon}{2}$.

Iterating this argument one obtains the last inequality for all $p \in [1,2]$ and small $|\xi|$. Using this estimate for $\xi = 0$ and the fact that

$$(H_{2,\mu} - w)^{-1} = (H_{p,\mu} - w)^{-1} \quad \text{on } L_p \cap L_2(Y)$$

one obtains $K \subset \rho(H_{p,\mu})$. ∎

1.4. Remarks. (a) A slightly different situation has been treated in [ScV]. In this paper the perturbation μ is the sum of a form small distributional part μ_0 (cf. [HS]) and $\mu_+ \in M_0$. This implies that the semigroup $(e^{-tH_\mu};\ t \geq 0)$ acts as a strongly continuous semigroup on $L_p(Y)$ for $p_0 \leq p \leq p_0'$ where $p_0 \in [1,2)$ depends on the form bound of μ_0 (cf. [BS]). It is then shown that $\sigma(H_{p,\mu}) = \sigma(H_{2,\mu})$ for all $p \in (p_0, p_0')$.

(b) The p-independence of the L_p-spectrum of elliptic operators on certain Riemannian manifolds was shown in [Stu]. In a similar context the p-independence for $1 < p < \infty$ was shown in [Sh; Proposition 2.6].

(c) The p-independence of spectra has been shown in [Al] for perturbations of certain translation invariant operators.

(d) If $U(\cdot)$ is a strongly continuous semigroup on $L_2(\Omega)$ (where $\Omega \subset \mathbf{R}^d$) satisfying a Gaussian estimate, then it was shown in [Ar] that the spectra of the generators of the corresponding semigroups on $L_p(\Omega)$ are p-independent.

2. The spectrum of $-\frac{1}{2}\Delta + \mu$ in spaces of continuous functions

We want to show that under suitable hypotheses the spectrum of $-\frac{1}{2}\Delta + \mu$ in

$$C_0(\mathbf{R}^d) = \{f \in C(\mathbf{R}^d);\ f(x) \to 0\ (|x| \to \infty)\}$$

(or in other spaces of bounded continuous functions) is the same as the L_p-spectrum.

It turns out that the main point which is specific about this situation is the question whether $(e^{-tH_\mu};\ t \geq 0)$ acts as a strongly continuous semigroup on $C_0(\mathbf{R}^d)$. The fact that then coincidence of spectra can be concluded will follow from very general considerations presented next.

Let X be a Banach space, $(U(t); t \geq 0)$ a strongly continuous semigroup on X, and T its generator. The semigroup dual of X is then defined by

$$X^\odot := \{x^* \in X^*;\ T(t)^* x^* \to x^*\ (t \to 0)\};$$

see, e.g., [HP; Chap. XIV], [BB; Sec. 1.4] (where $X^\odot$ is denoted by X_0^*), [Ne]. (We use the adjoint space X^* of continuous conjugate linear functionals on X in order to stay consistent with duality in L_2.)

2.1. Theorem. *Let $Y \subset X^\odot$ be a closed subspace which is invariant under $U^*(t)$ $(t \geq 0)$. Denote by $U_Y(\cdot)$ the part of the semigroup $U^*(\cdot)$ in Y, and by T_Y the generator of $U_Y(\cdot)$.*

(a) Then T_Y is the part of T^* in Y,

$$\begin{aligned} D(T_Y) &= \{x^* \in Y \cap D(T^*);\ T^*x^* \in Y\}, \\ T_Y &= T^*|D(T_Y). \end{aligned}$$

(b) $\rho_\infty(T) \subset \rho_\infty(T_Y)$, and $(\lambda - T_Y)^{-1}$ is the part of $((\lambda - T)^{-1})^*$ in Y, for $\lambda \in \rho_\infty(T)$. (Here $\rho_\infty(T)$ denotes the component of $\rho(T)$ containing a right half plane; and similarly for T_Y.)

(c) If additionally Y is equi-norming for X, i.e., the norm

$$\|x\|_Y := \sup\{| < x^*, x > |;\ x^* \in Y,\ \|x^*\| \le 1\} \quad (x \in X)$$

is equivalent to the original norm in X, then

$$\rho_\infty(T) = \rho_\infty(T_Y).$$

Proof. (a) This is known for $Y = X^\odot$, and the proof carries over to our case (cf. [BB; p. 51], [Ne; Theorem 1.3.3]).

(b) For $\lambda \in \mathbf{C}$ with Reλ larger than the type of $U(\cdot)$, the resolvents of T and T_Y are given by the Laplace transform of $U(\cdot)$ and $U_Y(\cdot)$, respectively, and therefore

$$< x^*, (\lambda - T)^{-1}x > \quad = \quad < (\lambda - T_Y)^{-1}x^*, x >$$

for all $x \in X$, $x^* \in Y$. Therefore $(\lambda - T_Y)^{-1}$ is the part of $((\lambda - T)^{-1})^*$ in Y. This implies that $((\lambda - T)^{-1})^*$ maps Y to Y for all $\lambda \in \rho_\infty(T)$. By uniqueness we obtain the claimed assertions.

(c) The equivalence of $\|\cdot\|$ and $\|\cdot\|_Y$ implies that there exists a constant c such that

$$\|(\lambda - T)^{-1}\| \le c\|(\lambda - T_Y)^{-1}\| \quad \text{for all} \quad \lambda \in \rho_\infty(T).$$

This implies $\partial(\rho_\infty(T)) \subset \sigma(T_Y)$, and therefore $\rho_\infty(T) = \rho_\infty(T_Y)$. ∎

2.2. Remark. The assymptions made in the previous theorem are satisfied, in particular, for $Y = X^\odot$. For this case, however, one has $\rho(T^\odot) = \rho(T)$; cf. [Ne; Theorem 1.4.2].

2.3. Corollary. *Assume that μ satisfies the hypotheses of Theorem 1.1. Let Y be a closed subspace of L_∞ which is equi-norming for L_1, invariant under $(e^{-tH_{1,\mu}})^*$ $(t \ge 0)$ and such that*

$$\|(e^{-tH_{1,\mu}})^* f - f\|_\infty \to 0 \quad (t \to 0)$$

for all $f \in Y$. Denote by $-H_{Y,\mu}$ the generator of the strongly continuous semigroup on Y induced by $((e^{-tH_{1,\mu}})^*;\ t \ge 0)$. Then

$$\sigma(H_{Y,\mu}) = \sigma(H_{2,\mu}).$$

2.4. Remarks. (a) The semigroup dual of $L_1(\mathbf{R}^d)$ for the unperturbed Schrödinger semigroup $(e^{-tH_1};\ t \geq 0)$ is

$$C_{b,u}(\mathbf{R}^d) = \{f \in C(\mathbf{R}^d);\ f \text{ bounded and uniformly continuous}\}.$$

The generator is then the part of $-H_\infty$ in $C_{b,u}$,

$$D(H_{C_{b,u}}) = \{f \in C_{b,u}(\mathbf{R}^d);\ H_{C_{b,u}}f = -\frac{1}{2}\Delta f \in C_{b,u}\}.$$

For $V \in C_{b,u}(\mathbf{R}^d)$, the multiplication operator by V is a bounded operator in $C_{b,u}(\mathbf{R}^d)$, and therefore Theorem 2.3 is applicable to $H+V$ with $Y = C_{b,u}(\mathbf{R}^d)$.

(b) The space $C_0(\mathbf{R}^d)$ is invariant under the unperturbed Schrödinger semigroup, and

$$D(H_{C_0}) = \{f \in C_0(\mathbf{R}^d);\ H_{C_0}f = -\frac{1}{2}\Delta f \in C_0\}.$$

For bounded $V \in C(\mathbf{R}^d)$ the multiplication by V is a bounded operator on $C_0(\mathbf{R}^d)$. Therefore Theorem 2.3 is applicable to $H+V$ with $Y = C_0(\mathbf{R}^d)$.

(c) For $V = V_+ - V_-$, $V_\pm \geq 0$, $V_- \in K_d$, $V_+ \in K_{d,\mathrm{loc}}$ it is shown in [S; Theorem B.3.1] that e^{-tH_V} maps L_∞-functions to continuous functions, for $t > 0$. As a consequence,

$$Y := L_1(\mathbf{R}^d)^\odot$$

consists of continuous functions, in this case.

3. An application of Shnol's method.

In order to establish a connection with the PDE-world, we will now discuss an alternative proof of the inclusions

$$\sigma(H_{p,V}) \supset \sigma(H_{2,V}), \qquad \sigma(H_{C_0,V}) \supset \sigma(H_{2,V}). \tag{3.1}$$

To this end, we will produce rather explicit "Weyl sequences" in L_p and also in C_0 which are obtained by applying suitably chosen cut-offs to generalized eigenfunctions associated with the expansion theorem for $H_{2,V}$ ([B], [S], [PStW]); this requires some mild modifications of Shnol's method (cf. [Shn], [S; Section C.4], and [HSt]). Therefore, we learn that properties of the Schrödinger operator in Hilbert space L_2 fully determine the spectra in L_p and even in C_0: while estimates for the resolvent kernel $(H_{2,V} - z)^{-1}(x,y)$ give the inclusion $\varrho(H_{p,V}) \supset \varrho(H_{2,V})$, the converse inclusion will now be a consequence of the eigenfunction expansion theorem for $H_{2,V}$. Related ideas are also discussed in [Sh].

It should be stressed, however, that the approach proposed here requires more restrictive assumptions on the potential V, as compared with the "duality and interpolation"-proof described in Section 2. In the following, we will restrict the discussion to the case $V \in L_\infty(\mathbf{R}^d)$ where it is easy to obtain L_p-bounds for the gradient of a generalized eigenfunction.

We first collect a few facts (where we always assume that V is bounded):

(1) For $1 \le p \le \infty$, we have ([HV1])

$$D(H_{p,V}) = D(H_p) = \{u \in L_p;\ \Delta u \in L_p\}. \tag{3.2}$$

If, more strongly, V is bounded and continuous, then (cf. Section 2)

$$D(H_{C_0,V}) = D(H_{C_0}) = \{u \in C_0;\ \Delta u \in C_0\}. \tag{3.3}$$

(2) From the generalized eigenfunction expansion theorem for $H_{2,V}$ ([B], [S], [PStW]), we can draw the following conclusion: for any $\mu \in \sigma(H_{2,V})$ and any $\varepsilon > 0$, there exists a $\lambda \in (\mu - \varepsilon, \mu + \varepsilon)$ and a (non-trivial) distributional solution u of the PDE

$$-\frac{1}{2}\Delta u + Vu = \lambda u, \tag{3.4}$$

satisfying a polynomial growth bound

$$|u(x)| \le c_1(1 + |x|)^K, \tag{3.5}$$

with some constants $c_1 > 0$ and $K \in \mathbf{N}$. For V bounded, it is also known that u is (equivalent to) a continuous function (cf., e.g., [S]).

(3) To control the cut-off errors, we need an L_p-bound on ∇u, for u satisfying (3.4), (3.5). Note that there is no L_p-analogue of the L_2-gradient bound given in [S; Lemma C.2.1]. Here we proceed as in [HV1], using an argument of L. Schwartz, to obtain the following lemma.

3.1. Lemma. *Let $p \in [1, \infty]$, and suppose that $\Omega \subset \Omega'$ are open sets in $\mathbf{R}^d$ with the property that* $\operatorname{dist}(\Omega, \partial\Omega') \ge 1$. *Then there exists a constant $C = C(p)$, which is independent of both Ω and Ω', such that*

$$\|\nabla u\|_{L_p(\Omega)} \le C\left(\|u\|_{L_p(\Omega')} + \|\Delta u\|_{L_p(\Omega')}\right), \tag{3.6}$$

for all $u \in L_p(\Omega')$ with the property that $\Delta u \in L_p(\Omega')$.

Proof. We proceed as in [HV1]: letting T denote the usual fundamental solution for $-\Delta$, and picking some $\chi \in C_c^\infty(\mathbf{R}^d)$ with support in the unit ball and $\chi(x) = 1$ for $|x| \le 1/2$, we have

$$\nabla u = (\nabla(\chi T)) * \Delta u - \nabla\zeta * u, \tag{3.7}$$

(where $\zeta = (\Delta\chi)T + 2\nabla\chi \cdot \nabla T \in C_c^\infty(\mathbf{R}^d)$), and the required estimate follows from Young's inequality ([RS]). Furthermore, it is clear from eq. (3.7) that ∇u is continuous, provided u and Δu are continuous functions. ∎

Now let u be a (continuous) generalized eigenfunction of $H_{2,V}$ and $\varphi \in C_c^\infty(\mathbf{R}^d)$. Then it follows from Lemma 3.1 and $\Delta(\varphi u) = \varphi\Delta u + 2\nabla\varphi\nabla u + (\Delta\varphi)u$ that φu

will belong to the domain of $H_{p,V}$, for $1 \leq p \leq \infty$. Similarly, if V is bounded and continuous, then φu will belong to the domain of $H_{C_0,V}$.

(4) Central to Shnol's method is the observation that the growth bound (3.5) implies that the L_2-norm of u, considered on a suitable sequence of balls, will not grow too rapidly (cf. [S]). While the exposition given in [S; Section C.4] can directly be carried over to the L_p-case for $1 \leq p < \infty$, it has to be modified for $p = \infty$ and, similarly, also for the space C_0. We therefore change the scenario used in [S] and consider

$$\mathcal{E}_n = \{x \in \mathbf{R}^d;\ |x| < 2^n\}, \qquad \mathcal{F}_n = \mathcal{E}_{n+1} \setminus \mathcal{E}_n \qquad (n \in \mathbf{N}). \tag{3.8}$$

We then have the following lemma.

3.2. Lemma. *Let $1 \leq p \leq \infty$, and let u be as in (3.5). Let $a > 2$ and set $c_2 = c_2(p) = a^{K+\frac{d}{p}}$. Then there exists a sequence $(n_j)_{j\in\mathbf{N}} \subset \mathbf{N}$, $n_j \to \infty$, such that*

$$\left\|u|_{\mathcal{F}_{n_j}}\right\|_p \leq c_2 \left\|u|_{\mathcal{E}_{n_j}}\right\|_p \qquad (j \in \mathbf{N}). \tag{3.9}$$

Proof. If the statement of the lemma were not true, there would exist some n_0 such that

$$\|u|_{\mathcal{F}_n}\|_p \geq c_2\ \|u|_{\mathcal{E}_n}\|_p > 0 \qquad (n \geq n_0), \tag{3.10}$$

so that

$$\|u|_{\mathcal{E}_n}\|_p \geq \left\|u|_{\mathcal{F}_{n-1}}\right\|_p \geq c_2 \left\|u|_{\mathcal{E}_{n-1}}\right\|_p \quad (n > n_0). \tag{3.11}$$

This leads to

$$\|u|_{\mathcal{E}_n}\|_p \geq c_2^{n-n_0} \|u|_{\mathcal{E}_{n_0}}\|_p \quad (n \geq n_0), \tag{3.12}$$

in contradiction with the polynomial growth bound of u. ∎

With these preparations, it is now easy to prove the inclusions stated in eq. (3.1).

Proposition 3.3. *Let $V \in L_\infty(\mathbf{R}^d)$. Then $\sigma(H_{p,V}) \supset \sigma(H_{2,V})$, for all $p \in [1,\infty]$. If, moreover, V is (bounded and) continuous, then $\sigma(H_{C_0,V}) \supset \sigma(H_{2,V})$.*

Proof. We first choose a function $\varphi \in C_c^\infty(-2,2)$ with the property that $\varphi(x) = 1$, for $|x| \leq 4/3$, and $\varphi(x) = 0$, for $|x| \geq 5/3$, and we define

$$\varphi_n(x) = \varphi(2^{-n}|x|)\,, \qquad x \in \mathbf{R}^d.$$

Then $\mathcal{G}_n := \operatorname{supp}(\nabla\varphi_n) \subset \mathcal{F}_n$ and $\operatorname{dist}(\mathcal{G}_n, \partial\mathcal{F}_n) \geq 1$, for $n \geq 2$. Furthermore, we have $\|\nabla\varphi_n\|_\infty \leq c_3 2^{-n}$ and $\|\Delta\varphi_n\|_\infty \leq c_4 2^{-2n}$.

Now let $\mu \in \sigma(H_{2,V})$ be given, and let $\varepsilon > 0$. By what was said in point (2), there exists some $\lambda \in (\mu - \varepsilon, \mu + \varepsilon)$ and a (non-trivial) generalized eigenfunction u of $H_{2,V}$ that satisfies (3.4), (3.5). For given $p \in [1,\infty]$, we will prove that there exists a sequence $(n_j) \subset \mathbf{N}$ so that

$$\left\|(H_{p,V} - \lambda)(\varphi_{n_j} u)\right\|_p / \left\|\varphi_{n_j} u\right\|_p \to 0, \quad j \to \infty. \tag{3.13}$$

Therefore, $H_{p,V}-\lambda$ does not have a bounded inverse, whence $\lambda \in \sigma(H_{p,V})$. Taking $\varepsilon \to 0$ then gives $\mu \in \sigma(H_{p,V})$.

Applying Lemma 3.2 to u, we find a constant c_2 and a sequence (n_j) such that (3.9) holds. As $\varphi_{n_j} u \in \mathcal{D}(H_{p,V})$ and $(H_{p,V}-\lambda)(\varphi_{n_j} u) = -(\nabla \varphi_{n_j})\nabla u - \frac{1}{2}(\Delta \varphi_{n_j})u$, we have

$$\begin{aligned} \left\|(H_{p,V}-\lambda)(\varphi_{n_j} u)\right\|_p &\le \left\|\nabla \varphi_{n_j}\right\|_\infty \left\|\nabla u|_{\mathcal{G}_{n_j}}\right\|_p + \left\|\Delta \varphi_{n_j}\right\|_\infty \left\|u|_{\mathcal{G}_{n_j}}\right\|_p \\ &\le c_5\, 2^{-n_j} \left(\left\|u|_{\mathcal{F}_{n_j}}\right\|_p + \left\|\Delta u|_{\mathcal{F}_{n_j}}\right\|_p\right), \end{aligned}$$

by Lemma 3.1. From $V \in L_\infty$ and $\frac{1}{2}\Delta u = (V-\lambda)u$ we now conclude that

$$\left\|(H_{p,V}-\lambda)(\varphi_{n_j} u)\right\|_p \le c_6 2^{-n_j} \left\|u|_{\mathcal{F}_{n_j}}\right\|_p \le c_7\, 2^{-n_j} \left\|u|_{\mathcal{E}_{n_j}}\right\|_p \le c_8 2^{-n_j} \left\|\varphi_{n_j} u\right\|_p ,$$

and the result follows.

The proof in the case of the space C_0 is essentially identical with the $p=\infty$ proof and omitted. ∎

Acknowledgements. R. Hempel would like to thank T. Hoffmann-Ostenhof for the kind invitation to the Erwin Schrödinger Institute at Vienna.

References.

[Al] D. ALBOTH: *Closable translation invariant operators and perturbations by potentials.* Dissertation, Kiel 1991.

[Ar] W. ARENDT: *Gaussian estimates and p-independence of the spectrum in L^p.* Manuscript 1993.

[BS] A. G. BELYI, YU. A. SEMENOV: *L^p-theory of Schrödinger semigroups II.* Sibirskii Matematicheskii Zhurnal **31**, 16 - 26 (1990) (russian). Translation: Siberian Math. J. **31**, 540 - 549 (1990).

[B] J. M. BEREZANSKI: *Expansions in eigenfunctions of selfadjoint operators.* Transl. Math. Monogr., vol. **17**, Amer. Math. Soc., Providence, 1968.

[BB] P. L.BUTZER, H. BERENS: *Semi-groups of operators and approximation.* Springer-Verlag, Berlin, 1967.

[HV] R. HEMPEL, J. VOIGT: *The spectrum of a Schrödinger operator in $L_p(\mathbf{R}^N)$ is p-independent.* Commun. Math. Phys. **104**, 243 - 250 (1986).

[HV1] R. HEMPEL, J. VOIGT: *On the L_p-spectrum of Schrödinger operators.* J. Math. Anal. Appl. **121**, 138 - 159 (1987).

[HS] I. W. HERBST, A. D. SLOAN: *Perturbation of translation invariant positivity preserving semigroups on $L^2(\mathrm{R}^N)$.* Transactions Amer. Math. Soc. **236**, 325 - 360 (1978).

[HP] E. HILLE, R. S. PHILLIPS: *Functional analysis and semi-groups.* Amer. Math. Soc., Providence, 1957.

[HSt] A. M. HINZ, G. STOLZ: *Polynomial boundedness of eigensolutions and the spectrum of Schrödinger operators.* Math. Ann. **294**, 195 - 211 (1992).

[Ne] J. VAN NEERVEN: *The adjoint of a semigroup of linear operators.* Lecture Notes in Math. **1529**, Springer-Verlag, Berlin, 1992.

[PStW] TH. POERSCHKE, G. STOLZ, J. WEIDMANN: *Expansions in generalized eigenfunctions of selfadjoint operators.* Math. Z. **202**, 397 - 408 (1989).

[RS] M. REED, B. SIMON: *Methods of modern mathematical physics II: Fourier analysis, self-adjointness.* Academic Press, New York, 1975.

[ScV] G. SCHREIECK, J. VOIGT: *Stability of the L_p-spectrum of Schrödinger operators with form small negative part of the potential.* In: "Functional Analysis", Proc. Essen 1991, Bierstedt, Pietsch, Ruess, Vogt eds., Marcel Dekker, to appear.

[Shn] I. EH. SHNOL': *Ob ogranichennykh resheniyakh uravneniya vtorogo poryadka v chastnykh proizvodnykh.* Dokl. Akad. Nauk SSSR **89**, 411 - 413 (1953).

[Sh] M. A. SHUBIN: *Spectral theory of elliptic operators on non-compact manifolds.* In: "Méthodes semi-classiques, vol. 1", Astérisque **207**, 37 - 108 (1992).

[S] B. SIMON: *Schrödinger semigroups.* Bull. (N. S.) Amer. Math. Soc. **7**, 447 - 526 (1982).

[StV] P. STOLLMANN, J. VOIGT: *Perturbation of Dirichlet forms by measures.* Preprint 1992.

[Stu] K.-TH. STURM: *On the L^p-spectrum of uniformly elliptic operators on Riemannian manifolds.* J. Funct. Anal., to appear.

Rainer Hempel, Department of Mathematics, University of Alabama in Birmingham, Birmingham, AL 35294, USA

Jürgen Voigt, Technische Universität Dresden, Abteilung Mathematik, D-01062 Dresden, Germany

Operator Theory:
Advances and Applications, Vol. 70

On the spectral properties of generalized Schrödinger operators

J.F.Brasche

Let X be a locally compact separable metric space, m a positive Radon measure on X such that $m(U) > 0$ for each open set $U \subset X$ and H a self-adjoint operator in $L^2(X,m)$ which is uniquely associated to some regular Dirichlet form $\mathcal{E}$ in the sense that

$$\mathcal{E}(f,g) = \langle Hf, g\rangle \quad \forall f \in D(H) \subset D(\mathcal{E}) \quad \forall g \in D(\mathcal{E}).$$

We refer to [1] and [2] for the notions from the theory of Dirichlet forms.

We are mainly interested in the following

Example 1: $X = \mathbf{R}^d$, $dm = dx$ =Lebesgue measure, $H = -\Delta$. ♮

More generally we are interested in the following

Example 2: Let $\phi \in L^2_{\text{loc}}(\mathbf{R}^d)$. Suppose that for each compact set $K \subset \mathbf{R}^d$ there exists a constant $c_K > 0$ such that $\phi \geq c_K$ on K. Let

$$\mathcal{E}_\phi(f,g) := \int \overline{\nabla f} \cdot \nabla g \, \phi^2 dx \quad \forall f, g \in D(\mathcal{E}_\phi) := C_0^\infty(\mathbf{R}^d).$$

$\mathcal{E}_\phi$ is closable in $L^2(\mathbf{R}^d, \phi^2 dx)$ (cf. [3]) and its closure $\overline{\mathcal{E}_\phi}$ is a regular Dirichlet form in $L^2(\mathbf{R}^d, \phi^2 dx)$. Thus we may put $X = \mathbf{R}^d$, $dm = \phi^2 dx$ and $H = H_\phi$ where H_ϕ is the self-adjoint operator uniquely associated with $\overline{\mathcal{E}_\phi}$. ♮

Let μ_+ and μ_- be positive Radon measures on X charging no set with $\mathcal{E}$-capacity zero. We put $\mu := \mu_+ - \mu_-$ and $|\mu| := \mu_+ + \mu_-$. We suppose that there exist a constant $a < 1$ and a finite constant b such that

$$\int |\tilde{f}|^2 \, d|\mu| \leq a \, \mathcal{E}_b(f,f) \quad \forall f \in D(\mathcal{E}). \tag{1}$$

Here $\tilde{f}$ denotes any quasi-continuous representative of f and $\mathcal{E}_b := \mathcal{E} + b\langle \cdot, \cdot \rangle$. Let

$$(\mathcal{E} + \mu)(f,g) := \mathcal{E}(f,g) + \int \overline{\tilde{f}} \tilde{g} d\mu \quad \forall f, g \in D(\mathcal{E} + \mu) := D(\mathcal{E}).$$

By the inequality (1) and the KLMN–Theorem, $\mathcal{E}+\mu$ is lower semi–bounded and closed in $L^2(X,m)$. We shall denote by $H+\mu$ the self–adjoint operator uniquely associated with $\mathcal{E}+\mu$.

Example 1': If $H=-\Delta$ and the measure μ is supported on some closed set Γ then the operator $-\Delta+\mu$ describes the interaction of a quantum mechanical particle with a potential which is concentrated on the set Γ. ♮

Example 2': In the situation of the Example 2 consider the special case that $\phi = G_\alpha * \nu$ for some $\alpha>0$ and some positive Radon measure ν supported by some closed set Γ. Here $G_\alpha(x-y)=(-\Delta+\alpha)^{-1}(x,y)$. Suppose that the measure μ is also supported on Γ. Then the operator

$$\phi(H_\phi+\mu)\frac{1}{\phi}-\alpha$$

in $L^2(\mathbf{R}^d,dx)$ describes the interaction of a quantum mechanical particle with a potential concentrated on Γ (cf. [4]). It is noteworthy that, in contradistinction to the Example 1', the classical capacity of the set Γ may be equal to zero. ♮

Let $\nu,\rho\in\{m,|\mu|\}$, $\alpha>0$ and $|m|:=m$. Let $h\in L^2(X,\nu)$. By (1) and Schwarz' inequality, the linear functional $g\mapsto\int\bar{h}\tilde{g}\,d\nu$ is bounded on the Hilbert space $(D(\mathcal{E}),\mathcal{E}_\alpha)$. Thus there exists a unique $U_\alpha(h\nu)\in D(\mathcal{E})$ such that

$$\mathcal{E}_\alpha(U_\alpha(h\nu),g)=\int\bar{h}\tilde{g}d\nu \quad \forall g\in D(\mathcal{E}).$$

We define the mapping $R_{\nu\rho\alpha}:L^2(X,\nu)\longrightarrow L^2(X,\rho)$ by

$$R_{\nu\rho\alpha}h:=\widetilde{U_\alpha(h\nu)} \quad \rho\text{–a.e.} \quad \forall h\in L^2(X,\nu).$$

By (1), there exist an $a<1$ and an $\alpha_0>0$ such that for all $\alpha\geq\alpha_0$ and all $f\in D(\mathcal{E})$

$$\int|\tilde{f}|^2\,d\rho\leq a\,\mathcal{E}_\alpha(f,f) \quad\text{and}\int|\tilde{f}|^2\,d\nu\leq a\,\mathcal{E}_\alpha(f,f).$$

Let $\alpha\geq\alpha_0$ and $S_\alpha:=\{f\in D(\mathcal{E}):\mathcal{E}_\alpha(f,f)=1\}$. For each $h\in L^2(X,\nu)$ we have

$$\begin{array}{rlcl} & \int|\widetilde{U_\alpha(h\nu)}|^2\,d\rho & \leq & a\mathcal{E}_\alpha(U_\alpha(h\nu),U_\alpha(h\nu)) \\ = & a\sup_{f\in S_\alpha}|\mathcal{E}_\alpha(U_\alpha(h\nu),f)|^2 & = & a\sup_{f\in S_\alpha}|\int\bar{h}\tilde{f}\,d\nu|^2 \\ \leq & a\int|h|^2\,d\nu\cdot\sup_{f\in S_\alpha}\int|\tilde{f}|^2\,d\nu & \leq & a^2\int|h|^2\,d\nu. \end{array}$$

Thus the operator $R_{\nu\rho\alpha}$ is bounded with operator norm less than one.

Theorem 1: *There exists an α_0 such that for all $\alpha \geq \alpha_0$ the number $-\alpha$ is in the resolvent set of the operator $H + \mu$ and we have*

$$\begin{aligned} & (H+\mu+\alpha)^{-1} - (H+\alpha)^{-1} \\ = & -R_{|\mu| m\alpha}[I + \gamma R_{|\mu||\mu|\alpha}]^{-1}\gamma R_{m|\mu|\alpha}. \end{aligned}$$

Here the function γ is chosen such that $\gamma|\mu| = \mu$.

Scetch of the proof: Let $f \in L^2(X, m)$. Let $\alpha > 0$ be such that $\|R_{|\mu||\mu|\alpha}\| < 1$. We put $h := [I + \gamma R_{|\mu||\mu|\alpha}]^{-1}\gamma R_{m|\mu|\alpha} f$. It suffices to show that

$$(\mathcal{E} + \mu)_\alpha((H+\alpha)^{-1} f - R_{|\mu| m\alpha} h, g) = \langle f, g\rangle \quad \forall g \in D(\mathcal{E}).$$

A short computation gives that the expression on the left hand side equals

$$\begin{aligned} \langle f, g\rangle \quad & + \quad \int \overline{\widetilde{(H+\alpha)^{-1}} f}\tilde{g} d\mu - \int \bar{h}\tilde{g} d|\mu| \\ & - \quad \int \overline{U_\alpha \widetilde{(h|\mu|)}}\tilde{g} d\mu. \end{aligned}$$

We have

$$\widetilde{(H+\alpha)^{-1}} f = U_\alpha \widetilde{(fm)} = R_{m|\mu|\alpha} f \quad |\mu|\text{-a.e.}$$

and, since $\mathcal{E}$ is regular, for each $g \in C_0(X)$ there exists a sequence $\{g_n\}$ in $D(\mathcal{E}) \cap C_0(X)$ converging to g uniformly such that $\operatorname{supp}(g_n) \subset \operatorname{supp}(g)$ for each $n \in \mathbf{N}$. Thus we have only to show that

$$\gamma R_{m|\mu|\alpha} f - h - \gamma U_\alpha \widetilde{(h|\mu|)} = 0 \quad |\mu|\text{-a.e.}$$

This equality follows from a simple computation. □

A straightforward computation gives the following lemma which is useful in order to derive Birman–Schwinger bounds for the number of negative eigenvalues of the operator $H + \mu$.

Lemma: $\forall \alpha > 0 : \dim \ker(H + \mu + \alpha) = \dim \ker[I + \gamma R_{|\mu||\mu|\alpha}]$.

In the special case that $H = -\Delta$ we have that

$$U_\alpha \widetilde{(h\nu)} = G_\alpha * (h\nu) \quad \rho\text{-a.e.}$$

(cf.[5]). By this equation and the above theorem one has an explicit representation of the resolvent of the operator $-\Delta + \mu$. In [5] this explicit representation has been the starting point for a detailed investigation of the spectral properties of the operator $-\Delta+\mu$. In particular, in [5] there have been derived the following results:

Theorem 2: *If* $|\mu|(\mathbf{R}^d)<\infty$ *then* $\sigma_{ess}(-\Delta+\mu)=[0,\infty)$.

Scetch of the proof: For sufficiently large α the operators $R_{|\mu|m\alpha}$ and $[I+\gamma R_{|\mu||\mu|\alpha}]^{-1}$ are bounded and the Schur test gives that $R_{m|\mu|\alpha}$ is compact. Thus by Weyl's essential spectrum theorem and Theorem 1 the operator $-\Delta+\mu$ has the same essential spectrum as the operator $-\Delta$. □

Theorem 3: *If the measure* μ *has compact support and belongs to the Kato class then* $\sigma_{ac}(-\Delta+\mu)=[0,\infty)$, $\sigma_{sc}(-\Delta+\mu)=\emptyset$ *and the set of positive eigenvalues of* $-\Delta+\mu$ *is discrete.*

Scetch of the proof: Let $\nu,\rho\in\{m,|\mu|\}$ and $(\nu,\rho)\neq(m,m)$. Let $\mathbf{C}^+$ be the set of all complex numbers with positive real part. First one shows that the mapping $z\mapsto R_{\nu\rho}(z):=R_{\nu\rho-z}$ from $(-\infty,0)$ to the Banach space of bounded everywhere defined operators from $L^2(X,\nu)$ to $L^2(X,\rho)$ has an analytic continuation to the set $D:=(\{z\in\mathbf{C}:\mathrm{Re}(z)\leq 0,\mathrm{Im}(z)\geq 0\}\cup\mathbf{C}^+)\setminus\{0\}$, that for each z of the extended definition domain D the operator $R_{\nu\rho}(z)$ is compact and that for each z in D with $\mathrm{Im}(z)>0$ the following equation holds:

$$(-\Delta+\mu-z)^{-1}-(-\Delta-z)^{-1}=-R_{|\mu|m}(z)[I+\gamma R_{|\mu||\mu|}(z)]^{-1}\gamma R_{|\mu|m}(z).$$

By the analytic Fredholm theorem and Theorem 1, it follows that the mapping $z\mapsto(-\Delta+\mu-z)^{-1}$ from the set $\{z\in\mathbf{C}:\mathrm{Im}(z)>0\}$ to the Banach space of everywhere defined bounded operators on $L^2(\mathbf{R}^d,dx)$ has an analytic continuation to $D\setminus S$ for some discrete set S. By the limiting absorption principle, the theorem is proved. □

Theorem 4: *Let* $1<q\leq 2$. *There exist the following bounds for the number (counting multiplicities)* $N(-\alpha)$ *of eigenvalues of the operator* $-\Delta+\mu$ *below* $-\alpha$:

$$N(-\alpha)\leq\int\gamma_-(x)m(dx)\left\{\sup_{x\in\mathbf{R}^d}\int G_\alpha(x-y)^q\gamma_-(y)m(dy)\right\}^{\frac{1}{q-1}}\tag{2}$$

Scetch of the proof: One uses the Birman–Schwinger method and the lemma in order to show that $N(-\alpha)\leq\|R_{|\mu||\mu|\alpha}\|_p^p$ for each $1\leq p<\infty$. Here $\|\cdot\|_p$ denotes the norm in the trace ideal of order p (with the convention that $\|A\|_p=\infty$ if the operator A does not belong to this ideal). Then estimates on the $\|\cdot\|_p$–norm of an integral operator due to Solom'jak (cf. [6]) give the above inequalities. □

Of course one is mainly interested in the number (counting multiplicities) $N(0)$ of negative eigenvalues of the operator $-\Delta+\mu$. An upper bound for $N(0)$

can be obtained by taking the limit as α tends to zero on the right hand side of the inequality (2). This method, however, gives the trivial upper bound $+\infty$ in the cases $d = 1$ and $d = 2$ since then $G_\alpha(x) \to \infty$, as $\alpha \to 0$ for all $x \in \mathbf{R}^d$. Fortunately the method can be modified in the same way as in [7], [8], [9] because the singularity does not depend effectively on the spectral parameter and corresponds therefore to just one bound state which can be taken into account separately.

Theorem 5: *Suppose that* $\int \gamma_- dm > 0$. *Then we have the following bounds for the number* $N(0)$ *of negative eigenvalues of* $-\Delta + \mu$:
$d = 1$:

$$N(0) \leq 1 + \frac{\frac{1}{2}\int\int |x-y|\gamma_-(x)\gamma_-(y)m(dy)m(dx)}{\int \gamma_-(x)m(dx)}.$$

$d = 2$:

$$\begin{aligned} N(0) \quad \leq 1 \quad &+ \left\{\int \gamma_-(x)m(dx)\right\}^{-2} \int\int\int\int \gamma_-(x)\gamma_-(y)\gamma_-(z)\gamma_-(u) \\ &\ln|x-y|\ln\left(\frac{|x-y|}{|x-z|}\frac{|z-u|}{|y-u|}\right) m(dx)m(dy)m(dz)m(du). \end{aligned}$$

References

[**1**] Fukushima, M.: Dirichlet Forms and Markov Processes. North–Holland/ Kodansha, Amsterdam–Oxford–New York (1980).

[**2**] Ma, Z.; Röckner, M.: Introduction to the Theory of (Non–Symmetric) Dirichlet Forms. Springer, Berlin (1991).

[**3**] Röckner, M.; Wielens, N.: Dirichlet forms–Closability and change of speed measure. In Infinite dimensional analysis and stochastic processes, Research Notes in Mathematics 124, Pitman, Boston–London–Melbourne, 1985.

[**4**] Brasche, J. F.: Dirichlet forms and non–standard Schrödinger operators. In Schrödinger operators–standard and non–standard, World Scientific, Singapore–New Jersey–London–Hong Kong, 1989.

[**5**] Brasche, J. F.; Exner, P.; Kuperin, Y.; Šeba, P.: Schrödinger operators with singular interactions. Will appear in Journ. Math. Anal. Appl.

[**6**] Solomjak, M. Z.: On estimates of the singular numbers of integral operators IV, Vestnik Leningrad Univ. Vol.3 (1976), 83–96.

[**7**] Klaus, N.: On the bound state of Schrödinger operators in one dimension, Ann. Phys. 108 (1977), 288–300.

[8] Newton, R. G.: Bounds on the number of bound states for the Schrödinger equation in one and two dimensions, J. Operator Theory 10 (1983), 119–125.

[9] Seto, N.: Bargmann's inequalities in spaces of arbitrary dimension, Publ. R.I.M.S. 9 (1974),429–461.

Operator Theory:
Advances and Applications, Vol. 70

A Fermi-type rule for contact embedded-eigenvalue perturbations

J-P. Antoine, P. Exner, P.Šeba and J.Shabani

Abstract

A perturbation theory of embedded eigenvalues is constructed for a class of models with a contact interaction which are inspired by heavy-quarkonia mesonic decays.

1 Introduction

Contact–type interactions have been studied intensively in the last decade — see [2, 3, 14, 15] and the papers [5, 7–10, 16–18]; a more complete bibliography can be found in a journal version of this paper [4]. Recently they have been shown to yield solvable models of some decay and resonance–scattering processes [12, 13]; the importance of this observation stems from the fact that there are only a few situations where the embedded–eigenvalue perturbation problem can be solved — see [11], [21, Sec.XII.6] and references therein.

The aim of this talk is to present another model of this type; in distinction to [13] the embedded eigenvalues will correspond here to a potential interaction rather than to boundary conditions. On the other hand, comparing to [12], the perturbation–theory parameter will be now contained in the boundary condition as the strength of the contact interaction. This allows us to prove a Fermi–type rule for the considered class of contact–interaction decays.

To link the mathematical problem with a physical situation, we shall consider excited states of heavy quarkonia decaying into mesons. In reality, this is a complicated process governed by the QCD Lagrangian, however, it is known that both the quark–antiquark pair and the mesons resulting from the decay are non–relativistic with a reasonable degree of accuracy [6, 19], so we can model this unstable system coupling the two dynamics directly. Moreover, we know that the transition between the quark and meson states can occur only at small distances. Hence it is reasonable to employ a contact–type interaction: we shall suppose that the quarks can annihilate when they hit each other giving rise to a meson pair at a distance R (this parameter plays essentially no role in the model).

2 Description of the model

For simplicity, we remove the centre–of–mass motion in both the quark and meson channels. Moreover, we shall assume that the interaction is rotationally invariant so one can perform the partial–wave decomposition. Following the standard contact–interaction ideology, we construct then the model Hamiltonian as

$$H = \bigoplus_{\ell=0}^{\infty} W^{-1} H^{(\ell)} W \otimes I_\ell \tag{1}$$

where I_ℓ is the unit operator on $L^2(S_2)$, $W := W_1 \oplus W_2$ with $(W_j\psi_j)(r) := r\psi(r)$, and furthermore, $H^{(\ell)}$ denotes a self–adjoint extension of the operator $H_0^{(\ell)} := H_{0,1}^{(\ell)} \oplus H_{0,2}^{(\ell)}$ on $L^2(R_1,\infty) \oplus L^2(R_2,\infty)$ defined by

$$H_{0,j}^{(\ell)} := -\frac{1}{m_j}\frac{d^2}{dr_j^2} + V_j(r_j) + \frac{\ell(\ell+1)}{m_j r_j^2} + 2(m_j - m_2)c^2 \tag{2}$$

with $D(H_{0,j}^{(\ell)}) := C_0^\infty(R_j,\infty)$. In other words, $H_{0,j}^{(\ell)}$ are the partial–wave Hamiltonians in the quark and meson channels, $j = 1,2$, respectively, restricted to function supported out of the interaction region; a possible difference of the rest energies has been added to the quark potential.

We set in the following $R_1 = 0$ and $R_2 := R \geq 0$. Furthermore, we have to specify requirements on the potentials. For simplicity, we suppose that (apart from the interaction with the quark channel) the mesons are free, *i.e.*,

$$V_2 = 0 \tag{3}$$

(this is not quite realistic in case of the decay to charged mesons but the model will be easier to solve with this assumption). On the other hand, for the quark potential we adopt the rather weak assumption that

$$V_1 \text{ is locally integrable} \quad \text{and} \quad \lim_{r\to 0+} V_1(r) \quad \text{exists and is finite}. \tag{4}$$

With the standard picture of a quarkonium state in mind, we shall also suppose that the quarks are confined,

$$\lim_{r\to\infty} V_1(r) = \infty, \tag{5}$$

even if the model works without this hypothesis as long as the first–channel free Hamiltonian has eigenvalues embedded in the continuous spectrum of the other channel (in conclusion, we shall mention the case when (4) is modified by adding a Coulomb potential in the quark channel). To be able to construct the self–adjoint extensions, one has to know first whether it is possible and how many there are. Using [21, Thm.X.10] we check easily

2.1 Proposition: Under the assumptions (3)–(5), the deficiency indices of the operators $H_0^{(\ell)}$ are:

(a) $(2,2)$ if $\ell=0$,

(b) $(1,1)$ if $\ell\geq 1$ and $R=0$,

(c) $H_0^{(\ell)}$ is *e.s.a.* if $\ell\geq 1$ and $R=0$.

Thus the two channels can be coupled in the case (a) only, since otherwise at least one of the operators $H_{0,j}^{(\ell)}$ is *e.s.a.* This means, in particular, that the decay into higher partial waves, $\ell=1,2,\dots$, is forbidden, hence we shall put $\ell=0$ and drop the index ℓ in the following.

3 Boundary conditions

The most straightforward way to couple the operators $H_{0,j}$ is to subject them to suitable boundary conditions at the points $r_1=0$ and $r_2=R$. Using the standard argument, one can check that the adjoint H_0^* acts as the same differential operator as H_0 and its domain consists of all $f:=\begin{pmatrix} f_1 \\ f_2 \end{pmatrix}$ with f_j, f_j' absolutely continuous, $f_j''\in L^2_{loc}$ and $f_j''-m_jV_jf_j\in L^2(R_j,\infty)$ — see, *e.g.*, [21, Appendix to Sec.X.1]; the self-adjoint extension are then obtained by suitable restrictions of this domain. The most general form of the boundary conditions is

$$f_1(0) = a_{11}f_1'(0)+a_{12}f_2'(R)\,,\quad f_2(R) = a_{21}f_1'(0)+a_{22}f_2'(R)\,. \tag{6}$$

Choosing appropriate boundary functionals, we find easily

3.1 Proposition: The conditions (6) specify a self-adjoint extension of $H_{0,j}$ *iff* the coefficients satisfy the relations $\overline{a_{jj}}=a_{jj}$, $j=1,2$, and $m_1\overline{a_{21}}=m_2a_{12}$.

We choose for our model a subclass among the extensions specified by the proposition above. The diagonal coefficients a_{jj} correspond to a point interaction in the j-th channel supported by the point $r_j=R_j$. There is no physical reason why such an interaction should be present in the considered system, hence we put $a_{jj}=0$. The channels are coupled through the off-diagonal coefficients, *i.e.*, the model boundary conditions read

$$f_1(0) = a\,f_2'(R)\,,\quad f_2(R) = \frac{m_2}{m_1}\,\bar{a}\,f_1'(0)\,, \tag{7}$$

where $a:=a_{12}$ is the coupling strength; it is a reasonable choice because in the non-interacting case, $a=0$, we get the Dirichlet boundary condition in both channels as expected. The extension corresponding to the boundary conditions (7) will be denoted as H_a in the following.

4 The resolvent

In the standard picture [21, Sec.XII.6], [11, Chap.3] the leading behaviour of the unstable states is determined by poles in the analytic continuation of the resolvent, in fact, the mere existence of decaying modes of the system is usually put into correspondence with existence of these resonance states. Hence, in order to solve the model, one has to find the resolvent of the Hamiltonian H_a. In view of the particular form of this operator, this can be done by means of the Krein formula [1, Sec.106]. Denote by ϕ, χ the solutions to the equation

$$\left(-\frac{1}{m_j}\frac{d^2}{dr^2} + V_1(r) + 2(m_1 - m_2)c^2 \right) f(r) = z f(r) \tag{8}$$

for $z \notin I\!R$ such that $\phi(0) = 0$ and χ is L^2 at infinity; because of the assumptions (4) and (5) they are unique up to a constant (the facts from the theory of ordinary differential equations we use here are usually formulated for the case of smooth coefficients but they hold for locally integrable coefficients as well [20]). They allow us to express the free quark–channel resolvent

$$G_1(r,s;z) = \frac{1}{W(\phi,\chi)} \begin{cases} \phi(r;z)\chi(s;z) & \ldots \quad r \le s \\ \chi(r;z)\phi(s;z) & \ldots \quad r \ge s \end{cases} \tag{9}$$

where $W(\phi,\chi) := \phi\chi' - \chi\phi'$ is the Wronskian of the two solutions. On the other hand, by the assumption (3) the free meson resolvent can be expressed explicitly as

$$G_2(r,s;z) = \frac{1}{2ik}\left(e^{ik(r+s-2R)} - e^{ik|r-s|}\right), \tag{10}$$

where $k := \sqrt{m_2 z}$ is the meson momentum with the cut conventionally chosen along the positive real axis. Now we have the following result.

4.1 Theorem: Under the stated assumptions, the resolvent of H_a is given, for z outside the two spectra, by the formulae

$$(H_a - z)^{-1} = (H_0 - z)^{-1} + \sum_{j,k=1}^{2} \lambda_{jk}(z)(F_k^{\bar z}, \cdot)F_j^z, \tag{11}$$

where

$$F_1^z(r) := \begin{pmatrix} \chi(r;z) \\ 0 \end{pmatrix}, \quad F_2^z(r) := \begin{pmatrix} 0 \\ e^{ikr} \end{pmatrix} \tag{12}$$

and the coefficient functions are

$$\lambda_{11}(z) := \frac{-ikm_2|a|^2}{m_1\chi(0;z)D(\chi,a;z)}, \quad \lambda_{21}(z) := \frac{-m_2\bar a e^{-ikR}}{m_1 D(\chi,a;z)},$$
$$\lambda_{12}(z) := \frac{a e^{-ikR}}{D(\chi,a;z)}, \quad \lambda_{22}(z) := \frac{m_2|a|^2\chi'(0;z)e^{-2ikR}}{m_1 D(\chi,a;z)}, \tag{13}$$

where

$$D(\chi,a;z) := \chi(0;z) - ik|a|^2\frac{m_2}{m_1}\,\chi'(0;z)\,. \tag{14}$$

Proof: The component functions in (12) solve the equations $(H^*_{0,j}-z)f=0$ for $j=1,2$, respectively. Furthermore, χ is L^2 at infinity by definition, and the same is true for e^{ikr} if $z\notin I\!R_+$ as long as k belongs to the upper complex halfplane; it remains therefore to find the coefficients $\lambda_{jk}(z)$. To this end, we use the fact that for any $g\in L^2(I\!R_+)\oplus L^2(R,\infty)$, the vector $f:=(H_a-z)^{-1}g$ has to belong to $D(H_a)$, *i.e.*, its components must satisfy the boundary conditions (7). This requirement yields a system of four linear equations for the coefficients which is solved by (13). ∎

Hence we are able to determine the analytic structure of the resolvent, in particular, of its projection on the quark–channel discrete spectral subspace which is essential for determining the decaying–state poles [11, Sec.3.1]. The behaviour of $D(\chi,a;\cdot)$ plays a decisive role. The first term in this function is zero at the quark bound–state energies, where the solutions ϕ,χ become linearly dependent, however, the corresponding singularities are easily seen to cancel with those of $G_1(r,s;\cdot)$ so the only poles in the analytically continued resolvent come from the lower–halfplane zeros of (14); they are given by the equation

$$\chi(0;z) - ik|a|^2\frac{m_2}{m_1}\,\chi'(0;z) = 0 \tag{15}$$

which has to be solved with respect to z.

5 The Fermi rule

The equation (15) can be solved for particular potentials; we refer to [4] for the examples of square–well, linear and harmonic confinement. The results motivate the following general claim.

5.1 Theorem: Under the assumptions (3)–(5), the quarkonium decay width is given for the n–th s–wave state by

$$\Gamma_n(a) = 8\pi k_n\,\frac{m_2}{m_1^2}\,|a|^2|\psi_n(0)|^2 + \mathcal{O}(|a|^6)\,,\quad k_n := \sqrt{m_1E_n}\,, \tag{16}$$

provided $E_n>0$, where E_n is the bound–state energy (scaled by the difference of the rest energies — *cf.* (8)) and $\psi_n(0)$ is the value of the corresponding wavefunction at the origin.

5.2 Remark: Though it is not the matter of our interest here, the present analysis yields also an expression for perturbation of the quarkonium energies in the case where the unperturbed eigenvalues are isolated, $E_n < 0$. One has

$$E_n(a) = E_n + 4\pi\kappa_n \frac{m_2}{m_1^2} |a|^2 |\psi_n(0)|^2 + \mathcal{O}(|a|^4), \tag{17}$$

where $\kappa_n := \sqrt{-m_1 E_n}$; the fourth–order term can be computed easily from the formula (18) below.

Proof of Theorem 5.1: For simplicity, we put $m_1 = 1$ and define $g := m_2|a|^2$; the extension to the general case is straightforward. Equation (15) takes then the form

$$D(k,g) := \chi(0,k^2) - ikg\chi_r(0,k^2) = 0,$$

where χ_r, χ_{k^2} *etc.* denote the partial derivatives with respect to the indicated variable. One has $D(k_n,0) = 0$ for $n = 0,1,\dots$, so the implicit function theorem may be applied provided $D_k(k_n,0) = 2k_n\chi_{k^2}(0,k_n^2) \neq 0$. It yields

$$\frac{dk}{dg}(k_n,0) = \frac{i\chi_r(0,k_n^2)}{2\chi_{k^2}(0,k_n^2)}.$$

The higher derivatives can be computed in the same way; we restrict ourselves to the second one obtaining

$$\begin{aligned} k_n(g) = k_n &+ \frac{ig}{2} \frac{\chi_r(0,k_n^2)}{\chi_{k^2}(0,k_n^2)} \\ &- \frac{g^2}{2} \left[\frac{3}{4k_n} \left(\frac{\chi_r}{\chi_{k^2}} \right)^2 + \frac{k_n}{2\chi_{k^2}^3} (2\chi_r\chi_{k^2}\chi_{rk^2} + \chi_{k^2k^2}) \right] (0,k_n^2) + \mathcal{O}(g^3). \end{aligned} \tag{18}$$

Hence to prove the formulae (16) and (17), one has to check that

$$2i\frac{dk}{dg}(k_n,0) = 4\pi|\psi_n(0)|^2 = (f_n'(0))^2 = \frac{\chi_r(0,k_n^2)^2}{\int_0^\infty \chi(s,k_n^2)^2 ds},$$

where f_n is the normalized reduced wavefunction, *i.e.*, to establish the identity

$$\chi_{k^2}(0,k_n^2) = -\frac{1}{\chi_r(0,k_n^2)} \int_0^\infty \chi(s,k_n^2)^2\, ds. \tag{19}$$

To do that one has realize that the function $\chi(\cdot,k'^2)$ satisfies the equation

$$-f'' + (V-k^2)f = (k'^2 - k^2)f. \tag{20}$$

In order to solve this equation, we interpret its *lhs* as $(H_N - k^2)f$, where H_N is quark Hamiltonian of the equation (8), but with the *Neumann* boundary condition at the origin. Introducing the corresponding regular solution $\tilde{\phi}$ which verifies the boundary conditions $\tilde{\phi}_r(0,k^2)=0$ and $\tilde{\phi}(0,k^2)=1$, we may solve the equation (20) immediately as

$$\chi(r,k'^2) = c_1(k'^2)\chi(r,k^2) + c_2(k'^2)\tilde{\phi}(r,k^2) + (k^2-k'^2)g(r,k^2), \tag{21}$$

with

$$\begin{aligned} g(r,k^2) &:= ((H_N-k^2)^{-1}\chi)(r,k^2) \\ &= \frac{\chi(r,k^2)}{\chi_r(0,k^2)}\int_0^r \tilde{\phi}(s,k^2)\chi(s,k^2)\,ds + \frac{\tilde{\phi}(r,k^2)}{\chi_r(0,k^2)}\int_r^\infty \chi(s,k^2)^2 ds\,. \end{aligned} \tag{22}$$

Since the *rhs* of (21) is L^2 at infinity by definition, it suffices to check that $g(\cdot,k^2)$ is square integrable; indeed, this implies $c_2(k'^2)=0$ for any k' and $\lim_{k'\to k} c_1(k'^2)=1$ so that

$$\chi_{k^2}(r,k^2) = c_1'(k^2)\chi(r,k^2) - g(r,k^2)\,. \tag{23}$$

However, at the points $k=k_n^2$ the solutions χ and ϕ are linearly dependent and equal up to a multiplicative factor to the bound-state (reduced) wavefunction which satisfies, of course, the Dirichlet boundary condition at the origin. Hence (23) gives $\chi_{k^2}(0,k_n^2)=-g(0,k_n^2)$, which is nothing but the identity (19).

Finally, since $\chi(\cdot,k^2)$ belongs to $L^2(0,\infty)$ for any k, the function $g(\cdot,k^2)$ is square integrable provided k^2 does not belong to the spectrum of H_N. This is true, of course, since the free quark Hamiltonian H_D corresponds to the Dirichlet condition, and therefore $\sigma(H_D)\cap\sigma(H_N)=\emptyset$ (this is a reason why we have considered the operator H_N). Notice also that the *rhs* of (19) is non-zero so $\chi_{k^2}(0,k_n^2)\neq 0$; this justifies the use of the implicit-function theorem. ∎

5.3 Remark: In fact the regularity assumption in (4) is not very realistic; since quarks are charged particles, they should have in addition to the confining potential V_1 also an attractive Coulomb potential $V_C(r)=-\gamma r^{-1}$, $\gamma>0$ [19]. Fortunately, the above analysis can be easily modified to cover this case. It suffices to replace in the boundary conditions (7) the value $f_1'(0)$ which becomes now singular by

$$f'_{1,reg}(0) := \lim_{r\to 0+} r^{-1}\left[f_1(r) - f_1(0+)(1+\gamma r\log|\gamma|r)\right], \tag{24}$$

where $f_1(0+)$ is the boundary value of the function itself, which remains well defined. Replacing now $V_1(r)$ in (8) by $V_1(r)-\gamma/r$, we may repeat the discussion of Sec.4. First, the particular solutions ϕ and ψ are replaced by their Coulombic counterparts ϕ_C and χ_C with $\phi_C(0)=0$ and χ_C being L^2 at infinity. As before, these solutions are unique up to a multiplicative constant. The main difference is that (as in the pure Coulombic case, $V_1=0$), χ' has a logarithmic singularity at the origin for $k\neq -i\gamma/n$. The free quark-channel resolvent reads, of course

$$G_1^C(r,s;z) = \frac{1}{W(\phi_C,\chi_C)}\begin{cases}\phi_C(r;z)\chi_C(s;z) & \dots \quad r\le s\\ \chi_C(r;z)\phi_C(s;z) & \dots \quad r\ge s\end{cases} \tag{25}$$

Thus the analogue of Theorem 4.1 can be proved in exactly the same way; the statement is unchanged except that the denominator function D takes the form

$$D(\chi_C, a; z) := \chi_C(0; z) - ik|a|^2 \frac{m_2}{m_1} \chi'_{C,reg}(0; z) . \tag{26}$$

As for Theorem 5.1 and Remark 5.2, they remain valid too; we refer to [4] for details.

References

[1] N.I.Akhiezer, I.M.Glazman. *Theory of Linear Operators in Hilbert space*, 3rd edition, Viša Škola, Kharkov 1978 (in Russian; English translation of the 1st ed.: F.Ungar, New York 1963).

[2] S.Albeverio, J.Fenstad, H.Holden, T.Lindstrøm, eds.: *Ideas and Methods in Mathematical Physics*, vol.2, Cambridge University Press 1992.

[3] S.Albeverio, F.Gesztesy, R.Høegh-Krohn, H.Holden: *Solvable Models in Quantum Mechanics*, Springer, Heidelberg 1988.

[4] J.-P.Antoine, P.Exner, P.Šeba, J.Shabani: A mathematical model of heavy–quarkonia mesonic decays, *preprint UCL–IPT–93–03*, Louvain–la–Neuve.

[5] J.-P.Antoine, F.Gesztesy, J.Shabani: Exactly solvable models of sphere interaction in quantum mechanics, *J.Phys.* **A20** (1987), 3627-3712.

[6] T.Appelquist, H.D.Politzer: Heavy quarks and long–lived hadrons, *Phys. Rev.* **D12** (1975), 1404–1414.

[7] J.Brasche, P.Exner, Yu.A.Kuperin, P.Šeba: Schrödinger operators with singular interactions, *J.Math.Anal.Appl.*, to appear

[8] W.Bulla, T.Trenckler: The free Dirac operator on compact and non-compact graphs, *J.Math.Phys.* **31** (1990), 1157-1163.

[9] S.E.Cheremshantsev: Hamiltonians with zero–range interactions supported by a Brownian path, *Ann.Inst.H.Poincaré: Phys.théor.* **56** (1992), 1–25.

[10] J.Dittrich, P.Exner, P.Šeba: Dirac operators with a spherically symmetric δ-shell interaction, *J.Math.Phys.* **30** (1989), 2875-2882.

[11] P.Exner: *Open Quantum Systems and Feynman Integrals*, D.Reidel, Dordrecht 1985.

[12] P.Exner: A model of resonance scattering on curved quantum wires, *Ann.Physik* **47** (1990), 123-138.

[13] P.Exner: A solvable model of two–channel scattering, *Helv.Phys.Acta* **64** (1991), 593–609.

[14] P.Exner, P.Šeba, eds.: *Applications of Self-Adjoint Extensions in Quantum Physics*, Lecture Notes in Physics, vol.324, Springer, Heidelberg 1989.

[15] P.Exner, P.Šeba, eds.: *Schrödinger Operators, Standard and Non–Standard*, World Scientific, Singapore 1989.

[16] P.Exner, P.Šeba, P.Šťovíček: Quantum interference on graphs controlled by an external electric field, *J.Phys.* **A21** (1988),4009-4019.

[17] N.I.Gerasimenko, B.S.Pavlov: Scattering problem on non-compact graphs, *Teor. Mat.Fiz.* **74** (1988), 345-359 (in Russian).

[18] F.Gesztesy, P.Šeba: New exactly solvable models of relativistic point interactions, *Lett.Math.Phys.* **13** (1987), 345-358.

[19] W.Lucha, F.Schoberl, D.Gromes: Bound states of quarks, *Phys.Rep.* **200** (1991), 127–240.

[20] M.A.Naimark: *Linear Differential Operators*, 2nd edition, Nauka, Moscow 1968 (in Russian; English transl. of the 1st ed.: F.Ungar, New York 1967).

[21] M.Reed, B.Simon: *Methods of Modern Mathematical Physics, I.Functional Analysis, II. Fourier Analysis. Self-Adjointness, III. Scattering Theory, IV. Analysis of operators*, Academic Press, New York 1972-1979.

[22] R.Van Royen, V.F.Weiskopf: Hadron decay processes and the quark model, *N.Cim.* **50A** (1967), 617–645.

J-P. Antoine[a], Institut de Physique Théorique, Université Catholique de Louvain, B-1348 Louvain-la-Neuve, Belgium;
e-mail antoine@fyma.ucl.ac.be

P. Exner, P. Šeba, Theory Division, Nuclear Physics Institute, AS CR, CZ–25068 Řež near Prague, Czechia;
e-mail exner *and* seba@ujf.cas.cz

J.Shabani, Faculté des Sciences, Université du Burundi, Bujumbura, Burundi

Operator Theory:
Advances and Applications, Vol. 70

A Simple Model for Predissociation

P. Duclos and B. Meller

Abstract

We analyse a very simple class of one dimensional two by two matrix Schrödinger operators. Their diagonal part has embedded eigenvalues in the continuous spectrum which become resonances when the off-diagonal part is turned on. Our analysis is semiclassical and contains a regular perturbative calculus of these resonances, asymptotics of the Fermi rule contribution to the width of these as well as lower bounds on the corresponding life time.

1 Introduction

We start directly exposing the model; the discussion about the content of this paper is split in several parts which are put at the end of their relevent sections.

1.1 The Model

Let

$$H := H^d + W, \quad H^d := \begin{pmatrix} H^1 & 0 \\ 0 & H^2 \end{pmatrix}, \quad W := \begin{pmatrix} 0 & V^{1,2} \\ V^{2,1} & 0 \end{pmatrix}$$

be a matrix Schrödinger operator acting on $L^2(\mathbb{R}) \oplus L^2(\mathbb{R}) =: \mathcal{H}$, where

$$\begin{aligned} H^k &:= D^2 + V^k, \quad k = 1, 2, \quad D := \frac{\hbar}{i}\frac{d}{dx}, \\ V^2(x) &:= x^2, \quad V^{1,2} := \hbar(tD + Dt) = -V^{2,1}. \end{aligned}$$

Dilation and translation analyticity of the potentials play here an important role; we assume

V^1 and it are bounded analytic multiplication operators in $\Sigma_{\beta_0,\eta_0} := \{z \in \mathbb{C}, |\operatorname{Arg} z| < \beta_0 \text{ or } |\operatorname{Im} z| < \eta_0\}$, β_0, η_0 *being both strictly positive; V^1 and it are real on $\mathbb{R}$.* **(H1)**

The images of all operators under the scaling $x \to e^\theta x$ will be denoted when necessary by a subcript θ. With these assumptions we have the following

Theorem 1. H_θ, H_θ^i, i=1,2 are selfadjoint analytic families of type A for all θ such that $|\mathrm{Im}\,\theta| < \hat{\beta}_0 := \min\{\beta_0, \frac{\pi}{4}\}$ with domains

$$\begin{aligned} \mathcal{D}(H_\theta) &= \mathcal{D}(H_\theta^1) \oplus \mathcal{D}(H_\theta^2), \\ \mathcal{D}(H_\theta^1) &= \mathcal{H}^2(I\!R), \ \ \mathcal{D}(H_\theta^2) = \mathcal{H}^2 \cap \hat{\mathcal{H}}^2(I\!R). \end{aligned}$$

The definition of "type A family" may be found in [Ka, Ch.VII §2.1]. $\mathcal{H}^n(I\!R)$ denotes the usual Sobolev space and $\hat{\mathcal{H}}^n$ its Fourier image. The proof of this theorem is rather standard. For H_θ^1 it is obvious, since V_θ^1 is bounded analytic; for H_θ^2 a detailed proof can be found in [BCD2]. H_θ can be treated perturbatively as in [DEŠ]; in the form sense on $\mathcal{D}(H_\theta^d)$ one has

$$|e^\theta W_\theta|^2 \le \hbar^2\{8\|t_\theta\|^2 D^2 + 2\hbar^2\|t'_\theta\|^2\} \begin{pmatrix} 1 & 0 \\ 0 & 1 \end{pmatrix}$$

which shows that W_θ is H^d-bounded with relative bound zero.

In order to have a reasonable spectrum for H^1 we assume

$\exists\varepsilon > 0,\ \exists v_\infty < 0,\ \forall 0 < |\mathrm{Im}\,\theta| < \beta_0,\ \ V_\theta^1 = v_\infty + \mathcal{O}(x^{-1-\varepsilon})$, *as* $|x|$ *tends to infinity.* **(H2)**

This implies in particular that the essential spectrum of H_θ^1 is simply: $\sigma_{ess}(H_\theta^1) = v_\infty + e^{-2\theta} I\!R_+$.

1.2 Discussion and Further Hypothesis

H^2 has only discrete spectrum, $\sigma(H^2) = (2I\!N + 1)\hbar$. When θ equals zero, these eigenvalues are embedded in the continuous spectrum of H^1. The effect of the perturbation W is usually to couple these bound states to the scattering states of H^1. If the quantum system is initially prepared in a state $0 \oplus \varphi^2$, φ^2 being a bound state of H^2, it will eventually turn into a scattering state. The mechanism behind this effect is very similar to the one which causes the existence of shape resonances, viz. tunneling through the potential barrier (see e.g. [CDKS, HeSj]).

If the two potentials V^1 and V^2 cross one can reduce the problem for the lowest energies to a shape resonance situation [K]; the relevent effective potential is $\min\{V^1, V^2\}$ which possesses a well, separated from the escaping regions by a potential barrier.

Here we want to address explicitly the case with no crossing of V^1 and V^2: we assume

$$\max V^1 < \min V^2 = 0. \qquad \textbf{(H3)}$$

Now, at first glance, one would say there is no barrier. But classically one can see that for the energy $e_0 = (2n+1)\hbar$ (the nth quantum state of H^2) the allowed momenta for H^1 and H^2 are separated by a gap of size $\sqrt{-\max V^1}$ for $\hbar$ small enough. This gap indicates a classically forbidden region for the hamiltonian H^d in the momentum space rather than in the configuration space. So we can speak of a *dynamical barrier* being present and of *dynamical tunneling* as the reason for the escape of the bound state (see [AD] for the same discussion in the case of the reflection over a potential barrier). However assumption (H3) is much stronger than necessary for this phenomenon to take place. Wilkinson [W] and Martinez [Ma2] have remarked that it is sufficient to require that the energy shells do not cross: $\{H^1_{cl}(q,p) = H^2_{cl}(q,p) = E\}$[1] $= \emptyset$ to get such an effect which they called respectively *tunneling in phase space* and *microlocal tunneling.*

The dynamical tunneling manifests itself in the so-called resonances of the quantum system. According to the standard machinery ([AgC, RS4]) these resonances are recognized as complex eigenvalues of H_θ which are the perturbed eigenvalues of H^d_θ. Notice that these eigenvalues do not depend on θ. Since the essential spectrum of H^1_θ has turned down in the complex energy plane, the eigenvalues of H^d_θ to be perturbed are now isolated provided H^1_θ does not have eigenvalues too close. This last requirement is achieved by imposing a nontrapping condition on V^1:

$$\exists S < 0, \quad \forall 0 \le \beta < \beta_0 \quad \operatorname{Im} e^{i2\beta} V^1_{i\beta} \le \beta S \tag{\textbf{H4}}$$

The perturbation W_θ being H^d_θ-bounded with relative bound zero our problem falls into the category of regular perturbation theory. Section 2 is devoted to this perturbation theory. Once the resonances are shown to exist we shall give for a restricted model the asymptotics of the Fermi rule contribution to their width in section 3 and finally estimate this width in section 4.

Such a model with $V^1 = -1$ and $V^{1,2} = V^{2,1} = \hbar^2$ has been proposed by J. Asch [A] as a simple model to understand the predissociation phenomenon in diatomic molecules. V^1 and V^2 play the role of the electronic curves, $\hbar^2$ the inverse of the nucleus mass. We have chosen, here, more realistic coupling terms. Reducing the complete molecular hamiltonian to a two by two matrix of this type is the purpose of the Born-Oppenheimer approximation, see e.g. [CDS] and [Ma1]. In these articles the basic algebraic tool is a method bearing numerous names: Brilloin-Wigner, Feshbach, Grushin, Schur... not to forget the Livsic Matrix [Ho]. However by "putting" into H^1 all the electronic curves except the second one, it is conceivable to obtain the same result not using the energy dependent perturbation theory, see e.g. [DEŠ].

2 Perturbation Expansion of the Resonances

We first prove that under (H1,4) H^1_θ has no spectrum close to a given eigenvalue e_0 of H^2. It will be sufficient to have this property in the closed neighbourhood of e_0 bounded by the contour $\Gamma := \{z \in \mathbb{C}, |z - e_0| = \hbar\}$.

[1] H^i_{cl} denotes the classical Hamiltonian funtion associated to H^i.

2.1 Nontrapping Estimates

Lemma 2. Under hypothesis (H1,4) one has:

$$\forall 0<\beta<\beta_0,\ \exists c_\beta^1 := \frac{-2}{\beta S},\ \forall 0<\hbar<\frac{1}{c_\beta^1},\ \forall z\in\Gamma,\ \|R_{i\beta}^1(z)\|\le c_\beta^1.$$

Sketch of the proof: Such type of result is now rather standard (cf. [BCD1]). The condition on $\hbar$ (stronger than necessary), insures that

$$\Gamma\subset\nu := \{z\in\mathbb{C},\ \operatorname{Im} e^{i2\beta}z\ge\beta S\}.$$

ν is a set of complex energies that cannot be resonances of H^1 due to (H4). More precisely, since in the form sense on $\mathcal{D}(H^1)$ one has:

$$|H_{i\beta}^1 - z|\ge \operatorname{Im} e^{i2\beta}(z-H_{i\beta}^1) = \operatorname{Im} e^{i2\beta}(z-V_{i\beta}^1)\ge \operatorname{Im} e^{i2\beta}z-\beta S\ge -\beta\frac{S}{2},$$

which yields the a priori estimate we need to bound $R_{i\beta}^1(z)$. In the last step we have explictely used that z belongs to Γ and the condition on $\hbar$. ■

2.2 Stability of the Resolvent Set of H_θ^d

The previous lemma gives sufficient conditions to insure that Γ is included in $\rho(H_\theta^1)$, the resolvent set of H_θ^1. Since it is well known (see e.g. [BCD2]) that the spectrum of H_θ^2 is invariant with respect to θ as long as $|\operatorname{Im}\theta|<\frac{\pi}{4}$, we conclude that Γ is also in $\rho(H_\theta^d)$ under this extra condition on $\operatorname{Im}\theta$. Therefore, by standard perturbation theory, Γ will also be in $\rho(H_{i\beta})$ if in addition

$$\forall z\in\Gamma,\quad \|R_{i\beta}^1(z)V_{i\beta}^{1,2}R_{i\beta}^2(z)V_{i\beta}^{2,1}\|<1.$$

$R_{i\beta}^1(z)$ is already estimated by lemma 2. In the next lemma we shall estimate $V_{i\beta}^{1,2}R_{i\beta}^2(z)V_{i\beta}^{2,1}$ and other quantities needed in the sequel. As in Kato [Ka, Ch.II §2.1], we use the notation:

$$S_{i\beta}^{2\,(k)} := (\hat{R}_{i\beta}^2(e_0))^k,\ \text{if } k\ge 1 \text{ and } S_{i\beta}^{2\,(0)} = -P_{i\beta}^2$$

where $\hat{R}_{i\beta}^2(e_0)$ is the reduced resolvent of $H_{i\beta}^2$ at e_0 and $P_{i\beta}^2$ the corresponding spectral projection.

Lemma 3. For any $0<\beta<\hat{\beta}_0$ there exists c_β^2 such that for any $\hbar>0$,

$$\begin{aligned}\forall z\in\Gamma,\quad \|V_{i\beta}^{1,2}R_{i\beta}^2(z)V_{i\beta}^{2,1}\| &\le c_\beta^2\hbar^2,\\ \forall k\ge 0,\quad \|V_{i\beta}^{1,2}S_{i\beta}^{2(k)}V_{i\beta}^{2,1}\| &\le c_\beta^2\hbar^{3-k}.\end{aligned}$$

Sketch of proof: By the scaling $x \to \sqrt{\hbar}x$ $V^{1,2}_{i\beta}R^2_{i\beta}(z)V^{2,1}_{i\beta}$ is unitarily equivalent to:

$$\hbar^2(t_{i\beta}(\hbar^{\frac{1}{2}}x)\partial_x + c.)(-e^{-i2\beta}\partial_x^2 + e^{i2\beta}x^2 - \zeta)^{-1}(\partial_x t_{i\beta}(\hbar^{\frac{1}{2}}x) + c.)$$

where $c.$ means 'commutated term' and ζ belongs to the fixed compact set $\hbar^{-1}\Gamma := \{\zeta \in \mathbb{C}, |\zeta - (2n+1)| = 1\}$. The first statement follows easily by the continuity in ζ of the lhs of the formula above since $t_{i\beta}$ is bounded and ∂_x relatively bounded to $-e^{-i2\beta}\partial_x^2 + e^{i2\beta}x^2$. For the derivation of the second statement we use the Cauchy formula and the same scaling trick. ■

Thus we have obtained the stability of the resolvent set:

$$\left(0 < \beta < \hat{\beta}_0 \text{ and } 0 < \hbar < \frac{1}{c_\beta}\right) \Rightarrow \Gamma \subset \rho(H_{i\beta}), \quad c_\beta := \max\{c^1_\beta, c^2_\beta\}. \tag{1}$$

2.3 Stability of the Spectrum of H^2 and Existence of Resonances for H

The preceeding analysis proved that for $\hbar$ small enough $P_{i\beta}$, the eigenprojection of $H_{i\beta}$ associated to Γ, is well defined. This certainly remains true if one replaces W by αW with $0 \le \alpha \le 1$ thus defining a continuous family of projections interpolating between $P_{i\beta}$ $(\alpha = 1)$ and $P^2_{i\beta}$ $(\alpha = 0)$. Consequently

$$\forall 0 < \beta < \hat{\beta}_0,\ \forall 0 < \hbar < \frac{1}{c_\beta}, \quad \dim P_{i\beta} = 1.$$

Standard arguments on resonances ([RS4, Ch. XIII.10]) insure that the imaginary part of the eigenvalue associated to $P_{i\beta}$ cannot be positive. So we have proven the existence of a resonance of H close to each eigenvalue of H^2; however we cannot exclude a vanishing imaginary part of this resonance.

Remark 1. c^1_β depends only on β and S whereas c^2_β depends on β and the quantum number n of the eigenvalue e_0 of H^2. So the range of values of $\hbar$ for which the existence of the resonances is obtained depends on β, S and n.

2.4 Convergent Expansion of the Resonances

We denote by E the resonance obtained by perturbation of $e_0 = (2n+1)\hbar$. With the standard formula of regular perturbation theory [Ka, Ch.II§1] and noticing that W is off-diagonal we get:

$$E = \sum_{m=0}^{\infty} e_m, \tag{2}$$

$$e_m = \frac{1}{2m}\sum_{l=0}^{m-1}\sum_{\sigma} \operatorname{tr} V^{2,1}(R^1)^{k_1}V^{1,2}(S^2)^{l_1}\dots V^{2,1}(R^1)^{k_m}V^{1,2}(S^2)^{l_m},$$

$$\sigma = \{\sum l_i = l \wedge \sum k_i = 2m-1-l,\ l_i \geq 0,\ k_i \geq 1\},$$

where we have dropped the indices $i\beta$; all the resolvents in the above formula are evaluated at $z = e_0$. Straightforward combinatorics, lemmas 2 and 3, the definition (1) of c_β and the extra condition $\hbar c_\beta < 1$ to simplify the analysis yield

$$|e_m| \leq \frac{1}{mc_\beta}\binom{3m-2}{m-1}(c_\beta \hbar)^{2m+1}.$$

Using the d'Alembert criterium we arrive at

Theorem 4. Under hypothesis (H1,4) and for $\hbar$ small enough each eigenvalue of H^2 gives rise to a resonance of H of multiplicity one. Furthermore for all β in $(0, \hat{\beta}_0)$ let c_β be defined by (1). If $\hbar$ is in $(0, \hbar_0)$, where $\hbar_0 := \frac{2}{3\sqrt{3}}{c_\beta}^{-1}$, then the series (2) converges to this resonance.

2.5 Discussion

The method of this section follows tightly [DEŠ] with extra niceties due to the simple form of H and the fact that the perturbation is off diagonal. Also we have been able to give a critical value of $\hbar$ below which the convergence of the perturbation series is assured. c_β^1 is easily estimated in terms of V^1 but for c_β^2 we have only an existence result since we do not yet know how to estimate the resolvent of the harmonic oscillator scaled with a complex parameter θ and for a spectral parameter in the numerical range. Thus this critical value of $\hbar$ is for the moment merely theoretical. We stress that we do not give here expansions of the resonances in $\hbar$; Martinez [Ma1] has shown that such expansions are asymptotic, see also [CDS].

3 Fermi-Rule Contribution to the Width of the Resonance

3.1 Asymptotics of $\operatorname{Im} e_1$ as $\hbar$ Tends to Zero

Heuristic arguments [LL, §90] lead to the conclusion that the width of the resonances of H which we found in section 2 are in fact exponentially small as $\hbar$ tends to zero. As an indication of this property we compute below the asymptotics of the imaginary part of the first order coefficient in the expansion (2).

Definition 1. Each point x in $\sum_{\beta_0,\eta_0}$ which is a root of $V^1(x) = V^2(x)$ is called a *transition point*.

We shall only consider here the restricted model

$$V^1 = v_\infty, \quad v_\infty < 0. \tag{H5}$$

We then conclude immediately that there are two transition points

$$x_\star := i\sqrt{-v_\infty} \quad \text{and} \quad -x_\star.$$

The other important points are the singularities of $V^{1,2}$ or equivalently of t. We suppose

$\eta_0 > |x_\star|$ *which means that* t *is analytic beyond the transition points.* **(H6)**

Theorem 5. Under (H1,5,6) and for the resonance associated to $e_0 := (2n+1)\hbar$, $n \in I\!N$ one has

$$\operatorname{Im} e_1 = -\frac{2\sqrt{2\pi}}{n! e^{2n+1}} \hbar^2 \left(\frac{4d_\star}{\hbar}\right)^{n+\frac{1}{2}} \exp\left(-\frac{2d_\star}{\hbar}\right)\{|t(x_\star)|^2 + \mathcal{O}(\hbar)\} \tag{3}$$

where

$$d_\star := \frac{-v_\infty}{2} = |\operatorname{Im} \int_0^{x_\star} \sqrt{-V^2(y)}dy|.$$

Proof: Under (H5) H^1 obviously has a pair of generalized eigenvectors at energy e_0

$$H^1\varphi^{1,\nu} = e_0\varphi^{1,\nu}, \nu = \pm 1, \quad \varphi^{1,\nu} := (2\pi\hbar)^{-\frac{1}{2}} e^{\nu i \frac{k}{\hbar} x}, k := \sqrt{e_0 - v_\infty}.$$

e_1 which is constant with respect to θ may be computed for $\theta = 0$ using the boundary value of R^1 at $e_0 + i0$. We get [RS4, XII.6]

$$\operatorname{Im} e_1 = -\frac{\pi}{2k} \sum_{\nu=\pm 1} |(V^{2,1}\varphi^{1,\nu}, \varphi^2)|^2,$$

where

$$\varphi^2(x) := \frac{C_n}{\hbar^{\frac{1}{4}}} \mathcal{P}_n\left(\frac{x}{\sqrt{\hbar}}\right) e^{-\frac{x^2}{2\hbar}}, \quad C_n := \pi^{-\frac{1}{4}}(2^n n!)^{-\frac{1}{2}},$$

$\mathcal{P}_n$ being the nth Hermite polynomial. One derives easily

$$(V^{2,1}\varphi^{1,\nu}, \varphi^2) = -\hbar(f\varphi^{1,\nu}, \varphi^2), \quad f := 2\nu k t + \frac{\hbar}{i} t'.$$

Since by (H6) $I\!R + i\nu k$ is in $\sum_{\beta_0,\eta_0}$, we may take it as the "contour" of integration:

$$(V^{2,1}\varphi^{1,\nu}, \varphi^2) = -\frac{C_n}{\sqrt{2\pi}} \hbar^{\frac{1}{4}} e^{-\frac{k^2}{2\hbar}} \int_{I\!R} f(x + \nu i k) \mathcal{P}_n\left(\frac{x + \nu i k}{\sqrt{\hbar}}\right) e^{-\frac{x^2}{2\hbar}}.$$

It is now easy to compute the asymptotics of the last integral by first scaling x by $\sqrt{\hbar}$, then expanding

$$f(\nu ik+\sqrt{\hbar}x)\mathcal{P}_n(\frac{\nu ik}{\sqrt{\hbar}}+x)=\Big(f(\nu ik)+f'(\nu ik)\sqrt{\hbar}x\Big)\Big(\mathcal{P}_n(\frac{\nu ik}{\sqrt{\hbar}})+\mathcal{P}_n'(\frac{\nu ik}{\sqrt{\hbar}})x\Big)$$
$$+\,\mathcal{O}\left(\hbar^{-\frac{n}{2}+1}(1+|x|^{n+2})\right)\ \text{(as } \hbar\to 0 \text{ and uniformly in } x\text{)},$$

and finally integrating with $e^{-\frac{x^2}{2}}$. ■

3.2 Discussion

The idea of looking at the imaginary part of the first non real coefficient of the perturbation expansion (2) has a long history, see e.g. [RS4, Ya, GMS, DEŠ] not to mention the physics litterature. Methods to compute the aymptotics of $\operatorname{Im} e_1$ using the analyticity of the potential are all based more or less on the stationary phase or steepest descent methods.

However since it is expected that all coefficients of (2) will contribute to the imaginary part of the resonance E with the same exponential behavior, the prefactor of (3) has not the right asymptotics for $\operatorname{Im} E$. Such a phenomenon is discussed for example in [Be]. Below we give a classical interpretation of $d_\star$.

3.2.1 Exponential Decay of $\operatorname{Im} E$ and Classical Action of Instantons

The following interpretation of $d_\star$ is a well known heuristic fact (see [LL]); we would like to illustrate it in detail with our simple model.

On the complex phase space $\mathbb{C}\times\mathbb{C}$ we consider the two energy shells

$$\Sigma^{(i)}(e_0):=\{(q,p)\in\mathbb{C},\quad H^i_{cl}(q,p)=e_0\},\quad i=1,2,$$

and their trace on the real phase space $\Sigma^{(i)}_{\mathbb{R}}, i=1,2$. The two real energy shells do not intersect, but the complex ones do:

$$\begin{cases} p^2+v_\infty &= e_0\\ p^2+q^2 &= e_0\end{cases} \Longleftrightarrow \begin{cases} p &= \pm\sqrt{e_0-v_\infty}=\pm k\\ q &= \pm i\sqrt{-v_\infty}=\pm x_\star\end{cases}$$

These points will also be called *transition points.*

We want to endow the union of the two complex energy shells $\Sigma(e_0):=\Sigma^{(1)}(e_0)\cup\Sigma^{(2)}(e_0)$ with a (pseudo-) distance δ as follows. Let A and B be two points of $\Sigma(e_0)$, then

$$A,B\in\Sigma^{(i)}(e_0)\quad\Longrightarrow\quad \delta(A,B):=|\operatorname{Im} s(A,B)|,$$

where $s(A,B)$ is the minimal (complex) action to join A and B by a (complex) trajectory on the energy shell $\Sigma^{(i)}(e_0)$:

$$s(A,B):=\int_{A\to B} p\,dq=\int_{A\to B}\sqrt{e_0-V^i(q)}\,dq.$$

Such a complex trajectory is usually called an *instanton*. Otherwise,

$$\begin{aligned} A \in \Sigma^{(i)}(e_0),\ B \in \Sigma^{(j)}(e_0),\ i \neq j, \quad &\Longrightarrow \\ \delta(A,B) \ := \ \min\{|\mathrm{Im}\, s(A,T_\star) + \mathrm{Im}\, s(T_\star,B)|,\ & T_\star \in \Sigma^{(1)}(e_0) \cap \Sigma^{(2)}(e_0)\}. \end{aligned}$$

In other words the distance $\delta(A,B)$ is the minimal imaginary part of the action of all instantons joining A and B.

We want to compute the distance $\delta(\Sigma_{\mathbb{R}}^{(1)}(e_0), \Sigma_{\mathbb{R}}^{(2)}(e_0))$ between the two real energy shells. We remark that the distance between two points of the same connected component of $\Sigma_{\mathbb{R}}^{(i)}(e_0), i = 1,2$, of course vanishes. As consequence we may take any point in each one and compute their distance. So let $A = (0,k)$ be in $\Sigma_{\mathbb{R}}^{(1)}(e_0)$ and $B = (0, \sqrt{e_o})$ be in $\Sigma_{\mathbb{R}}^{(2)}(e_0)$. We choose $T_\star = (x_\star, k)$ among the transition points. Then we have:

$$\begin{aligned} s(A,T_\star) + s(T_\star,B) &= \int_0^{x_\star} k dq + \int_{x_\star}^0 \sqrt{e_0 - q^2} dq \\ &= kx_\star + \frac{1}{2}\left[q\sqrt{e_0 - q^2} + e_0 \arcsin(\frac{q}{\sqrt{e_0}})\right]_{x_\star}^0 \\ &= id_\star + \mathcal{O}(\hbar \ln \hbar) \end{aligned}$$

Performing the analoguous calculations for the other transition points, we see that the above result gives indeed the minimal contribution to the definition of $\delta(A,B)$. This shows that the distance between the two real energy shells is $d_\star$ in the limit $\hbar$ tending to zero. But on the other hand this is nothing else than the rate of exponential decay of the width of the resonance due to quantum dynamical tunneling between the two real energy shells. This tunneling takes place through

3.2.2 The Dynamical Barrier.

Looking at the two real energy shells in the real phase space we see three curves. $\Sigma_{\mathbb{R}}^{(2)}(e_0)$ is the circle centered at the origin with radius $\sqrt{e_0}$ and $\Sigma_{\mathbb{R}}^{(1)}(e_0)$ consists of the two straight lines $p = \pm k$. As already mentioned the curves do not intersect. More precisely their projections in the configuration space do intersect but not their projection in the momentum space. The latter are the classically allowed regions in the momentum space: $\{\pm k\}$ for H_{cl}^1 and $[-\sqrt{e_0}, \sqrt{e_0}]$ for H_{cl}^2. They are separated by the classically forbidden region: $(-k, -\sqrt{e_0}) \cup (\sqrt{e_0}, k)$. In analogy with tunneling in the configuration space we would like to say that associated to this classically forbidden region there is a *dynamical barrier*. And through this barrier the bound state of H^2 has to tunnel to become a scattering state of H^1. The strength of this tunneling, as in the configuration space, depends on the diameters of the dynamical barriers. These are measured by the length of each component of the classically forbidden region in the *instanton metric*:

$$\left((V^2)^{-1}(p^2 - e_0)\right)_+^2 dp^2 = (p^2 - e_0)_+ dp^2. \tag{4}$$

Computing this diameter for the barrier of the positive momenta we obtain

$$\int_{\sqrt{e_0}}^{k} \sqrt{(p^2 - e_0)_+}dp = d_\star + \mathcal{O}(\hbar \ln \hbar)$$

which is in agreement with the result of theorem 5. The reason why V^1 does not enter in formula (4) is due to the fact that the instanton joining A and $T_\star$ has a constant velocity.

4 Exponential Bounds on the Resonance Width

We recall that in section 2 we proved that H possesses a resonance E in the vicinity of each eigenvalue e_0 of H^2 for small enough $\hbar$. In section 3, for the restricted model $V^1 = v_\infty$ (see H5), we have shown that the imaginary part of the first term e_1 of the perturbation expansion (2) of $E - e_0$ is actually exponentially small as $\hbar$ tends to zero. The purpose of this section is to show that this exponential behavior is also true for the full width of the resonance E.

For technical reasons we have to require the following additional condition on the coupling term $V^{1,2}$:

$\exists\varepsilon > 0, \quad t(z) = \mathcal{O}(|z|^{-\varepsilon})$ *as* z *tends to infinity in* $\sum_{\beta_0,\eta_0}$, **(H7)**

to state our main

Theorem 6. Assume (H1,5,6,7). Then each eigenvalue e_0 of H^2 gives rise for $\hbar$ small enough to a resonance E of H and:

$$\forall 1 \geq \xi > 0, \quad 0 \geq \operatorname{Im} E = \mathcal{O}(\xi^{-2}\hbar \exp(-\frac{2d_\xi}{\hbar})) \quad \text{as } \hbar \to 0,$$

where

$$d_\xi := \int_0^{|x_\star|} \sqrt{(V^2(y) - e_0 - \hbar\xi)_+}dy.$$

Remark 2. Since the lifetime τ_E of the resonance E is defined as $\frac{2}{\operatorname{Im} E}$, the above theorem provides a lower bound on τ_E.

One can check easily that

$$\xi^{-2}\hbar e^{-2\hbar^{-1}d_\xi} = \mathcal{O}(\xi^{-2}\hbar^{-n+(1-\xi)/2}e^{-2\hbar^{-1}d_\star}).$$

This shows that the exponential behaviour of the bound of theorem 6 is the same as the one of $\operatorname{Im} e_1$ (see (3)). However the exponent in the prefactor differs by the quantity $-1 - \frac{\xi}{2}$. We do not know yet whether the $\hbar$ behavior of the prefactor of $\operatorname{Im} e_1$ differs from the one of $\operatorname{Im} E$ or if our upper bound is not optimal.

Since the proof is rather involved we only give a synopsis of it. The details of the proof will appear somewhere else together with a more general analysis including non constant V^1's.

4.1 Passing to the Fourier Image

As we explained in §3.2, the exponential behaviour of the width of the resonance E is due to tunneling through the barrier of classically forbidden momenta between the two energy shells $\Sigma_{\mathbb{R}}^{(1)}(e_0)$ and $\Sigma_{\mathbb{R}}^{(2)}(e_0)$.

A Fourier transformation of H causes the exchange of q and p in the classical picture. Consequently the tunneling takes now place in the configuration space, a situation we are more familiar with. We denote below by the same symbols the Fourier image of $H^i, i = 1, 2$ and $V^{1,2}$:

$$H^1 = x^2 + v_\infty, \quad H^2 = D^2 + x^2, \quad V^{1,2} = \hbar(t(D)x + xt(D)).$$

The price to pay is that we have to deal with non local operators as, e.g. $t(D)$ (see §4.5 below). We emphasize that this transformation is only for convenience and in principle not necessary.

4.2 Why Exterior Scaling?

Consider now (E, ϕ_θ) a resonance and its resonance function: $H_\theta \phi_\theta = E \phi_\theta$. By a simple algebraic manipulation one also has

$$(H_\theta^2 - B_\theta)\phi_\theta^2 = E\phi_\theta^2, \quad B_\theta := V_\theta^{2,1} R_\theta^1(E) V_\theta^{1,2},$$

where ϕ_θ^i denotes the ith component of ϕ_θ. Thus

$$\operatorname{Im} E \|\phi_\theta^2\|^2 = (\operatorname{Im}(H_\theta^2 - B_\theta)\phi_\theta^2, \phi_\theta^2). \tag{5}$$

In section 2 we have seen that ϕ_θ and therefore ϕ_θ^2 "converge" in norm to φ_θ^2, the corresponding eigenfunction of H_θ^2. The latter is of course known to decay exponentially as x goes to infinity and/or $\hbar$ tending to zero. So suppose that the operator $\operatorname{Im}(H_\theta^2 - B_\theta)$ is localized on Ω_e where

$$\Omega_i := (-\omega, \omega), \quad \Omega_e := \mathbb{R} \setminus \overline{\Omega}_i, \quad \omega^2 = -v_\infty,$$

we would obtain that $\operatorname{Im} E$ is roughly $\|\chi_e \phi_\theta^2\|^2 \sim e^{-\hbar^{-1}\omega^2}$ which is more or less what we are looking for. χ_a will denote the sharp characteristic function of $\Omega_a, a = i, e$. Ω_i is nothing but the dynamical barrier for $\hbar = 0$ (see §3.2.2). With the usual complex scaling $\operatorname{Im}(H_\theta^2 - B_\theta)$ is certainly not localized in Ω_e, whereas $\operatorname{Im} H_\theta^2$ would be with the exterior scaling

$$s_\theta(x) := \begin{cases} x & \text{if } x \in \Omega_i \\ \pm\omega + e^\theta(x \mp \omega) & \text{if } \pm x > \omega. \end{cases} \tag{6}$$

Thus in this section we choose the above complex exterior scaling to deform all our operators; a subscript θ will now mean the image under (6).

It is well known that the resonances of H do not depend on the choice of the complex deformation (see e.g. [Hu]). Here they will be considered as eigenvalues of H_θ for a certain complex θ where, as announced above, H_θ is now obtained by (6). Making sense out of all our scaled objects is rather standard (see e.g [CDKS]) except for the non local terms $V_\theta^{1,2}$ and B_θ. We shall explain briefly in §4.5 how we proceed. We have in particular the analogue of theorem 1:

Theorem 7. $H_\theta^i, i = 1,2$ and H_θ are selfadjoint analytic families for all θ such that $|\mathrm{Im}\,\theta| < \hat{\beta}_0$. H_θ^1 is of type A with domain $\mathcal{D}(H_\theta^1) = \mathcal{D}(x^2)$. The domain of H_θ^2 is given by

$$u \in \mathcal{D}(H_\theta^2) \iff \begin{array}{l} u \in \mathcal{H}^2(\Omega_i) \oplus \mathcal{H}^2(\Omega_e) \text{ and} \\ u(\pm\omega \pm 0) = e^{\theta/2} u(\pm\omega \mp 0) \\ u'(\pm\omega \pm 0) = e^{3\theta/2} u'(\pm\omega \mp 0). \end{array}$$

4.3 Spectral Stability Again

We shall need in the sequel information on the spectrum of H_θ^d and some bounds on its resolvent. Since H_θ^1 is a multiplication operator its spectrum is nothing but the range of the function $x \mapsto v_\infty + s_\theta^2(x)$. There are two facts to remark here. First the spectrum of H_θ^1 consists of the union of the interval $[v_\infty, 0]$ and a curve starting from zero, contained in the sector $\{|\mathrm{Im}\,\theta| < \mathrm{sgn}(\mathrm{Im}\,\theta)\arg(z) < 2|\mathrm{Im}\,\theta|\}$; secondly the essential spectrum of H^1 above zero turns up in the upper half plane, if $\mathrm{Im}\,\theta > 0$. In order to work with the resonances which have a negative imaginary part we shall from now on take only θ with negative imaginary part. The spectrum of H_θ^2 remains $(2\mathbb{N}+1)\hbar$ by standard arguments. So for small enough $\hbar$ the same contour Γ as in §2 is contained in the resolvent set of H_θ^d. To prove that it lies also in $\rho(H_\theta)$ we need the

Lemma 8. For any $0 < |\beta| < \hat{\beta}_0$ and for any z on Γ one has
i) $\|V_{i\beta}^{1,2} R_{i\beta}^2(z) V_{i\beta}^{2,1}\| = \mathcal{O}(\hbar^2)$.
ii) $\|R_{i\beta}^1(z)\|, \|xR_{i\beta}^1(z)\|, \|xR_{i\beta}^1(z)x\|$ are all $\mathcal{O}(\hbar^{-1})$; this is also valid for z inside Γ.

Thus stability of e_0 is assured for $\hbar$ small enough.

4.4 Estimate of $(\mathrm{Im}\,(H_\theta^2\phi_\theta^2, \phi_\theta^2)$

We want to show that $(\mathrm{Im}\,(H_\theta^2\phi_\theta^2, \phi_\theta^2)$ is exponentially small using the decay properties of ϕ_θ^2 and the localisation of $\mathrm{Im}\,H_\theta^2$. The following estimate is adapted from [Agm]. However there is a novelty with respect to the standard situation: the operator $H_\theta^2 - B_\theta$ contains the non local energy dependent "potential term" B_θ. Let

$\exp(-\hbar^{-1}\rho)$ be the expected decay behavior and let $\phi^2_{\theta,\rho} := \exp(\hbar^{-1}\rho)\phi^2_\theta$ be the boosted eigenfunction; more generally we denote all objects boosted by $\exp(\hbar^{-1}\rho)$ with the subscript ρ. $\phi^2_{\theta,\rho}$ is a solution of $(H^2_{\theta,\rho} - B_{\theta,\rho} - E)\phi^2_{\theta,\rho} = 0$ and therefore

$$\mathrm{Re}\,((H^2_{\theta,\rho} - B_{\theta,\rho} - E)\phi^2_{\theta,\rho}, \phi^2_{\theta,\rho}) = 0.$$

To cope with the $B_{\theta,\rho}$ problem we use the following property

Lemma 9. Assume (H6), then there exists $a, b > 0$ such that for any $0 < |\mathrm{Im}\,\theta| < \hat{\beta}_0$ and any real Lifschitz function ρ one has:

$$\rho'^2 \leq \omega^2\chi_i \quad \Longrightarrow \quad -\mathrm{Re}\,B_{\theta,\rho} \geq -a\hbar\mathrm{Re}\,x^2_\theta - b\hbar^3.$$

From this result we deduce that

$$\hbar^2\|s'^{-\frac{1}{2}}_\theta \partial_x\phi^2_{\theta,\rho}\|^2 + (((1-a\hbar)\mathrm{Re}\,x^2_\theta - b\hbar^3 - \mathrm{Re}\,E - \rho'^2)\phi^2_{\theta,\rho}, \phi^2_{\theta,\rho}) \leq 0.$$

This motivates the choice of ρ:

$$\rho'^2 := \left((1-a\hbar)x^2 - b\hbar^3 - \mathrm{Re}\,E - \hbar\xi\right)_+ \chi_i, \quad \xi > 0,$$

and we obtain the

Lemma 10. For any $0 < |\mathrm{Im}\,\theta| < \hat{\beta}_0$ one has as $\hbar$ tends to zero and for any $0 < \xi \leq 1$:

$$\|\phi^2_{\theta,\rho}\| = \mathcal{O}(\xi^{-\frac{1}{2}}), \quad \|\chi_e x\phi^2_{\theta,\rho}\| = \mathcal{O}(\hbar^{\frac{1}{2}}), \quad \hbar\|\chi_e\partial_x\phi^2_{\theta,\rho}\| = \mathcal{O}(\hbar^{\frac{1}{2}}).$$

We can now estimate $\mathrm{Im}\,(H^2_\theta\phi^2_\theta, \phi^2_\theta)$:

$$\begin{aligned}\mathrm{Im}\,(H^2_\theta\phi^2_\theta, \phi^2_\theta) &= \mathrm{Im}\,(e^{-\hbar^{-1}\rho}H^2_\theta e^{-\hbar^{-1}\rho}\phi^2_{\theta,\rho}, \phi^2_{\theta,\rho}) \\ &= e^{-2\hbar^{-1}\rho(\omega)}(\mathrm{Im}\,H^2_\theta\phi^2_{\theta,\rho}, \phi^2_{\theta,\rho}) = \mathcal{O}(\hbar e^{-2\hbar^{-1}\rho(\omega)}).\end{aligned}$$

An elementary calculus shows that $e^{-2\hbar^{-1}\rho(\omega)} = \mathcal{O}(e^{-2\hbar^{-1}d_\xi})$ so that this term has the announced behavior. It remains to do the

4.5 Estimate of $\mathrm{Im}\,(B_\theta\phi^2_\theta, \phi^2_\theta)$.

We start by explaining how we make sense out of B_θ. We may write formally

$$B_\theta = -\hbar^2(2t(D_\theta)x_\theta + i\hbar t'(D_\theta))R^1_\theta(E)(2x_\theta t(D_\theta) - i\hbar t'(D_\theta)).$$

The operator x_θ is controlled by $R^1_\theta(E)$ (see lemma 8) so it is sufficient to show that $t(D_\theta)$ and $t'(D_\theta)$ are bounded operators. D_θ is the image of D under our exterior scaling for which we know the

Theorem 11. $\{D_\theta, \theta \in \mathbb{C}\}$ is a selfadjoint analytic family of operators and:
i) the domain of D_θ is defined by

$$\mathcal{D}(D_\theta) \ni u \iff u \in \mathcal{H}^1(\Omega_i) \oplus \mathcal{H}^1(\Omega_e) \text{ and } u(\pm\omega \pm 0) = e^{\theta/2} u(\pm\omega \mp 0);$$

ii) the spectrum of D_θ is just: $\sigma(D_\theta) = e^{-\theta}\mathbb{R}$.
iii) The following resolvent estimate holds: let $\nu := \{z \in \mathbb{C}, \operatorname{Im} s'_\theta z > 0 \text{ or } \operatorname{Im} s'_\theta z < 0\}$; then for every z in ν, one has

$$\|(D_\theta - z)^{-1}\| \leq |\operatorname{Im} s'_\theta z|^{-1} = \operatorname{dist}(z, \mathbb{C} \setminus \nu)^{-1}.$$

To define $t(D_\theta)$ we use the Dunford-Taylor integral

$$t(D_\theta) := \frac{1}{2\pi i} \int_C t(\lambda)(D_\theta - \lambda)^{-1} d\lambda, \quad |\operatorname{Im}\theta| < \hat{\beta}_0$$

where the contour C is taken in $\nu \cap \sum_{\beta_0,\eta_0}$ enclosing $\sigma(D_\theta)$ and such that we have $\operatorname{dist}(C,\nu)^{-1} = \mathcal{O}(\lambda^{-1})$ as $|\lambda|$ tends to infinity. The convergence of the above integral is due to (H7) and lemma 11 iii). Furthermore we know that

Lemma 12. $t(D_\theta), t'(D_\theta)$ are bounded selfadjoint analytic families as long as $|\operatorname{Im}\theta| < \hat{\beta}_0$.

We now explain how one can define $B_{\theta,\rho}$. Using the same strategy as for B_θ and the fact that the boost $e^{\hbar^{-1}\rho}$ commutes with x_θ and $R^1_\theta(E)$, it suffices to show that $t(D_{\theta,\rho})$ and $t'(D_{\theta,\rho})$ are bounded. One can show the analogue of theorem 11 and lemma 12 for them. Moreover $t(D_{\theta,\rho})$ and $t'(D_{\theta,\rho})$ are uniformly bounded with respect to ρ as long as ρ obeys: $\rho'^2 \leq \omega^2 \chi_i$. Here one must use assumption (H6). By a straightforward though rather involved algebraic calculus one gets a reformulation of (5)

$$\operatorname{Im} E(\|\phi_\theta\|^2 + (1 - e^{-2\hbar^{-1}\rho(\omega)})\|\phi^1_{\theta,\rho}\|^2) = e^{-2\hbar^{-1}\rho(\omega)} \operatorname{Im}((H^2_\theta - B_{\theta,\rho})\phi^2_{\theta,\rho}, \phi^2_{\theta,\rho}). \quad (7)$$

One has the estimate

$$\|\phi^1_{\theta,\rho}\|^2 = \|R^1_\theta(E) V^{1,2}_{\theta,\rho} \phi^2_{\theta,\rho}\|^2 = \mathcal{O}(\xi^{-1}),$$

(use lemma 8, 12 and 10). Finally one also has

$$(B_{\theta,\rho}\phi^2_{\theta,\rho}, \phi^2_{\theta,\rho}) = \mathcal{O}(\hbar\xi^{-1}).$$

Inserting this estimate and the one of §4.4 in (7) completes the proof of theorem 6.

4.6 Discussion

The use of a complex deformation in the momentum space leaving invariant $\Omega_{i,\varepsilon} := \{p \in I\!R, |p| < \omega - \varepsilon\}, \varepsilon$ small, i.e. a region which contains approximately the dynamical barrier (see §3.2.2), was brought to our attention by M. Rouleux. In his notes [Ro] he combines this idea with the Green's formula

$$2\,\mathrm{Im}\,E \int_{\Omega_{i,\varepsilon}} |\phi(x)|^2 dx = -\hbar^2 \mathrm{Im}\,\phi'\overline{\phi}\Big|_{\partial\Omega_{i,\varepsilon}}; \tag{8}$$

as above ϕ stands here for the Fourier image of the original ϕ.

The last ingredient to get the asymptotics of $\mathrm{Im}\,E$ would be the knowledge of the behavior of the r.h.s. of (8) since obviously the norm of ϕ on $\Omega_{i,\varepsilon}$ is $1 + o(1)$ as $\hbar$ tends to zero.

The idea of using the Green's formula can be traced back to Herring [Her] through e.g. [HeSj, AsHa, W].

The first idea, the exterior complex deformation, is now rather standard when translated to the shape resonance framework (see e.g. [CDKS, HeSj]).

The choice of $\Omega_{i,\varepsilon}$ with a non zero ε is a consequence of the use of a smooth exterior distortion. Here we hope to get immediately the right exponential decay rate by employing an exterior scaling (§4.2).

Also instead of the Green's formula (8) we use a full L^2 calculus (5). With (5) we need only L^2-exponential decay estimates on ϕ (see §4.4,5) which are simpler to derive than the pointwise estimates needed in (8). However by using this L^2 approach we are forced to loose a little bit of the exponential decay rate; the correct one being given by d_ξ with $\xi = 0$ (see theorem 6).

Acknowledgement

The main part of this work has been done while both of us were paying a visit to Ph. Blanchard at BiBos, University of Bielefeld.

References

[A] J. Asch: Ein Modell für die Prädissoziation von diatomischen Molekülen, Diplomarbeit, TU-Berlin, FB Physik 1985

[AgC] J. Aguilar, J.M. Combes: A class of analytic perturbations for one-body Schrödinger hamiltonians, CMP **22**(1971) 269-279

[AD] J. Asch, P. Duclos: An elementary model of dynamical tunneling. Proc. of the Atlanta Conf. "Differential Equations with Applications to Mathematical Physics", W.F.Ames, E.M.Harrell II, J.V.Herod eds., Math. Science and Engineering **192** Academic Press 1993

[Agm] S. Agmon: Lectures on Exponential Decay of Solutions of Second Order Elliptic Operators, Princeton University Press 1982

[AsHa] M. Ashbaugh, E. Harrell II: Pertubation Theory for Shape Resonances and Large Barrier Potentials, CMP **83**(1982) 151-170

[BCD1] Ph. Briet, J.M.Combes, P.Duclos: On the location of resonances for Schrödinger operators in the semiclassical limit I: Resonance free domains, J.Math.Anal.Appl. **125**(1987) 90-99

[BCD2] Ph. Briet, J.M. Combes, P.Duclos: On the location of resonances for Schrödinger operators in the semiclassical limit II: Barrier top resonances, Comm. P. D. E. **12**(1987) 201-222

[Be] M. Berry: Semiclassical weak reflection above analytic and non-analytic potential barriers, J.Phys.A **15**(1982) 3693-3704

[CDS] J.M. Combes, P. Duclos, R. Seiler: The Born-Oppenheimer Approximation, Wightman, Velo (Eds),Rigorous Atomic and Molecular Physics Proceedings, Plenum Press (1981) 185-212

[CDKS] J.M. Combes, P. Duclos, M. Klein, R. Seiler: The Shape Resonance, CMP **110**(1987) 215-236

[DEŠ] P. Duclos, P. Exner, P. Šťovíček: Curvature induced resonances in a two-dimensional Dirichlet tube, CPT-93/P.2881 (1993)

[GMS] V. Grecci, M. Maioli, A. Saccetti: Wannier ladders and perturbation theory, J.Phys.A (Letters) **26** 1993

[Her] C. Herring: Critique of the Heitler-Landau Method of Calculating Spin Coupling at Large Distances, Rev.Mod.Phys. **34**(1962) 631-645

[HeSj] B. Helffer, J. Sjöstrand: Resonances dans la limite semiclassique, Suppl.Bull.SMF nouvelles séries **25/26**(1986)

[Ho] J. Howland: The Lisvic Matrix in Perturbation Theory, J. Math. Anal. Appl. **50**(1975) 415-437

[Hu] W. Hunziker: Dilation analyticity and molecular resonance curves, Ann.I.H.P. **45**(1986) 339-358

[K] M. Klein: On the mathematical theory of predissociation, Ann. Phys **178**(1987) 48-73

[Ka] T. Kato: Perturbation theory of linear operators, 2nd ed. Springer 1980

[LL] L. Landau, L. Lifschitz: Quantenmechanik, Theoretische Physik III, Akademie Verlag Berlin 1988

[Ma1] A. Martinez: Développements asymtotiques et effet tunnel dans l'approximation Born-Oppenheimer, Ann.I.H.P. **50**(1989) 239-257

[Ma2] A. Martinez: Estimates on complex interactions in phase space, Preprint Unversité Paris Nord 92-5 (1992)

[Ro] M. Rouleux, Private communication

[RS4] M. Reed, B. Simon: Analysis of Operators, Modern Methods in Mathematical Physics VI, Academic Press 1978

[W] M. Wilkinson: Tunneling between tories in phase space, Physica **21**D(1986) 341-354

[Ya] K. Yajima: Resonances for the AC-Stark effect, CMP **87**(1982/83) 331-352

P. Duclos, Centre de Physique Théorique, CNRS, Luminy, Case 907, F-13288 Marseille Cedex 9

B. Meller, PhyMat, Université de Toulon et du Var, BP 132, F-83957 La Garde Cedex

Operator Theory:
Advances and Applications, Vol. 70

Scattering on Several Solenoids

Pavel Šťovíček

Abstract

The problem is treated in an idealized setup and reduced to a scattering in the plane. The Hamiltonian $\hat{H}$ in $L^2(\mathbb{R}^2)$ is defined as a self-adjoint extension of the symmetric operator $X = -\Delta$, with $D(X) = C_o^\infty(\mathbb{R}^2_+) \oplus C_o^\infty(\mathbb{R}^2_-)$, determined by boundary conditions on the first coordinate axis. The wave operators $W_\pm(\hat{H}, \hat{H}_o)$ exist and are complete. Generalized eigen-functions of $\hat{H}$ giving a stationary picture of the scattering are constructed.

1 Introduction

The Aharonov–Bohm effect [1] is an exciting problem from the both physical and mathematical point of view. Here we are going to consider an idealized setup with infinitely thin and parallel solenoids. This means that we can treat this problem as a scattering in the plane. While the case of one solenoid enjoys the rotational symmetry and is solvable explicitly [1, 2] the case of two and more solenoids is considerably more complicated. It is also worth of noting that this problem differs from the usual potential scattering. In the gauge we have chosen the given Hamiltonian and the free Hamiltonian are not related by an additive potentional but, on the other hand, they are self-adjoint extensions of the same symmetric operator [3, 4]. This fact enables to apply the Krein's formula and to prove, in the framework of the Kato–Birman theory, that the wave operators exist and are complete. Using the theory of self-adjoint extensions rather than the Lippman–Schwinger equation, one can also construct the generalized eigen-functions giving a stationary picture of the scattering and thus obtain the S-matrix [5]. This result is only of theoretical importance. But in the case of two solenoids and in the asymptotic region $kd \gg 1$, with k being the length of the wave vector and d the distance between the solenoids, one can simplify the formula to allow also numerical evaluation [6]. It turns out that, alike the one-solenoid case, the differential cross section diverges for the forward scattering. The only exception is the configuration with equal but opposite fluxes. This situation is physically most consistent since the total flux is zero [7].

The aim of this paper is to treat this scattering problem in the framework of the regular mathematical theory. It contains only a brief survey of results; the complete proofs will appear elsewhere.

2 Boundary conditions defining the Hamiltonian

Suppose we are given a measurable real function $\nu(u)$ defined on $\mathbb{R}$. Let Q_ν be the closed subspace in the orthogonal sum of Sobolev spaces $\mathcal{H}^1(\mathbb{R}^2_+) \oplus \mathcal{H}^1(\mathbb{R}^2_-) \subset L^2(\mathbb{R}^2)$ determined by the boundary condition

$$\psi(u, 0_-) = e^{2\pi i\, \nu(u)} \psi(u, 0_+)\,. \tag{1}$$

The restriction to Q_ν of the scalar product in the sum of Sobolev spaces yields a closed form q_ν in $L^2(\mathbb{R}^2)$ fulfilling $q_\nu(\psi, \psi) \geq \|\psi\|^2$. Hence by a standard construction, one can relate to q_ν a positive self–adjoint operator $\hat{H}_\nu$. It is easy to see that the intersection of $D(\hat{H}_\nu)$ with the subspace $\mathcal{H}^2(\mathbb{R}^2_+) \oplus \mathcal{H}^2(\mathbb{R}^2_-) \subset L^2(\mathbb{R}^2)$ is formed by those functions ψ which fulfill in addition to (1) also

$$\partial_2\psi(u, 0_-) = e^{2\pi i\, \nu(u)} \partial_2\psi(u, 0_+)\,. \tag{2}$$

In this case, $\hat{H}_\nu\psi = -\Delta\psi$ where the generalized derivative on the RHS is taken in $\mathbb{R}^2_+ \cup \mathbb{R}^2_-$. In this sense, $\hat{H}_\nu$ is determined by the boundary conditions on the first coordinate axis (1), (2). It is also clear that for $\nu(u) \equiv 0$ one obtains the free Hamiltonian $\hat{H}_0 = -\Delta$ in $L^2(\mathbb{R}^2)$.

Proposition 1. *The point spectrum of $\hat{H}_\nu$ is empty.*

In our problem, the function ν is piecewise constant with finitely many discontinuities. So let $a_1, a_2, \ldots, a_N$ be a finite sequence of reals ordered increasingly and $\alpha_1, \alpha_2, \ldots, \alpha_N$ be some numbers from the interval $(0, 1)$. Denote $I_0 = (-\infty, a_1)$, $I_j = (a_j, a_{j+1})$ for $j = 1, \ldots, N-1$, $I_N = (a_N, +\infty)$, and set

$$\nu(u) = \nu_k := \sum_{j=1}^{k} \alpha_j \quad for\ u \in I_k\,. \tag{3}$$

The Hamiltonian is then denoted simply by $\hat{H}$.

The both Hamiltonians $\hat{H}$ and $\hat{H}_0$ are self–adjoint extensions of the same symmetric operator $X = -\Delta$ with the domain $D(X) = C_o^\infty(\mathbb{R}^2_+) \oplus C_o^\infty(\mathbb{R}^2_-)$. The resolvents are related by the Krein's formula. Denote by $\mathcal{N}(z)$ the deficiency subspace and let $P(z) : L^2(\mathbb{R}^2) \to \mathcal{N}(z)$ be the orthogonal projector. Furthermore, $V(z), V_0(z) : \mathcal{N}(z) \to \mathcal{N}(\bar{z})$ designate the unitary operators determining these self–adjoint extensions. The Krein's formula then reads

$$(\hat{H} - z)^{-1} = (\hat{H}_0 - z)^{-1} + P(z)^* \, \frac{V(\bar{z}) - V_0(\bar{z})}{\bar{z} - z} \, P(\bar{z}). \tag{4}$$

Denote by $\Lambda(z)$ the closed operator in $L^2(\mathbb{R}, d\kappa)$ acting by multiplication with $(\kappa^2 - z)^{1/2}$, $z \in \mathbb{C} \setminus [0, +\infty)$. Ω stands for an operator in $L^2(\mathbb{R}\,)$ defined by

$$\Omega\hat{\psi} := (e^{2\pi i \nu}\psi)^\wedge\,. \tag{5}$$

So $\Lambda(z)^{-1}$ is bouded and Ω is unitary.

Denote by $\mathfrak{p}_z$, with $z \notin [0, +\infty)$, the following strictly m-accreative form in $L^2(\mathbb{R})$,

$$\mathfrak{p}_z(\psi, \psi) = \int_{-\infty}^{+\infty} d\kappa \, \sqrt{\kappa^2 - z}\,(|\psi(\kappa)|^2 + |\Omega\psi(\kappa)|^2)\,,$$

and by $\Xi(z)$ the associated closed operator [8]. One finds easily that for $\psi \in D(\Xi(z))$,

$$\operatorname{Re}\langle\psi, \Xi(z)\psi\rangle \geq (2(|z| - \operatorname{Re} z))^{1/2}\langle\psi, \psi\rangle\,. \tag{6}$$

Consequently, $\Xi(z)^{-1}$ exists and is bounded. Since $\mathfrak{p}_z^* = \mathfrak{p}_{\bar{z}}$ it holds $\Xi(z)^* = \Xi(\bar{z})$. It follows that $\operatorname{Ran}\Xi(z) = D(\Xi(z)^{-1})$ is dense in $L^2(\mathbb{R})$ and so coincides with $L^2(\mathbb{R})$. One can show that $\Xi(z)$ is the closure of the operator $\Lambda(z) + \Omega^{-1}\Lambda(z)\Omega$. Consequently, $\operatorname{Ran}(\Lambda(z) + \Omega^{-1}\Lambda(z)\Omega)$ is dense in $L^2(\mathbb{R})$ and $\Xi(z)^{-1}$ is the closure of $(\Lambda(z) + \Omega^{-1}\Lambda(z)\Omega)^{-1}$.

The deficiency subspaces are easy to describe. $\mathcal{N}(z)$ is the range of the unitary mapping

$$U(z) : \mathbb{C}^2 \otimes L^2(\mathbb{R}) \rightarrow \mathcal{N}(z)$$

defined by $\psi = U(z)\varphi$,

$$\begin{aligned} &\psi(x) = \mathcal{F}^{-1}_{\kappa\to x_1}\,\tilde{\varphi}(\kappa, x_2)\ \text{, where} \\ &\tilde{\varphi}(\kappa, u) = \left(2\operatorname{Re}\sqrt{\kappa^2 - z}\right)^{1/2}\exp(-\sqrt{\kappa^2 - z}\,|u|)\,(\varphi_+(\kappa)\vartheta(u) + \varphi_-(\kappa)\vartheta(-u))\,. \end{aligned}$$

We have ascribed the symbols $\{e_+, e_-\}$ to the standard basis in $\mathbb{C}^2$ and ϑ stands for the Heaviside step function. With respect to the chosen basis in $\mathbb{C}^2$, $\tilde{V}(z) := U(\bar{z})^{-1}V(z)U(z)$ splits into a 2×2 matrix $(V_{\varepsilon\varrho}(z))$, ε, $\varrho = \pm$.

Proposition 2. *It holds*

$$\begin{aligned} \mathbf{I} + V_{++} &= (z - \bar{z})\,(2\operatorname{Re}\Lambda(z))^{-1/2}\Xi(\bar{z})^{-1}(2\operatorname{Re}\Lambda(z))^{-1/2}\,, \\ V_{+-} &= (z - \bar{z})\,(2\operatorname{Re}\Lambda(z))^{-1/2}\Xi(\bar{z})^{-1}\Omega^{-1}(2\operatorname{Re}\Lambda(z))^{-1/2}\,, \\ V_{-+} &= (z - \bar{z})\,(2\operatorname{Re}\Lambda(z))^{-1/2}\Omega\,\Xi(\bar{z})^{-1}(2\operatorname{Re}\Lambda(z))^{-1/2}\,, \\ \mathbf{I} + V_{--} &= (z - \bar{z})\,(2\operatorname{Re}\Lambda(z))^{-1/2}\Omega\,\Xi(\bar{z})^{-1}\Omega^{-1}(2\operatorname{Re}\Lambda(z))^{-1/2}\,. \end{aligned} \tag{7}$$

3 A perturbative formula

Let us relate to the jth solenoid the operator $R^{(j)}(z) = (R^{(j)}_{\varepsilon\varrho})_{\varepsilon,\varrho=\pm}$ in $\mathbb{C}^2 \otimes L^2(\mathbb{R})$ determined by the kernel

$$\begin{aligned} \langle\kappa|R^{(j)}_{\varepsilon\varrho}|\lambda\rangle \;=\; & -\frac{\sin\pi\alpha_j}{\pi}\,(\kappa^2 + w^2)^{-1/4}\exp(-ia_j\kappa) \\ & \times\ \frac{\exp[\alpha_j(\varepsilon\,\omega(\kappa) + \varrho\,\omega(\lambda)) - i\frac{\pi}{2}\alpha_j(\varepsilon - \varrho)]}{1 + \exp[\varepsilon\,\omega(\kappa) + \varrho\,\omega(\lambda) - i(\frac{\pi}{2} - 0)(\varepsilon - \varrho)]} \\ & \times\ \exp(ia_j\lambda)\,(\lambda^2 + w^2)^{-1/4}\,, \end{aligned} \tag{8}$$

with $w = \sqrt{-z}$ and $\omega(\kappa) := \operatorname{argsh}(w^{-1}\kappa)$. This definition makes sense directly only for $z < 0$ and hence $w > 0$. Nevertheless, one can show that $R^{(j)}(z)$ extends analytically to the domain $z \in \mathbb{C} \setminus [0, +\infty)$. Clearly, $R^{(j)}(z)^* = R^{(j)}(\bar{z})$.

We denote by $\{\epsilon_1, \ldots, \epsilon_N\}$ the standard basis in $\mathbb{C}^N$. The following formula was derived in a form of an infinite series in [4]. It can be also extracted from the results given in [9].

Theorem 3. *For any $z \in \mathbb{C} \setminus [0, +\infty)$, it holds*

$$\Xi(z)^{-1} = \frac{1}{2}\Lambda(z)^{-1} + \frac{1}{2}\Lambda(z)^{-1/2}\left(\mathbf{f}^t \otimes e_-^t \,(\mathbf{I} - \mathcal{D}\mathcal{J})^{-1}\mathcal{D}\,\mathbf{f} \otimes e_-\right)\Lambda(z)^{-1/2}, \quad (9)$$

where

$$\mathcal{D} = \operatorname{diag}(R^{(1)}(z), \ldots, R^{(N)}(z))$$

is an operator in $\mathbb{C}^N \otimes \mathbb{C}^2 \otimes L^2(\mathbb{R})$,

$$\mathcal{J} = \begin{pmatrix} 0 & K & \ldots & K \\ K^t & 0 & \ldots & K \\ \ldots & \ldots & \ldots & \ldots \\ \ldots & \ldots & \ldots & \ldots \\ K^t & K^t & \ldots & 0 \end{pmatrix}, \qquad \textit{with} \quad K = \begin{pmatrix} 0 & 1 \\ 0 & 0 \end{pmatrix},$$

is a $2N \times 2N$ matrix and $\mathbf{f} = \epsilon_1 + \epsilon_2 + \ldots + \epsilon_N$.

One can deduce from the following Lemma and from the analytic Fredholm theorem [10] that $(\mathbf{I} - \mathcal{D}\mathcal{J})^{-1}$ exists. For z negative, a perturbative expansion into a geometric series is possible if the numbers characterizing the strength of the magnetic fluxes, $\sin \pi\alpha_j$, are small enough or, on the other hand, if $|z|$ is large enough.

Lemma 4. *For $z < 0$, the operator $(\mathcal{D}\mathcal{J})^{2N}$ belongs to the trace class and its norm can be estimated by*

$$\|(\mathcal{D}\mathcal{J})^{2N}\| \leq C\,(\max_j \sin(\pi\alpha_j))\,K_0(wd), \quad (10)$$

where $w = \sqrt{-z} > 0$, $d = \min_{j \neq k} |a_j - a_k|$ and $K_\tau(x)$ is the Macdonald function.

4 Wave operators, generalized eigen–functions

Using the Krein's formula (4), the explicit expressions (7) and the Kato–Birman theory [11] one can prove

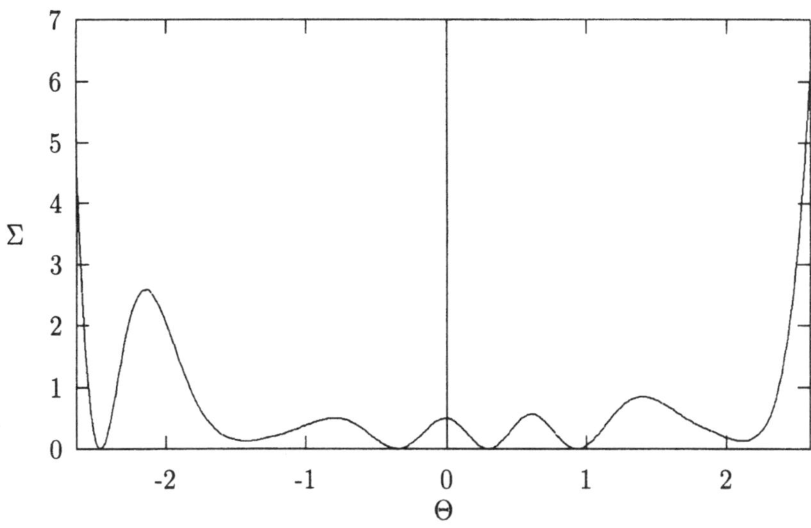

Figure 1:

Theorem 5. *The wave operators $W_{\pm}(\hat{H}, \hat{H}_0)$ exist and are complete.*

Assume that $k > 0$. Instead by means of the Lippman-Schwinger equation one can gain generalized eigen–functions of the Hamiltonian $\hat{H}$ as follows. Suppose that one can choose $\varphi(z) \in \mathcal{N}(z)$ depending on the spectral parameter $z \in \mathbb{C} \setminus [0, +\infty)$ in such a way that the function $\psi_+(z) = \varphi(z) + V(z)\varphi(z)$ has a limit when $z = k^2 - i\sigma$, $\sigma \downarrow 0$. Since $\hat{H}(\varphi(z) + V(z)\varphi(z)) = z\,\varphi(z) + \bar{z}\,V(z)\varphi(z)$ the limiting function $\psi_+(k^2 - i0)$ is expected to be a generalized eigen–function.

To proceed more formally, choose $h \in C_o^\infty((-k, k))$ and define

$$\psi_+(z; x) := \mathcal{F}^{-1}_{\kappa \to x_1} f(z; \kappa, x_2)\,. \tag{11}$$

where

$$\begin{aligned} f(z; \kappa, u) &= (\exp(-\sqrt{\kappa^2 - z}\, u) - \exp(-\sqrt{\kappa^2 - \bar{z}}\, u))\, h(\kappa) \\ &\quad + \exp(-\sqrt{\kappa^2 - \bar{z}}\, u)\, \Xi(\bar{z})^{-1}(\Lambda(\bar{z}) - \Lambda(z))\, h(\kappa)\,, \qquad \textit{for } u > 0\,, \\ &= \exp(\sqrt{\kappa^2 - \bar{z}}\, u)\, \Omega\, \Xi(\bar{z})^{-1}(\Lambda(\bar{z}) - \Lambda(z))\, h(\kappa)\,, \qquad \textit{for } u < 0\,. \end{aligned}$$

Furthermore, set

$$\psi_0(k; x) := \frac{1}{\sqrt{2\pi}} \int_{|\kappa| < k} d\kappa\, \exp(i\kappa\, x_1 - i\sqrt{k^2 - \kappa^2}\, x_2)\, h(\kappa)\,. \tag{12}$$

Theorem 6. *The functions $\psi_+(k^2 - i0; x)$ given in (11), with $k > k_o$ large enough, are generalized eigen–functions of the Hamiltonian $\hat{H}$ corresponding to the value k^2. Moreover, it holds*

$$W_- \psi_0(k; x) = \psi_+(k^2 - i0; x) \tag{13}$$

in the usual sense.

Finally we note that it is possible to extract from the eigen–function ψ_+ the S-matrix $S(\theta, \theta_o)$ [5]. Here θ_o is the angle of the incoming plane wave, $\theta_o \in (\pi, 2\pi)$, and θ is the angle of the scattered wave. Figure 1 depicts $\Sigma := 2\pi\, |S(\theta, \theta_o)|^2$ in the case of two solenoids. For the sake of convenience, it depends on $\Theta \equiv \theta - \theta_o + \pi$ (mod 2π), $\Theta \in (-\pi, \pi)$, rather than on θ. The values of parameters are $\alpha_1 = 0.4$, $\alpha_2 = 0.6$, $kd = 10$ ($d = |a_1 - a_2|$) and $\theta_o = 1.4\,\pi$.

References

[1] Aharonov A., Bohm D.: Phys. Rev. **115**, 485 (1959)

[2] Ruijsenaars S. N. M.: Ann. Phys. (NY) **146**, 1 (1983)

[3] Šťovíček P.: Phys. Lett. A **142**, 5–10 (1989)

[4] Šťovíček P.: J. Math. Phys. **32**, 2114–2122 (1991)

[5] Šťovíček P.: Phys. Lett. A **161**, 13–20 (1991)

[6] Šťovíček P.: in preparation

[7] Peshkin M., Talmi I., Tassie L. J.: Ann. Phys. (N.Y.) **12**, 426 (1961)

[8] Kato T.: Perturbation theory for linear operators. Heidelberg, New York: Springer 1966

[9] Sato M., Miwa T., Jimbo M.: Holonomic quantum fields III, Publ. RIMS, Kyoto Univ. **15**, 577–629 (1979)

[10] Reed M., Simon B.: Methods of modern mathematical physics I: Functional analysis. New York: Academic Press 1972

[11] Reed M., Simon B.: Methods of modern mathematical physics III: Scattering theory. New York: Academic Press 1979

Pavel Šťovíček, Department of Mathematics, Faculty of Nuclear Science, CTU, Trojanova 13, 120 00 Prague, Czech Republic

Operator Theory:
Advances and Applications, Vol. 70

Hall conductance of Riemann surfaces

Markus Klein

We shall briefly describe some spectral results on Schrödinger operators with a constant magnetic field on Riemann surfaces of finite volume. These essentially go back to some old results on automorphic forms of arbitrary real weight [14]. Their physical interpretation in the context of magnetic fields, however, is new and leads to some natural generalizations. We refer to our joint paper [3]for background and results. Detailed proofs shall be presented elsewhere [4].

We consider a Riemann surface M of genus g, with r cusps. For simplicity we do not consider elliptic points. By the uniformization theorem [7], each such surface can be represented up to conformal equivalence as the quotient of the complex upper half plane $\mathbf{H}$, with the Poincare metric $ds^2 = y^{-2}(dx^2 + dy^2)$, by some discrete subgroup Γ of $PSL(2, \mathbf{R})$, provided $\chi = 2g - 2 + r > 0$. Thus the surface is represented by a geodesic polygon in $\mathbf{H}$ (the boundary of a fundamental domain) with appropriate identifications of the sides. Recall that the geodesics are semicircles centered on the boundary of $\mathbf{H}$, $\partial\mathbf{H} = \{y = 0\} \cup \{\infty\}$, and the sides of the polygon can be taken to be circular arcs. Since the group Γ is isomorphic to the fundamental group of M punctured at the r cusps, the canonical polygon of Γ is obtained (topologically) from the canonical polygon of a compact genus g surface by filling in adjoining edges leading to the cusps, and the sidepairing Moebius transformations, subject to the obvious circuit relation, generate Γ.

If we equip M with a Riemannian metric of constant negative curvature -1 by pulling back the hyperbolic metric on the upper half plane, it follows from the Gauss-Bonnet theorem that $vol(M) = vol(\mathbf{H}/\Gamma) = 2\pi(2g + r - 2)$. Because of the negative curvature, the geodesic flow on such surfaces is chaotic. The Laplacian associated with compact multihandle tori has been studied in the context of chaology [9],[5], [6], [8]. In classical dynamics at low energies, $[0, B^2)$, Lorentz force dominates and one finds closed orbits, while at high energies, (B^2, ∞), the negative curvature dominates and the dynamic is (presumably) chaotic [1]. For our purpose, however, this aspect is of minor importance. Instead we observe that, if M is noncompact (the case we are interested in here), it can be deformation retracted to a bouquett, and thus its second cohomology group $H^2(M, G)$ with values in any abelian group G is zero.

These remarks serve to simplify the introduction of constant magnetic fields on M and the corresponding Schrödinger operator. A constant magnetic field on M is given by a constant multiple of the Riemannian volume form on M, which in the noncompact case we may write as

$$By^{-2}dx \wedge dy = dA, \tag{1}$$

where the vector potential A is a globally defined 1-form on M. We identify A with a Γ-periodic vector potential on $\mathbf{H}$. Up to gauge equivalence, A is characterized by its holonomy around the homology cycles of M: Denoting by $\beta_j(\epsilon)$ loops on M shrinking to the cusps (in the sense that the enclosed volume tends to zero as ϵ tends to zero), we may prescribe the Aharonov-Bohm fluxes threading the cusps

$$\phi_j = \lim_{\epsilon \to 0} \int_{\beta_j(\epsilon)} A, \quad j = 1, \ldots, r \tag{2}$$

and the $2g$ handle fluxes

$$\left(\int_{a_1} A, \ldots, \int_{b_g} A\right) = (\phi_{r+1}, \ldots, \phi_{r+2g}) \tag{3}$$

where $a_1, \ldots b_g$ are loops based at some arbitrary reference point which generate the homology classes associated with the g handles of M, provided the magnetic field and the fluxes fulfil the Dirac quantization condition

$$Bvol(M) = \sum_{j=1}^{r} \phi_j. \tag{4}$$

In the noncompact case this simply follows by integrating A along a fundamental polygon of M. The Dirac quantization condition remains true in the compact case: there it restricts the allowed values of B, while in our case it imposes a consistency condition on the value of the magnetic field and the fluxes through handles and cusps. We remark that by adding an exact 1-form to A (corresponding to a gauge transformation) we may assume that $\phi_j \in [0, 2\pi)$.

The Schrödinger operator with constant magnetic field B on M (or equivalently on the fundamental domain $F \subset \mathbf{H}$) is given by

$$\tilde{H}(B, \phi) = (-id - A)^*(-id - A), \tag{5}$$

which is self adjoint on the domain

$$\mathcal{D}(\tilde{H}) = \{f \in H^2(F, y^{-2}dxdy); f(\gamma z) = f(z) \forall \gamma \in \Gamma\},$$

where * denotes the Hilbert space adjoint on forms and $H^2(F)$ is the second Sobolev space with respect to the hyperbolic measure. Since this Γ-periodic vector potential is not explicit, it is often convenient to work instead with a fixed, not Γ-periodic, connection $A_0 = By^{-1}dx$ on the fundamental domain F and put, via gauge transformation, all geometric complications in the boundary conditions of an appropriate Schrödinger operator. In this form one finds a Hamiltonian with constant magnetic field in the literature on automorphic forms (see in particular [16]). More precisely, letting

$$U(z, z_0) = \exp i \int_{z_0}^{z} (A - A_0),$$

where z_0 is an arbitrary reference point in F, we obtain by conjugation

$$H(B, \phi) = U^{-1} \tilde{H}(B, \phi) U = y^2(-\partial_x^2 - \partial_y^2) + 2iBy\partial_x + B^2, \tag{6}$$

acting on functions which satisfy

$$\psi(\gamma z) = u(\gamma, z)\psi(z), \qquad u(\gamma, z) \equiv \nu_\phi(\gamma) \frac{(cz+d)^{2B}}{|cz+d|^{2B}}. \tag{7}$$

Here $\gamma = \begin{pmatrix} a & b \\ c & d \end{pmatrix} \in \Gamma$ acts on $\mathbf{H}$ as a Möbius transformation, and $\nu_\phi(\gamma)$, a complex number of modulus one, is a *multiplier system* on Γ associated to the fluxes. The properties of a multiplier system are the consistency conditions ensuring univaluedness of the wave function on the universal covering space and are easily computed from the definition of the gauge transformation $U(z, z_0)$:

$$\nu_\phi(-1) = e^{-i2\pi B} \tag{8}$$

and

$$u(\gamma_1\gamma_2, z) = u(\gamma_1, \gamma_2 z)\, u(\gamma_2, z), \qquad \gamma_1, \gamma_2 \in \Gamma. \tag{9}$$

Equation (8) is the consistency condition to lift the definition of ν from the group of Moebius transformations to its lift in $SL(2, \mathbf{R})$, and equation (9) is a cocycle condition on the covering space.

We remark that the notion of multiplier system is particularly useful in the compact case, since it allows to discuss operators on nontrivial bundles on a manifold as long as only they pull back to trivial bundles on the covering space.

We fix the relation of multipliers to fluxes by setting

$$u(\gamma_j, z_j) = e^{i\phi_j}, \quad 1 \le j \le 2g + r,$$

where, for $1 \le j \le r$, z_j is the site of the j-th cusp, and γ_j generates the subgroup of Γ that leaves z_j fixed. For $r+1 \le j \le r+2g$, γ_j is a transformation identifying sides in the fundamental polygon which dissect the g handles, and z_j is an *arbitrary* reference point on such a side. (There is no distinguished reference point on the $2g$-dimensional torus of handle fluxes which naturally corresponds to zero flux). A piercing flux ϕ_0 at z_0 can be added via the usual vector potential, singular at z_0. We remark that this discussion in particular proves the existence of multiplier systems on Γ and replaces the arguments in [14] or [10].

Note that the fluxes through the handles do not enter this relation, while the cusp and piercing fluxes do.

The spectral analysis of Schrödinger operators with magnetic fields and flux tubes on leaky tori has a very long history in the theory of automorphic forms, where it is known as the spectral analysis of the Maass-Selberg Laplacian for non-classical automorphic forms of real weight with multipliers [14]. Below we list some key facts; see [16],[14],[8]. Towards the end of this note we shall indicate how they can be derived.

One distinguishes *four* energy ranges: $(-\infty, B)$ which is outside the spectrum; low energies $[B, B^2)$ where the spectrum is reminiscent of the usual *Landau levels* in the plane; intermediate energies $(B^2, B^2 + \frac{1}{4})$ where except for being discrete little is known about the spectrum; and high energies $(B^2 + \frac{1}{4}, \infty)$ which admit scattering states if at least one cusp flux is zero.

Scattering states: Each cusp which is threaded by a flux tube carrying an *integer number of flux quanta* is an open scattering channel. Each such scattering channel contributes the interval of energies $[\frac{1}{4} + B^2, \infty)$, with multiplicity 1, to the absolutely continuous spectrum. Cusps that carry fluxes which are not integral are in some sense plugged, and a particle can not leak through such cusps to infinity. If all the cusps are plugged the spectrum in $[\frac{1}{4} + B^2, \infty)$ is discrete.

Maass supersymmetry : For $B \geq 1$ and fixed multiplier system $\nu_\phi(\gamma)$, the spectrum of $H(B)$ coincides with the spectrum of $H(B+1) - (2B+1)$ with the ground state removed, counting multiplicity.

Spectrum of Landau levels when $\phi_0 = 0$: In the interval of energies $[B, B^2]$, $B \geq 1$, for $\phi_0 = 0$, the spectrum has $[B - \frac{1}{2}]$ points at energies:

$$E_n(B, \phi, \phi_0 = 0) \equiv B(2n+1) - n(n+1), \quad n = 0, 1, \ldots, [B - \frac{3}{2}], \tag{10}$$

where $[x]$ denotes the integer part of x. E_n depends explicitly on B only, and implicitly on the cusp fluxes through the Dirac quantization condition. It is completely *independent of the fluxes through the handles.*

Degeneracy of Landau levels when $\phi_0 = 0$: Like the *Landau levels* on the flat torus, the energies in equation 10 are in general degenerate. Unlike them their degeneracy decreases with energy. Let $\lfloor x \rfloor$ be the greatest integer strictly smaller than x and set $\{x\} \equiv x - \lfloor x \rfloor \in (0, 1]$. The degeneracy of the n-th *Landau level* is then given by:

$$\mathrm{D}(n, \phi_0 = 0) = (B - n)(2g - 2 + r) - \sum_{j=1}^{r} \{\frac{\phi_j}{2\pi}\} - (g - 1) + \delta_{B,1}\delta_{\nu,1} \ . \tag{11}$$

By Dirac quantization, the right hand side of equation (11) is an integer. (11) and (4) are invariant under deformations of the leaky torus within the moduli space of topologically equivalent Riemann surfaces. Since generically the spectrum and degeneracies are sensitive to deformations, this invariance is remarkable, especially since the dimension of the moduli space can be large. In particular, it will follow from this invariance that the transport properties we shall calculate are constant on the moduli space.

Adiabatic charge transport: Suppose that initially all cusp fluxes are non-integer and so are plugged. Now vary two of the cusp fluxes e.g. along the line $\phi_i + \phi_j = const \neq 0 \bmod 2\pi$ by decreasing ϕ_i by 2π. The initial and final Hamiltonians are unitarily equivalent: Up to a gauge transformation, the Schrödinger operator underwent a closed cycle. In particular, the initial and final spectra, counting multiplicity, coincide. As in Laughlin's original argument, this cycle can transport net charge, and indeed it does: As ϕ_i passes through an integral flux quantum, cusp i opens briefly and, according to equation 11, one state per Landau level is sucked in from (spatial) infinity. As ϕ_j passes through an integral flux quantum, these additional states disappear at (spatial) infinity via cusp j. We remark that these states correspond to classical holomorphic forms with real weight B (or $2B$, depending on the definition of weight) for Γ and the fixed multiplier system, which at the critical value of ϕ_j just fail an integrability condition in the cusp. If N Landau levels are occupied, N charges will be transported. The cycle describes a quantum charge pump which transports integer charges. It is noteworthy that it gives integral adiabatic charge transport for systems whose area is finite. In the Hall effect and in the Niu charge pump [13] precise integers require the thermodynamic limit.

Hall conductance: For the plane Laughlin defines the Hall conductance as the charge transported to infinity by increasing the piercing flux by 2π. In the present context the charge can be transported to infinity along any of the r cusps, and moreover, by Dirac quantization, the piercing flux can not be varied independently. We therefore define the j-th Hall conductance as the charge Q_j transported to infinity along the j-th cusp, increasing ϕ_0 by 2π along the path $\phi_0 + \phi_j = const$. All these r Hall conductances turn out to be identical. Unfortunately, since equations (10, 11) hold for $\phi_0 = 0$, we can not directly follow the charge transport along the path $\phi_0 + \phi_j = const$. To compute the charge transport we therefore deform the path: First, the ϕ_0 increase is compensated by an increase in B by $2\pi/area$. This changes the degeneracy of all *Landau levels* by one. Then B decreases to its original value at the expense of the j-th cusp flux. This sends one particle per *Landau level* to infinity along the j-th cusp, while the Hamiltonian returns to its initial form up to unitary equivalence. We see that *the Hall conductance of each Landau level (for non interacting electrons) is unity, for all leaky tori*, if the magnetic field is large enough, i.e. $B \geq 1$.

These results for the Hall conductances generalize what one knows for *Landau levels* in the plane where, interestingly enough, the condition on the strength of B does not enter [11].

The energy and degeneracies are independent of the fluxes through any of the handles, and so is Dirac quantization. It follows that manipulating handle fluxes only does not transport any charge from infinity even if the cusps are open. For a proof of (10) via supersymmetric arguments we refer to the classical paper of [16]. We shall now briefly outline a derivation of (11).

We remark that both the Dirac quantization condition and the formula for the degeneracy are due to Petersson, and go back to 1938! His proof of the degeneracy formula is based on the Riemann-Roch theorem. Alternatively the degeneracy can be derived from the Atiyah-Patodi- Singer index theorem [2]. The basic tool is Maass supersymmetry which identifies the ground state degeneracy with an index which can be computed from the small time asymptotics of the heat kernel. Because of the punctures the manifold is non-compact, and one needs to compute corrections to the heat kernel from the boundaries. Like most higher order corrections to the heat kernel, this is a computational effort. The boundary terms lead to the flux dependent terms in equation (11) (the handle fluxes drop because they are not boundary terms). The fluxes may be thought of as a 1-dimensional version of the 3-dimensional η-invariant for the signature [2]. In this way the degeneracy formula can be generalized to operators with a piercing flux ϕ_0 at some point $z_0 \in M$ on a manifold of nonconstant negative curvature, as long as the magnetic field stays constant (see [15] for the basic idea).

Finally we remark that there are additional $\binom{2g}{2}$ interesting adiabatic transport coefficients associated with transport of charges around the handles due to the fluxes through the handles. Such transport coefficients are related to first Chern classes [19].

¿From the spectral results in this paper it follows that, since variation of the handle fluxes never gives level crossing, all these first Chern classes as well as higher Chern classes are constant on the moduli space. Using the Schottky uniformization and a deformation argument, they can actually be computed and turn out to be 1 (for "intersecting" fluxes) or 0 (for "nonintersecting" fluxes), see [4].

References

[1] M. Antoine, A. Comtet, and S. Ouvry, J. Phys. A **23**, 3699, (1990).

[2] M. F. Atiyah, V. K. Patodi, I. M. Singer, Math. Proc. Camb. Phil. Soc. **77**, 43, (1975); **78**, 405, (1975); **79**, 71, (1976).

[3] J. E. Avron, M. Klein, A. Pnueli and L. Sadun, Phys.Rev.Lett. ,**69**, 128, (1992).

[4] J. E. Avron, M. Klein, A. Pnueli and L. Sadun, in preparation.

[5] N. L. Balasz and A. Voros, Phys. Rep. **143**, 109, (1986).

[6] M. Berry, *Some quantum to classical asymptotics*, in: Les Houches school on chaos and quantum physics (1989)

[7] H. M. Farkas, I. Kra, *Riemann surfaces*, Springer (1980).

[8] L. D. Faddeev, Trans. Moscow Math. Soc. **17**, 357, (1967)

[9] M. Gutzwiller, *Chaos in classical and quantum mechanics*, Springer (1990).

[10] D. Hejhal, *The Selberg Trace Formula for PSL(2,R)*, Vol. 1 and 2, LN Math. **548, 1001**, Springer (1976,1983).

[11] R. G. Laughlin, in: *The Quantum Hall Effect*, R.E. Prange and S.M. Girvin, Eds., Springer (1987) .

[12] P. D. Lax and R. S. Phillips, *Scattering theory for automorphic functions*, Ann. Math. Stud., PU Press (1976).

[13] Q. Niu, Phys. Rev. Lett. **64**, 1812, (1990) .

[14] H. Petersson, Math. Ann. **115**, 23-67, 175-204, (1938);

[15] A. Pnueli, *Spinors and scalars on Riemann surfaces*, to appear in Journ. Phys. A

[16] W. Rölcke, Math. Ann. **167**, 292, (1966), and **168**, 261, (1967);

[17] J.Zak, in *Solid State Physics* **27**, F. Seitz, D. Turnbull and H. Ehrenreich, 59, Acad. Press (1972).

[18] J. E. Avron, R. Seiler and B. Simon, Phys. Rev. Lett. **65**, 2185, (1990).

[19] D. J. Thouless, M. Kohmoto, P. Nightingale and M. den Nijs, Phys. Rev. Lett. **49**, 40, (1982).

Markus Klein, Fachbereich Mathematik MA 7-2, Technische Universität Berlin, D-1000 Berlin 12, FRG

Chapter 3

Stochastic spectral analysis

Operator Theory:
Advances and Applications, Vol. 70

Framework and Results of Stochastic Spectral Analysis

Michael Demuth and Jan van Casteren

Abstract

The framework of stochastic spectral analysis is explained. The central and initial magnitude is the transition density function in a Hausdorff space. Free and perturbed Feller operators are introduced. Spectral theoretical results can be obtained by compactness, continuity in Kato-Feller norms, semiclassical and large coupling estimates. A collection of results illustrates each possibility.

1 The framework of stochastic spectral analysis

The centre of this theory is a function

$$p : (0, \infty) \times E \times E \to [0, \infty)$$

(E - second countable locally compact Hausdorff space). This function has different names depending on the field of mathematics which is studied. In stochastic analysis it is a transition density function of a Markov process, in the theory of partial differential equations it is called fundamental solution. In operator theory it is an integral kernel of a semigroup. The following scheme shows that $p(t, x, y)$, $t \in (0, \infty), x, y \in E$, is the main link between operator theory and stochastic analysis. The consequence is that one can use the theory of stochastic processes to study the spectral behaviour of large classes of operators. On the other hand it directs the interest in the theory of Markov processes to spectral analytic properties of their generators.

Of course the whole theory is only interesting if $p(t, x, y)$ can not be estimated by the Wiener density. On the other hand the assumptions on $p(t, x, y)$ have to admit the use of stochastic analysis. For that we (Demuth, van Casteren, 1989 and 1992) established the following Basic Assumptions on Stochastic Spectral Analysis, shortly denoted as BASSA:

BASSA

1.Existence and Symmetry

Let $(E, \mathcal{E})$ be a second countable locally compact Hausdorff space with Borel field $\mathcal{E}$. A non-negative Radon measure is assumed on E and denoted by dx. Let p

Framework

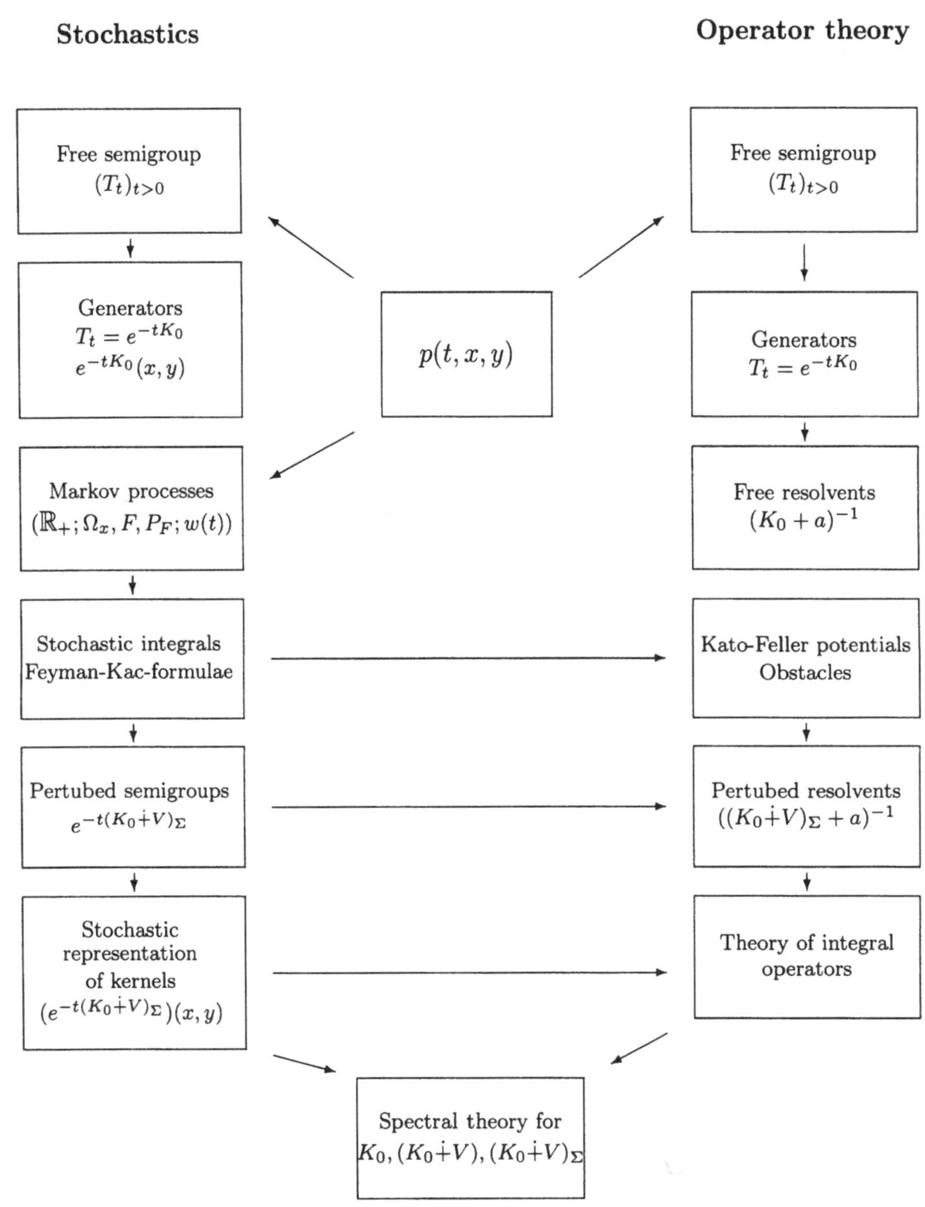
Stochastics
Operator theory
Free semigroup $(T_t)_{t>0}$
Generators $T_t = e^{-tK_0}$ $e^{-tK_0}(x,y)$
$p(t,x,y)$
Free semigroup $(T_t)_{t>0}$
Generators $T_t = e^{-tK_0}$
Markov processes $(\mathbb{R}_+; \Omega_x, F, P_F; w(t))$
Free resolvents $(K_0 + a)^{-1}$
Stochastic integrals Feyman-Kac-formulae
Kato-Feller potentials Obstacles
Pertubed semigroups $e^{-t(K_0 \dot{+} V)_\Sigma}$
Pertubed resolvents $((K_0 \dot{+} V)_\Sigma + a)^{-1}$
Stochastic representation of kernels $(e^{-t(K_0 \dot{+} V)_\Sigma})(x,y)$
Theory of integral operators
Spectral theory for $K_0, (K_0 \dot{+} V), (K_0 \dot{+} V)_\Sigma$

be a continuous function mapping $(0, \infty) \times E \times E \to [0, \infty)$ *with*

$$\int_A p(t, x, y)\, dy \leq 1\,,$$

$t > 0, x \in E, A \subset \mathcal{E}$ *and*

$$\int_E p(s, x, u)\, p(t, u, y)\, du = p(s + t, x, y)$$

Moreover p *is assumed to be symmetric, i.e.*

$$p(t, x, y) = p(t, y, x)$$

for all $t>0$ *and all* $x,\ y \in E$.

2. Continuity

Let C_∞ *be the set of continuous functions vanishing at infinity. For any* $f \in C_\infty$ *and any*
$x \in E$ *we assume*

$$\lim_{t \to 0} \int f(y)\, p(t, x, y)\, dy = f(x)$$

3. Feller property

For any $f \in C_\infty$ *we assume that the function*

$$x \mapsto \int_E f(y)\, p(t, x, y)\, dy \ \in C_\infty(E)$$

Under these assumptions exists a strong Markov process $(\mathbb{R}_+, \Omega_x, \mathcal{F}, P_{\mathcal{F}}, w(t))$ with the following properties:

The one-dimensional distribution is

$$P_F(w(t) \in B) = \int_B p(t, x, y)\, dy$$

$t > 0$, B Borel subset of E. Its sample paths are P_F-almost surely right continuous and possess P_F-almost surely left hand limits in E on their lifetime, and they start in $w(0) = x$. The free Feller operator K_0 is then the L^2-generator of the Feller semigroup determined by $p(t, x, y)$, i.e.

$$(e^{-tK_0} f)(x) = E_x\{f(w(t))\} = \int p(t, x, y) f(y)\, dy$$

and the free resolvents are given by

$$[(K_0 + a1)^{-1} f](x) = \int_0^\infty e^{-as} E_x\{f(w(s))\}\, ds$$

where a is strictly positive. The class of free Feller operators contains a variety of operators: second order elliptic differential operators with variable unbounded coefficients, Laplace Beltrami operators on locally finite Riemannian manifolds, pseudo-differential operators, relativistic Hamiltonians of quantum mechanics. Feller operators are free Feller operators together with a regular or singular perturbation. They can be introduced naturally by studying the properties of

$$E_x\{e^{-\int_0^t V(w(s))\,ds} f(w(t))\} =: (P_V(t)f)(x)$$

where V is a real-valued function on E. $P_V(t)$ is a strongly continuous, quasi-bounded semigroup on $L^2(E)$ with the selfadjoint generator $K_0 \dot{+} V$ if V is a Kato-Feller potential, i.e. if $V = V_+ - V_-$ satisfies

$$\lim_{\tau \to 0} \sup_x \int_0^\tau ds\, E_x\{V_-(w(s)) + \chi_B(w(s)V_+(w(s)))\} = 0$$

where B is a compact subset of E. Moreover, $P_V(t)$ is an integral operator and its kernel has the explicit representation

$$(e^{-t(K_0 \dot{+} V)})(x,y) = E_x^{y,t}\{e^{-\int_0^t V(w(s))\,ds}\}$$

where $E_x^{y,t}\{\}$ is the conditional Feller expectation. Instead of finite V_+ one can also include infinitely high parts of V_+. Let $V_+(x) = 1_\Gamma(x)\beta$ where Γ is some closed subset of E; β is a positive parameter tending to infinity. Let $S_\Gamma = S_\Gamma(w)$ be the penetration time of w in Γ, i.e.

$$S_\Gamma := \inf\{\tau > 0 : \int_0^\tau 1_\Gamma(w(s))\,ds > 0\}$$

Then

$$E_x\{e^{-\int_0^t V_-(w(s))\,ds} \chi\{w : S_\Gamma > t\} f(w(t))\}$$

restricted to $L^2(\Sigma)$, $\Sigma = E \setminus \Gamma$, is a Feller semigroup. Its generator is denoted by $(K_0 \dot{+} V_-)_\Sigma$. Alltogether we have the following integral kernel of regularly and singularly perturbed Feller semigroups:

$$(e^{-t(K_0 \dot{+} V_-)_\Sigma})(x,y) = E_x^{y,t}\{e^{-\int_0^t V_-(w(s))\,ds}, S_\Gamma > t\}$$

and

$$s-\lim_{\beta \to \infty} e^{-t(K_0 \dot{+} V_- + \beta 1_\Gamma)} f = e^{-t(K_0 \dot{+} V_-)_\Sigma} f$$

$f \in L^2(\Sigma)$. Coming back to our framework $p(x,t,y)$, given by BASSA, determines the free Feller semigroup, the class of free Feller operators, the corresponding Markov process. The expectation of the process provides perturbations of K_0. In all the cases the semigroups and resolvents are integral operators. Their kernels have explicit representations in terms of conditional Feller measures.

2 Principle spectral theoretical results

Assume always BASSA, two Kato-Feller potentials V and W, and the singularity region Γ as described above. Then there are several possibilities to study the spectral data of the Feller operators determined by the investigation of resolvent or semigroup differences.

Compactness: It is possible to find conditions on p, V, W or Γ such that the differences

$$e^{-t(K_0 \dot{+} V)} - e^{-t(K_0 \dot{+} W)},$$

$$(K_0 \dot{+} V + a)^{-1} - (K_0 \dot{+} W + a)^{-1},$$

$$Je^{-t(K_0 \dot{+} V)} - e^{-t(K_0 \dot{+} V)_\Sigma} J,$$

$$J(K_0 \dot{+} V + a)^{-1} - (K_0 \dot{+} W + a)_\Sigma^{-1} J$$

(where J is defined by $Jf := f \upharpoonright_\Sigma$, $\Sigma = E \backslash \Gamma$, Γ singularity region) are trace class, Hilbert-Schmidt, or compact operators. The conditions link always the density function $p(t, x, y)$ with V, W, or Γ. In order to verify these conditions one needs more information on p. Very often it is sufficient to have L^1-L^∞ smoothing, i.e.

$$\sup_{x,y} p(t, x, y) < \infty .$$

Moreover it is often very useful that the perturbed kernels satisfy

$$e^{-t(K_0 \dot{+} V)}(x, y) \leq c\, e^{ct}\, p^{1/2}(t, x, y) \sup_{x,y \in E} p^{1/2}(t, x, y)$$

Examples of results are given in the next section.

Continuity in V: For any Kato-Feller potential the Kato-Feller norm

$$\|V\|_{KF} = \sup_x \int_0^1 ds\, E_x \{|V(w(s))|\}$$

exists. Then the resolvent difference for regular resolvent values a, a large enough, can be estimated by this Kato-Feller-norm

$$\|(K_0 \dot{+} V + a)^{-1} - (K_0 \dot{+} W + a)^{-1}\| \leq c\, \|V - W\|_{KF}$$

For applications it is important that we treat here the operator norm. That can be used to study also the behaviour of these resolvents in the limiting absorption case.

Let φ be a nonvanishing continuous function mapping E into $\mathbb{R}_+$ with $\varphi^{-1} \leq 1$. For special real positive values λ it turns out that

$$\sup_{\varepsilon \in [0,1]} \| \varphi^{-1}[(K_0 \dot{+} V - \lambda + i\varepsilon)^{-1} - (K_0 \dot{+} W - \lambda + i\varepsilon)^{-1}]\varphi^{-1} \| \leq c \| (V - W)\varphi^2 \|_{KF} .$$

Again the operator norm (in weighted L^2-spaces) is studied. That implies consequences for any spectral property depending on the resolvents near the real axis.

Semiclassical limits: As explained in section 1 one has explicit representations for the kernels of the semigroups $e^{-t(K_0 \dot{+} V)}$. That remains true if we introduce a parameter $\hbar^2$, i.e. if we study generators of the form $\hbar^2 K_0 \dot{+} V$. For certain potentials the behaviour of

$$e^{-t(\hbar^2 K_0 \dot{+} V)} - e^{-t(\hbar^2 K_0 \dot{+} W)}$$

for small $\hbar^2$ can be studied.

Large coupling behaviour: The singularly perturbed semigroup $e^{-t(K_0 \dot{+} V_-)_\Sigma}$ was obtained by limits of semigroups the generators of which have finite potential heights

$$e^{-t(K_0 \dot{+} V_- + \beta 1_\Gamma)} .$$

The operator resolvent norm is

$$\| J(K_0 \dot{+} V_- + \beta 1_\Gamma)^{-1} - ((K_0 \dot{+} V)_\Sigma + a)^{-1} J \| =: f(\beta)$$

$f(\beta)$ is mainly determined by

$$\sup_{x \in \Sigma} \left[E_x \left\{ e^{-\beta \int_0^1 1_\Gamma (w(s)) \, ds} , S_\Gamma < 1 \right\} \right]$$

where S_Γ is the penetration time of Γ. For certain boundaries $\delta \Gamma$ the last term can be estimated uniformly in x.

3 Collection of results

In order to illustrate the kind of conditions typical in stochastic spectral analysis we collect some results concerning the principles mentioned in the preceding section. We always assume BASSA, Kato-Feller potentials and closed singularity regions Γ. Proofs are omitted. They are given in the articles referred. Hints are not given because it seems to be unmodest to mention always our names.

Compactness

Proposition 1 : *The semigroup difference*

$$e^{-t(K_0 \dot{+} V)} - e^{-tK_0}$$

is a Hilbert-Schmidt-operator if

$$\sup_{x,y} p(t,x,y) < \infty$$

and if

$$\int_0^{2t} d\lambda\, \lambda \int dx \int dy\, |V(x)|\, |V(y)|\, p(\lambda,x,y) < \infty$$

Proposition 2 : *The resolvent difference*

$$(K_0 \dot{+} V + a)^{-1} - (K_0 + a)^{-1}$$

is a trace class operator if

$$\int_0^{\infty} d\lambda\, \lambda\, e^{-a\lambda} \int dy\, E_y^{y,\lambda}\{e^{-\int_0^{\lambda} V(w(s))\, ds}\} |V(y)| < \infty$$

Proposition 3 : *For singular pertubations the difference*

$$e^{-t(K_0)_{E\setminus\Gamma}} - e^{-tK_0}$$

is Hilbert-Schmidt if

$$\sup_{x,y} p(t,x,y) < \infty$$

and if

$$\int dx\, [P_F\{S_\Gamma < t;\, w(0) = x\}]^2 < \infty$$

The singular semigroup difference is a trace class operator if

$$\int dx\, [P_F\{S_\Gamma < t\,,\, w(0) = x\}]^{1/2} < \infty$$

(see also Stollmann 1992).

Continuity in V

Proposition 4 : *$K_0\dot{+}V$ and $K_0\dot{+}W$ are selfadjoint operators in the Hilbert space $L^2(E)$. Let $E_V(.)$, $E_W(.)$ denote its spectral measures. Let φ be a non-vanishing Borel-function (typically
$\varphi(x) = (1+|x|^2)^\alpha$, $\alpha > 0$). Let*

$$[(K_0+a)^{-1}|\varphi|^2](x) \le c|\varphi(x)|^2$$

for all $x \in E$. For one of the potentials, take V, we assume

$$\sup_{\lambda\in\Delta} \|\varphi^{-1}(K_0\dot{+}V+\lambda+i0)^{-1}\varphi^{-1}\| < \infty$$

where $\Delta = (\alpha,\beta)$ is an interval in $\mathbb{R}_+$, α, β no eigenvalues of $K_0\dot{+}V$ or $K_0\dot{+}W$. Let $\|(V-W)\varphi^2\|_{KF}$ be sufficiently small, such that the last estimate holds also for $K_0\dot{+}W$. Then for $\lambda_0 \in \Delta$ we get

$$\left\|\varphi^{-1}\left[\frac{dE_V(\lambda)}{d\lambda} - \frac{dE_W(\lambda)}{d\lambda}\right]\varphi^{-1}\right\|_{\lambda=\lambda_0} \le c_{(\lambda_0,a)} \left\|(V-W)\varphi^2\right\|_{KF}$$

The constant $c_{(\lambda_0,\,a)}$ can be estimated quantitatively.

Proposition 5 : *Let V and W be Kato-Feller potentials in $L^1(E)$. Assume*

$$\int_0^\infty d\lambda\,\lambda e^{-a\lambda}\sup_x p(\lambda,x,x) < \infty.$$

Then the wave operators

$$\Omega_\pm(K_0\dot{+}V,K_0) := s-\lim_{t\to\pm\infty} e^{it(K_0\dot{+}V)}e^{-itK_0}P_{ac}(K_0)$$

and $\Omega_\pm(K_0\dot{+}W,\,K_0)$ exist. ($P_{ac}(K_0)$-projection operator onto the absolutely continuous subspace of K_0). Define the scattering operator by

$$S_V := \Omega_+^*(K_0\dot{+}V,K_0)\Omega_-(K_0\dot{+}V,K_0)$$

Both S_V and S_W commutes with K_0, providing that the corresponding scattering matrices $S_V(\lambda)$, $S_W(\lambda)$ are well defined. Assume that for some λ_0

$$\|V\varphi^{-1}(K_0\dot{+}W-\lambda_0-i0)^{-1}\varphi^{-1}\| < 1$$

Let $\sup_x|\varphi(x)V(x)| < \infty$ and $\sup_x|\varphi(x)W(x)| < \infty$. The operator norm of the scattering matrices is a norm in the fiber of the spectral resolution of the absolutely continuous subspace of K_0. This norm can be estimated as

$$\|S_V(\lambda_0) - S_W(\lambda_0)\| :\le c(\lambda_0)\left\|(V-W)\varphi^2\right\|_{KF}$$

Semiclassical limits

Proposition 6 : *Let B be a compact set in E and $(h^2K_0+V)_B$ the Feller operator with Dirichlet boundary conditions on δB. Assume positive V such that*

$$V = V1_{B\setminus\Gamma} + V1_\Gamma$$

$$V1_\Gamma \geq \gamma 1_\Gamma$$

$\Gamma \subset B$, i.e. V is larger than a constant γ on Γ. Let ψ_{h^2} be the ground state of $(h^2K_0+V)_B$, i.e.

$$((h^2K_0+V)_B\,\psi_{h^2})(x) = E_{h^2}\psi_{h^2}(x)$$

Let $\sup_{x,y} p(t,x,y) < \infty$ and $E_{h^2} = h^2E$. Then

$$|\psi_{h^2}(x)| \leq e^E\, E_x\{e^{-h^{-2}\gamma T_{1,\Gamma}(w)}\}$$

where $T_{t,\Gamma}(w) := meas\{s, s \leq t, w(s) \in \Gamma\}$ is the spending time of the trajectory w in Γ. If we consider x in a subset $\tilde{\Gamma} \subset \Gamma$ with $dist(\tilde{\Gamma}, B\setminus\Gamma) \geq r$, a uniform estimate is possible:

$$|\psi_{h^2}(x)| \leq e^E\, E_0\{e^{-h^{-2}\gamma T_{1,B(r)}(w)}\}$$

where $B(r)$ is a ball of radius r with centre in the origin. The right hand side tends to zero as $h \to 0$. A rate of convergence can be given for special K_0.

Large coupling limits

Proposition 7 : *Let $V \equiv 0$ and compare $K_\beta = K_0 + \beta 1_\Gamma$ with $(K_0)_\Sigma$, $\Sigma = E\setminus\Gamma$ for large parameters β. Denote again $Jf = f\restriction_\Sigma$. Then*

$$\|Je^{-tK_\beta} - e^{-t(K_0)_\Sigma}J\| \leq \sup_{x\in\Sigma} E_x\{e^{-\beta T_{t,\Gamma}}, T_{t,\Gamma} > 0\}$$

($T_{t,\Gamma}$ is the spending time defined in Proposition 6)

Remark: To estimate the Laplace transform of the spending time (or occupation time) is a difficult problem. If $K_0 = -\Delta$ in $L^2(\mathbb{R}^n)$ it is done recently by Demuth, Kirsch, Mc Gillivray (1993)and explained in another contribution of these proceedings.

References

[1] Demuth, M.; van Casteren, J. (1989), On spectral theory for selfadjoint Feller generators, Rev. Math. Phys.1, 325 - 414

[2] Demuth, M. (1992), Perturbations of spectral measures for Feller operators, Operator theory: Advances and Applications, Vol. **57**, 27 - 37

[3] Demuth, M.; van Casteren, J. (1992), A Hilbert-Schmidt property of resolvent differences of singularly perturbed generalized Schrödinger operators, Preprint Max-Planck-Institut Bonn, Nr. 92-48.

[4] Demuth, M.; van Casteren, J. (1992), Perturbation of generalized Schrödinger operators in stochastic spectral analysis, Lect. Notes in Physics **403**, 1-15

[5] Demuth, M.; Jeske, F.; Kirsch, W. (1992), On the rate of convergence for large coupling limits in quantum mechanics, Preprint Max - Planck - Institut Bonn, Nr.92 - 29. To be published in Ann. Inst. H. Poincaré.

[6] Demuth, M.; Kirsch, W.; Mc Gillivray, J.(1993), Schrödinger semigroups - geometric estimates in terms of the occupation time. To be published in the Preprint series of the Max-Planck-Institute, Bonn.

[7] Stollmann, P. (1992), Scattering by obstacles, Preprint Univ. of Frankfurt/Main, to be published in J. Functional analysis.

Author's addresses:

Michael Demuth
Max-Planck-Arbeitsgruppe
Universität Potsdam
FB Mathematik
Am Neuen Palais 10, PF 60 15 53
14415 Potsdam
Germany

Jan van Casteren
Department of Mathematics
and Computer Sciences
Univ. Instelling Antwerp
Universitetsplein 1
2610 Wilrijk/Antwerp
Belgium

Operator Theory:
Advances and Applications, Vol. 70

Occupation time asymptotics with an application to the decay of eigenfunctions

Michael Demuth, W. Kirsch and I. McGillivray

1 Main results

We aim to study the large coupling limit for the difference of Schrödinger semigroups with a view to an application in spectral theory, concerning the behaviour of eigenfunctions in special Dirichlet problems.
We start by detailing our set-up. We are interested in perturbed operators H_M of the positive Laplacian $H_0 := -\frac{1}{2}\Delta$ acting in $L^2\left(\mathbb{R}^d\right)$, $d \geq 2$ of the form

$$H_M := H_0 + V + MU$$

where V is a uniformly bounded potential and $U = 1_\Gamma$ for Γ a closed subset of $\mathbb{R}^d$, called the singularity region, and M is an arbitrary positive parameter. H_Σ, $\Sigma := \mathbb{R}^d - \Gamma$ denotes the operator $H_0 + V$ in $L^2(\Sigma, dx)$ with Dirichlet boundary conditions on $\partial\Sigma$. We study the asymptotic behaviour of the semigroup difference

$$\left\| Je^{-tH_M} - e^{-tH_\Sigma}J \right\| \, , \, t > 0$$

as $M \to \infty$. Here J is restriction $J : L^2\left(\mathbb{R}^d; dx\right) \to L^2\left(\Sigma, dx\right)$. It turns out that these asymptotics are linked to the geometry of the singularity region Γ. We assume that Γ is an uniform Lipschitz set, a condition slightly stronger than just Lipschitz. An important quantity in our estimates will be the so called cone constant. To explain this, a cone C in $\mathbb{R}^d$ will be a set of the form

$$C := \left\{ x \in \mathbb{R}^d : \langle x, e_1 \rangle \geq p\|x\| \right\} \, , \, -1 < p < 1$$

where $<,>$ is the inner product in $\mathbb{R}^d$ and e_1 is the unit vector in the x_1 direction. Let F be the closed subset of S^{d-1}, the unit sphere centered at the origin, given by $F := C \cap S^{d-1}$. The cone constant is defined by

$$A = A(C) := \frac{\lambda_1\left(F^C\right)}{\left(v^2 + 2\lambda_1\left(F^C\right)\right)^{1/2} + v}$$

where $v := \frac{d}{2} - 1$ and $\lambda_1\left(F^C\right)$ is the lowest eigenvalue of the Laplace-Beltrami operator $-\frac{1}{2}\Delta$ on $F^C := S^{d-1} - F$ with Dirichlet boundary conditions on ∂F^C. Given any uniform Lipschitz set in our sense, there exists $C = C_{int}$ such that

$$\forall\ y \in \partial\Gamma\ \exists \text{ rigid transformation } S_y \text{ of } \mathbb{R}^d$$
$$\text{such that } S_y \subset B(y,r) \cap \Gamma \text{ and } S_y(0) = y$$

Our main result is

Theorem. *Suppose that* Γ *is a uniform Lipschitz set and that* $A = A(C) \leq 1/2$, $C = C_{int}$. *Then for each* $0 < \gamma < 1$, $t > 0$ *there exists a constant* $c_\gamma(t)$ *such that*

$$\left\| Je^{-tH_M} - e^{-tH_\Sigma} J \right\| \leq c_\gamma(t) M^{-\gamma A}\,, \quad M\,large$$

We also have upper bounds for the resolvent differences of the same form. Under slightly different assumptions we have obtained similar lower bounds on the semigroup and resolvent differences.

2 Occupation time asymptotics

The key to these results lies in certain Brownian motion estimates. We have the following theorem, interesting in its own right.

Theorem. *Let* C *be a cone and*

$$T_{t,C} := |\{s \in [0,t] : X_s \in C\}|$$

the total occupation time up to time t *in* C. *Then there exist constants* $c, c' > 0$ *such that*

$$c'(Mt)^{-A} \leq E_0 e^{-MT_{t,C}} \leq c \left(\frac{\log Mt}{Mt}\right)^A, \quad A = A(C)$$

for $MT > e$.
This result rests on the next Lemma, whose proof we sketch to give the flavour of the methods used in the article.

Lemma. *Let* C *be as above. Then*

$$P_0\left(T_{t,C} \leq 1\right) \leq c \left(\frac{\log t}{t}\right)^A, \quad t \geq e$$

Proof. Let $v \in S^{d-1} \cap int\, C$. Then for each $a > 0$ there is a cube Q_a of side-length λ_a (λ independent of a) centered at the origin such that

$$\eta + Q_a \subset C - (2av + C)\,,\ \forall \eta \in av + \partial C$$

Write

$$P_0\left(T_{t,C} \leq 1\right) = P_0\left(T_{t,C} \leq 1, \sigma_{2a} < t\right) + P_0\left(T_{t,C} \leq 1, \sigma_{2a} > t\right)$$

when σ_a is the first hitting time of $av + C$. Since each tajectory in the first RHS term leaves $X_{\sigma_a} + Q_a$ after σ_a, before $\sigma_a + 1$

$$\leq P_0\left(\sigma\left(Q_a^c\right) \leq 1\right) + P_0\left(\sigma_{2a} > t\right) \leq ce^{-\lambda^2 a^2/2} + C\left(\frac{a^2}{t}\right)^A ,$$

the last line due to an estimate in [2] and standard Brownian motion estimates. Choosing $a^2 = \frac{2a}{\lambda^2} log\, t$ leads to the result. □

3 Applications to spectral theory

We are able to get estimates on the limiting absorption principle for the difference of Schrödinger and Dirichlet operators, but prefer here to describe an application to the decay of eigenfunctions. Let B be an open ball in $\mathbb{R}^d$, $d \geq 2$ centered at the origin and C be a cone. Let H_M be the operator $H_M := -\frac{1}{2}\Delta + M1_C$, $M > 0$ in $L^2(B)$ with Dirichlet boundary conditions on ∂B. Then H_M has discrete spectrum. If Φ_M is a normalized L^2 eigenvector of H_M we have

$$|\Phi_M(0)| \leq c\left(\frac{log\, M}{M}\right)^A , \quad M > e, A = A(C)$$

References

[1] M. Demuth, W. Kirsch, I. McGillivray. Schrödinger semigroups – estimates in terms of the occupation time. To appear as preprint of the Max-Planck-Institute for mathematics in Bonn.

[2] T. Meyre. Étude asymptotique du temps passé par le mouvement brownien dans un cône. Ann. Inst. Henri Poincaré, Vol.27, no.1 ,1991, 107–124

Michael Demuth, Max-Planck-Arbeitsgruppe, Fachbereich Mathematik
Universität Potsdam, Postfach 601 553, D-14415 Potsdam
W. Kirsch,Fakultät für Mathematik, Ruhr-Universität, Postfach 102 148
D-44801 Bochum
I. McGillivray, Fachbereich Mathematik, TU Berlin, SFB 288
Straße des 17.Juni 136, D-10623 Berlin

Operator Theory:
Advances and Applications, Vol. 70

Holomorphic semigroups and Schrödinger equations

El Maati Ouhabaz

1 L^p - holomorphy

We report on some recent results in [Ou] on holomorphy of semigroups. We also discuss the application of these results to the study of the general Schrödinger equations of the type $\begin{cases} u'(t) &= i(\Delta - V)u \\ u(0) &= x \end{cases}$ on $L^p(\mathbb{R}^n)$ $1 \leq p < \infty$, or $C_0(\mathbb{R}^N)$. Here $V = V_+ - V_-$ is a potential satisfying $V_+ \in L^1_{loc}(\mathbb{R}^N)$ and V_- is in the Kato class.

Let Ω be an open set of $\mathbb{R}^N$ (with the Lebesgue measure) and consider a self-adjoint semigroup $T = (T(t)_{t\geq 0})$ on $L^2(\Omega)$ with generator A. It is known that T is a *bounded holomorphic semigroup of angle* $\frac{\pi}{2}$ i.e. T has an extension to the domain $D = \{z; Re\, z > 0\}$ s.t.

1. $T(z + z') = T(z)T(z'), \quad z, z' \in D$
2. $z \to T(z)$ is holomorphic from D to $\mathcal{L}(L^2(\Omega))$.
3. $\lim_{z\to 0, z\in D} T(z)f = f$ for all $f \in L^2(\Omega)$
4. For each $\theta \in (0, \frac{\pi}{2})$ there exists a constant M (depending on θ) s.t.

$$|T(z)|_{\mathcal{L}(L^2(\Omega))} \leq M$$

for all $z \in \Sigma(\theta) = \{\lambda \in \mathbb{C}, |arg\lambda| < \theta\}$

Assume now that T interpolates on $L^p(\Omega)$ $1 \leq p < \infty$, that is, there exists for each p , a strongly continuous semigroup T_p on $L^p(\Omega)$ with $T_2 = T$ and satisfying $T_p(t)f = T_2(t)f$ $(t \geq 0)$ for $f \in L^p(\Omega) \cap L^2(\Omega)$. It follows from the Stein interpolation theorem that for $1 < p < \infty$ the semigroup T_p is holomorphic in the sector

$$\left\{z;\ |arg\, z| < \frac{\pi}{2}\left(1 - \left|\frac{2}{p} - 1\right|\right)\right\}$$

However, the case $p = 1$ is more delicate. The following result is shown in [Ou].

Theorem 1.1 *Assume that T has a* Gaussian estimate *i.e. there exist positive constants M and b s.t.*

$$(E) \quad |T(t)f| \leq MG(bt)\,|f| \quad for \quad 0 \leq t \leq 1,\ f \in L^2(\Omega)$$

where $G = (G(t))_{t\geq 0}$ *is the Gaussian semigroup on* $L^2(\mathbb{R}^N)$. *Then there exists* $w \geq 0$ *s.t. the semigroup* $e^{-w\cdot}T_p$ *is bounded holomorphic with angle* $\frac{\pi}{2}$ *on* $L^p(\Omega)$ $1 \leq p < \infty$.

Using the Gaussian estimates developed recently by Davies [Da](chap 3) for uniformly elliptic operators, this result is applicable in the following framework.

Assume that A is the associated operator with the following symmetric form

$$a(u,v) = \sum_{i,j=1}^{N} \int_\Omega a_{ij} D_i u D_j \bar{v}\, dx + \int_\Omega V u \bar{v}\, dx$$

with $a_{ij} = a_{ji} \in L^\infty(\Omega)$ satisfying the ellipticity condition $\sum a_{ij}(x)\xi_i\bar{\xi}_j \geq c|\xi|^2$ a.e. $x \in \Omega$ and all $\xi = (\xi_1, ..., \xi_N) \in \mathbb{C}^N$. The potential V is positive and in $L^1_{loc}(\Omega)$. The domain of the form a is given by $D(a) = W \cap \{u \in L^2(\Omega), \int_\Omega V|u|^2 < \infty\}$, where W is a closed subspace of the Sobolev space $H^1(\Omega)$ which contains $H^1_0(\Omega)$. We can apply Theorem 1.1 in the following situations:

1. $W = H^1_0(\Omega)$ for Ω any open set of $\mathbb{R}^N$ (this corresponds to the Dirichlet boundary conditions).

2. W satisfies the two following properties

   * $u \in W$ implies $|u| \in W$
   * If $0 \leq u \leq v$, $v \in W$ and $u \in H^1(\Omega)$ then $u \in W$.

In this case we assume that Ω has the extension property ($W = H^1(\Omega)$ corresponds to the Neumann boundary conditions).
If the boundary $\partial\Omega$ of Ω is regular, one can take as example of situation 2, $W = \{u \in H^1(\Omega);\ u_{|\Gamma} = 0\}$. Γ is a closed set of $\partial\Omega$. This corresponds to the mixed boundary conditions, Dirichlet on Γ and Neumann on $\partial\Omega\backslash\Gamma$. We recall that if the coefficients a_{ij} are smooth, bounded and of class C^2, it was shown by Amann [Am] that A generates a holomorphic semigroup on $L^1(\Omega)$. The result has been extended by Arendt and Batty [Ar.Ba] to arbitrary open set Ω under the Dirichlet boundary conditions. We recall that in these works [Am] and [Ar.Ba] it is shown that the holomorphy on $L^1(\Omega)$ holds with some "small" angle. More precisely, it is shown that the estimate

$$\|\lambda R(\lambda, A)\|_{\mathcal{L}(L^1(\Omega))} \leq M \tag{1}$$

holds for $Re\,\lambda > 0$. Here $R(\lambda, A)$ is the resolvent of A in $L^1(\Omega)$.
Our result gives that the estimate (1) holds in each sector
$\sum(\theta + \frac{\pi}{2}) := \{z; |arg\,z| < \theta + \frac{\pi}{2}\}$ for all $\theta \in (0, \frac{\pi}{2})$.

2 C_0-holomorphy

We denote by $C_0(\Omega)$ the space of continuous functions which vanish at infinity of Ω. We suppose that the assumptions of Theorem1.1 are satisfied and T acts as a semigroup on $C_0(\Omega)$ i.e. there exists a strongly continuous semigroup T_0 on $C_0(\Omega)$ s.t. $T_0(t)f = T(t)f$ for $t \geq 0$ and $f \in L^2(\Omega) \cap C_0(\Omega)$. We have

Theorem 2.1 *The semigroup $e^{-w\cdot}T_0$ is bounded holomorphic with angle $\frac{\pi}{2}$.*

To show this result one has to show that the estimate $\|\lambda R(\lambda, A_0)\|_{\mathcal{L}(C_0(\Omega))} \leq M$ holds in each sector $\sum\left(\theta + \frac{\pi}{2}\right)$, for all $\theta \in \left(0, \frac{\pi}{2}\right)$. Here A_0 is the generator of the semigroup $e^{-\omega\cdot}T_0$. The desired estimate follows by a duality argument and the result in $L^1(\Omega)$.

Corollary 2.1 *Let Ω be any open set of $\mathbb{R}^N$ and assume that there exists a realization A_0 of the Laplacian Δ s.t. A_0 is a generator of a semigroup T_0 on $C_0(\Omega)$. Then T_0 is bounded holomorphic with angle $\frac{\pi}{2}$.*
Moreover, $D(A_0) = \{u \in C_0(\Omega), \Delta u \in C_0(\Omega)\}$.

This corollary follows immediatly from Theorem 2.1 if we show that

$$T_0(t)f = T(t)f \quad f \in L^2(\Omega) \cap C_0(\Omega)\ (t \geq 0) \tag{2}$$

where T is the semigroup generated on $L^2(\Omega)$ by the Dirichlet Laplacian Δ_D. The equality (2) is easy if Ω is bounded and regular. Now if Ω is arbitrary we take a sequence $(\Omega_n)_n$ of bounded and regular open sets s.t. $\bar{\Omega}_n \subset \Omega_{n+1}$ and $\bigcup_n \Omega_n = \Omega$. We show that if $f \in C^\infty(\Omega)$ with compact support, the solution u_n of $(\lambda - \Delta)u = f$ on Ω_n converges to the solution of $(\lambda - A_0)u = f$ and to the solution of $(\lambda - \Delta_D)v = f$. This gives (2)

3 The general Schrödinger equation

We consider the general Schrödinger equation

$$(SE) \quad \begin{cases} u'(t) &= i(\Delta - V)u \qquad t \in \mathbb{R} \\ u(0) &= x \end{cases}$$

on $L^p\left(\mathbb{R}^N\right)$, $1 \leq p < \infty$. Here $V = V_+ - V_-$, $V_+ \in L^1_{loc}\left(\mathbb{R}^N\right)$ and V_- is in the Kato class.
On $L^2\left(\mathbb{R}^N\right)$ th equation (SE) is well-posed for all $x \in D(\Delta - V)$ by the Stone's theorem. But on $L^p\left(\mathbb{R}^N\right)$ for $p \neq 2$ it is known that $i\Delta$ is not a generator of a strongly continuous semigroup and then the equation (SE) is not easy in these spaces. We have the following

Theorem 3.1 *(see [B.dL] and [E.M]) let E be a Banach space. Assume that A generates a bounded holomorphic semigroup of angle $\frac{\pi}{2}$ on E. Assume that*

$$\|T(z)\|_{\mathcal{L}(E)} \leq c\left(\frac{|z|}{Re\, z}\right)^r \tag{3}$$

for all z, $Re\, z > 0$. Here C and r are positive constants. Then the evolution equation

$$\begin{cases} u'(t) &= iAu \qquad t \in \mathbb{R} \\ u(0) &= x \end{cases}$$

is well posed for $x \in D(A^\gamma)$ for all $\gamma > r$.

We apply this result to the operator $A = \Delta - V$ on $L^p(\mathbb{R}^N)$, $1 \leq p < \infty$ $(p \neq 2)$. It is known by the Feyman-Kac formula that the semigroup genereted by A has a Gaussian estimate (E) (see for example [Al.B.M] and [D.VC]. In [Al.B.M] it is shown that Gaussian estimates hold for $\Delta - \mu$ with $\mu = \mu_+ - \mu_-$ is a certain regular measure). It follows by theorem 1.1 that $a - \omega I$ is a generator of a bounded holomorphic semigroup with angle $\frac{\pi}{2}$ on $L^p(\mathbb{R}^N)$, $1 \leq p < \infty$. Moreover, the estimate (3) holds in $L^1(\mathbb{R}^N)$ with $r = N$ (see[Ou]). By the Riesz-Thorin interpolation theorem, the estimate (3) holds in $L^p(\mathbb{R}^N)$ with $r = N\left(\frac{2}{p} - 1\right)$, $1 \leq p \leq 2$.
We then use Theorem 1.1, Theorem 3.1 and these estimates to conclude that the equation (SE) has a unique classical solution in $L^p(\mathbb{R}^N)$ for $x \in D(A_p^\gamma)$, $\gamma > N\left(\frac{2}{p} - 1\right)$, where $A_p = \Delta - V$ considered in $L^p(\mathbb{R}^N)$. For $p > 2$ we use the duality. Finally, these arguments works also in $C_0(\mathbb{R}^N)$ if we show that $\Delta - V$ generates a semigroup on $C_0(\mathbb{R}^N)$.

References

[B.dL] K.Boyadzhiev and R. deLaubenfels. Boundary values of holomorphic semigroups (Preprint)

[Da] E. B. Davies. Heat Kernels and Spectral Theory. Cambridge Univer. Press 1989

[E.M] O. El-Mennaoui. Traces des semi-groupes holomorphes singuliers à l'origine et comportement asymptotic. Ph. D. Thesis. Besançon 1992.

[Am] H.Amann. Dual semigroups and second order linear elliptic boundary value problems. Israel J. Math. 45 (1983) 225-254.

[Ou] E. M. Ouhabaz. Gaussian estimates and holomorphy of semigroups (submitted).

[Al.B.M] S. Albeverio, P. Blanchard and Z. Ma. Feyman-Kac semigroups in terms of signed smooth measures. Preprint Bibos no.424 Univ. Bielefeld (1990).

[D.VC] M. Demuth and J. van Casteren. On Spectral theory of selfadjoint Feller generators. Rev. Math. Phys. Vol 1, no.3 (1989) 325-414.

[Ar.Ba] W. Arendt and C. Batty. Absorption semigroups and Dirichlet boundary conditions. Math. Ann. 295 (1993) 427-448

El Maati Ouhabaz, SFB 288, Technische Universität Berlin, Straße des 17. Juni 136,D-1000 Berlin 12,

and

Max-Planck Arbeitsgruppe, Fachbereich Mathematik, Universität Potsdam, Am Neuen Palais 10, 1571 Potsdam. Germany

Operator Theory:
Advances and Applications, Vol. 70

Some Problems on Submarkovian Semigroups

V. A. Liskevich*and Yu. A. Semenov

Abstract

We present a perturbation result for generators of submarkovian semigroups acting on L^p. It characterizes the domain of the generator of a submarkovian semigroup by means of the domain of the quadratic form. For the particular case of submarkovian perturbations the well known KLMN-theorem is extended to the L^p-spaces.

The object of this note is a generator of a submarkovian semigroup in L^p-space. We characterize the domain of the generator by means of the domain of its quadratic form in L^2. The main result concerns the non-symmetric form-bounded perturbations of the generator of the submarkovian semigroup. Under the assumption of form boundedness the form sum can be correctly defined by virtue of the well known KLMN- theorem. It is shown that this operator is the generator of a C_0-semigroup in L^p, where p depends on form bounds only. The domain of the perturbed generator is characterized and (L^p, L^q)-estimates of the resolvent are obtained.

Let (M, μ) be a measurable space with the σ-finite measure μ. We use the following notations: $L^p \equiv L^p(M, \mu)$, $\|\cdot\|_p$ is the norm in L^p, $\langle f, g\rangle \equiv \int_M f(x)\overline{g(x)}d\mu(x)$, $\mathcal{L}(L^p, L^q)$ is the space of bounded operators acting from L^p into L^q.

Let $A \geq 0$ be the self-adjoint operator acting in $L^2(M, \mu)$ with the domain $\mathcal{D}(A)$ and the form-domain $Q(A) = \mathcal{D}(A^{\frac{1}{2}})$. For $f \in Q(A)$ define $t_A[f] := \|A^{\frac{1}{2}} f\|_2^2$.

Definition. We say that A is a generator of a submarkovian semigroup (submarkovian generator) if the following conditions are satisfied:

(i) A is a non-negative self-adjoint operator in L^2.

(ii) $\|e^{-tA}f\|_\infty \leq \|f\|_\infty$, $\forall\, f \in L^1 \cap L^\infty$.

(iii) $0 \leq f \in L^2 \Longrightarrow e^{-tA}f \geq 0$ almost everywhere.

Now let A be a submarkovian generator. We can define the operator A_p as a generator of the contraction semigroup in L^p:

$$(e^{-tA}\restriction[L^2 \cap L^p])^{\sim}_{L^p \to L^p} = e^{-tA_p} \text{ (the closure in } L^p).$$

$$T^t_\infty =: (e^{-tA_1})^* \text{ (* denotes the adjoint operator).}$$

*Recipient of a Dov Biegun Postdoctoral Fellowship

By $-A_p$ we denote the generator of the semigroup T_p^t, so $T_p^t = e^{-tA_p}$, for all $t \geq 0$ and $p \in [1, \infty)$. Since T_p^t is a contraction the operator A_p is accretive with respect to the semi-inner product $[u, v] = \langle u, |v|^{p-1} sgn\ v \rangle \|v\|_p^{2-p}$ in L^p:

$Re\ [A_p u, u] \geq 0$ for all $u \in \mathcal{D}(A_p)$ and $[A_p u, u] \geq 0$ for all $u \in \mathcal{D}(A_p) \cap Re\ L^p$.

The following theorem makes more precise this property for submarkovian generators.

Theorem 1. *Let $f \in \mathcal{D}(A_p)$ for some $p \in (1, \infty)$. Then $f_p := f|f|^{\frac{p-2}{2}} \in Q(A)$, and*

$$4\frac{p-1}{p^2} t_A[f_p] \leq Re\ \langle A_p f, sgn\ f\ |f|^{p-1} \rangle \leq \varkappa(p) t_A[f_p] \tag{1}$$

where $\varkappa(p) = \sup\{(1 + s^{\frac{1}{p}})(1 + s^{\frac{1}{p'}})(1 + s^{\frac{1}{2}})^{-2};\ s \in (0, 1)\}$

$$|Im\ \langle A_p f, sgn\ f\ |f|^{p-1} \rangle| \leq \frac{|p-2|}{2\sqrt{p-1}} Re\ \langle A_p f, sgn\ f\ |f|^{p-1} \rangle \tag{2}$$

If $f \in \mathcal{D}(A_p) \cap L_+^p$, then $f^{\frac{p}{2}} \in Q(A)$, and

$$4\frac{p-1}{p^2} t_A[f^{\frac{p}{2}}] \leq \langle A_p f, f^{p-1} \rangle \leq t_A[f^{\frac{p}{2}}]. \tag{3}$$

Remark. (3) is the Stroock–Varopoulos inequality [S],[V],[CKS]. (1) with $f = Re\ f$ has been proved in [LSe]. From (2) the analyticity of the semigroup T_p^t follows (see [G]). (2) improves, in particular, the sector of analyticity which can be obtained from Stein's interpolation theorem [St].

Outline of the proof. Let $P(t, x, G) := (T_\infty^t 1\!\!1_G)(x)$, $G \in \mathcal{M}$, where $1\!\!1_G$ is the characteristic function of the set G. Then $P(t, x, \cdot)$ is finitely additive on $\mathcal{M}$, $P(t, x, M) \leq 1$ and $T_\infty^t f(x) = \int P(t, x, dy) f(y)$, $\forall f \in L^\infty$ (see, e.g. [D]). Using these properties of $P(t, x, G)$ and the following elementary inequalities

$$4\frac{p-1}{p^2}(s^{\frac{p}{2}} - t^{\frac{p}{2}})^2 \leq (s-t)(s^{p-1} - t^{p-1}) \leq (s^{\frac{p}{2}} - t^{\frac{p}{2}})^2$$

$$(s^{\frac{p}{2}} + t^{\frac{p}{2}})^2 \leq (s+t)(s^{p-1} + t^{p-1}) \leq \varkappa(p)(s^{\frac{p}{2}} + t^{\frac{p}{2}})^2$$

$$4\frac{p-1}{p^2}[s^p + t^p + 2b(st)^{\frac{p}{2}}] \leq s^p + t^p + b(st^{p-1} + ts^{p-1}),\quad \forall b \in [-1, 1]$$

$$|\sin\ \theta|\ |st^{p-1} - ts^{p-1}| \leq \frac{|p-2|}{2\sqrt{p-1}}(s^p + t^p - |\cos\theta|(st)^{\frac{p}{2}}),$$
$$\forall p \in (1, \infty),\ \forall \theta \in \mathbb{R}^1$$

$$s^p + t^p + b(st^{p-1} + ts^{p-1}) \leq \varkappa(p)[s^p + t^p + 2b(st)^{\frac{p}{2}}],\quad \forall p \in (1, \infty),\ \forall b \in [-1, 1]$$

(for all $0 < s, t < \infty$ and all $p \in [1, \infty)$) one can prove $\forall f \in L^p$

$$4\frac{p-1}{p^2}\langle (1 - T_2^t) f_p, f_p \rangle \leq Re\ \langle (1 - T_p^t) f, sgn\ f \cdot |f|^{p-1} \rangle \leq \varkappa(p) \langle (1 - T_2^t) f_p, f_p \rangle,$$

$$4\frac{p-1}{p^2}\langle (1 - T_2^t) f^{\frac{p}{2}}, f^{\frac{p}{2}} \rangle \leq \langle (1 - T_p^t) f, f^{p-1} \rangle \leq \langle (1 - T_2^t) f^{\frac{p}{2}}, f^{\frac{p}{2}} \rangle\ \ \forall f \in L_+^p$$

Since $Q(A) = \{g \in L^2 : \sup_{t>0} \frac{1}{t}\langle(1-T_2^t)g, g\rangle < \infty\}$ and $\lim_{t\downarrow 0} \frac{1}{t}\langle(1-T_2^t)g, g\rangle = t_A[g]$, $\forall g \in Q(A)$ the statements (1) and (3) follow from the spectral theorem for self-adjoint operators. (2) will follow from the inequality

$$|Im\ \langle(1-T_p^t)f, sgn\ f \cdot |f|^{p-1}\rangle| \leq \frac{|p-2|}{2\sqrt{p-1}} Re\ \langle(1-T_p^t)f, sgn\ f \cdot |f|^{p-1}\rangle.$$

□

Definition. We say that operator B is form-bounded relatively to A and write $B \in PK_\beta(A)$ if B is a self-adjoint operator in L^2, $\mathcal{D}(|B|^{\frac{1}{2}}) \supset \mathcal{D}(A^{\frac{1}{2}}) \equiv Q(A)$ and

$$\|\ |B|^{\frac{1}{2}}\varphi\|_2^2 \leq \beta\|A^{\frac{1}{2}}\varphi\|_2^2 + c_B(\beta)\|\varphi\|_2^2\ , \quad \forall\ \varphi \in \mathcal{D}(A^{\frac{1}{2}})$$

for some $\beta \in (0,1)$, $c_B(\beta) \geq 0$.

Theorem 2. Let A, B, C be submarkovian generators. Suppose that $B \in PK_{\beta_1}(A)$ for some $\beta_1 \in (0,1)$ and $C \in PK_{\beta_2}(A)$ for some $\beta_2 > 0$. Let t_1 and t_2 be the roots of the equation $(\beta_1 + \frac{|q-2|}{2\sqrt{q-1}}\beta_2)\varkappa(q) = 4\frac{q-1}{q^2}$, $1 < q < \infty$. [1] Let $H_2 \equiv H = A \dot{-} B \dot{+} iC$ be form-sum, $i := \sqrt{-1}$. Then for every $p \in [t_1, t_2]$ and $t \geq 0$

$$i) \qquad \|e^{-tH}f\|_p \leq e^{\varkappa(p)\{c_B(\beta_1) + \frac{|p-2|}{2\sqrt{p-1}}c_C(\beta_2)\}t}\|f\|_p, \quad f \in L^2 \cap L^p$$

and consequently this defines the family of operators $\{H_p\}_{t_1 \leq p \leq t_2}$ which generate quasi-contraction C_0-semigroups.

$$ii) \qquad H_p \supset A_p - B_p + iC_p$$

iii) the semigroup e^{-tH_p} is holomorphic with respect to t of the angle

$$\theta_1 = \frac{\pi}{2} - \arctan \mathfrak{U}, \quad 0 < \theta_1 < \frac{\pi}{2}, \quad p \in (t_1, t_2), \text{where}$$

$$\mathfrak{U} = \frac{\varkappa(p)[\frac{|p-2|}{2\sqrt{p-1}}(1+\beta_1) + \beta_2]}{4\frac{p-1}{p^2} - (\beta_1 + \frac{|p-2|}{2\sqrt{p-1}}\beta_2)\varkappa(p)}$$

$$iv) \qquad \forall p \in (t_1, t_2), \quad \forall f \in \mathcal{D}(H_p) \quad f_p = f|f|^{\frac{p-2}{2}} \in Q(A) \quad \text{and}$$

$$Re\ \langle H_p f, |f|^{p-1} sgn\ f\rangle \geq (4\frac{p-1}{p^2} - (\beta_1 + \frac{|p-2|}{2\sqrt{p-1}}\beta_2)\varkappa(p))\|A^{\frac{1}{2}}f_p\|_2^2$$
$$- \varkappa(p)\{c_B(\beta_1) + \frac{|p-2|}{2\sqrt{p-1}}c_C(\beta_2)\}\|f\|_p^p.$$

[1] The roots always exist and if $t_1 \in (1,2)$ then $t_2 = t_1' \in (2,\infty)$. This follows from the abovementioned properties of $\varkappa(p)$ and from $\frac{|q-2|}{\sqrt{q-1}} = \frac{|q'-2|}{\sqrt{q'-1}}$.

Proof. Let B_n and C_m be the Yosida approximations of B_p and C_p. It is enough to prove i) with B_n instead of B and with C_m instead of C, i.e. without loss we can assume that $B, C \in \mathcal{L}(L^2) \cap \mathcal{L}(L^p)$. Since $Re\ \langle \frac{d}{dt}u, u|u|^{p-2}\rangle = \frac{1}{p}\frac{d}{dt}\|u\|_p^p$, then

$$\begin{aligned}-\frac{1}{p}\frac{d}{dt}\|u\|_p^p &= Re\ \langle (A_p - B_p + iC_p)u, u|u|^{p-2}\rangle \\ &= Re\ \langle A_p u, u|u|^{p-2}\rangle - Re\ \langle B_p u, u|u|^{p-2}\rangle - Im\ \langle C_p u, u|u|^{p-2}\rangle \\ &\geq 4\frac{p-1}{p^2}\|A^{\frac{1}{2}}u_p\|_2^2 - \varkappa(p)\|B^{\frac{1}{2}}u_p\|_2^2 - \frac{|p-2|}{2\sqrt{p-1}}\varkappa(p)\|C^{\frac{1}{2}}u_p\|_2^2\end{aligned}$$

$u \in \mathcal{D}(A_p)$. (We used consecutively (1) and (2)). To finish the proof of i) it remains only to use the conditions $B \in PK_{\beta_1}$ and $C \in PK_{\beta_2}$. The proof of ii) is standard. In order to prove iii) let us estimate $|Im\ \langle \bar{H}_p u, sgn\ u \cdot |u|^{p-1}\rangle|$ by $Re\ \langle \bar{H}_p u, sgn\ u \cdot |u|^{p-1}\rangle$, where $\bar{H}_p = H_p + l$ for an appropriate l which will be chosen below. As before we assume B and C to be bounded operators in L^2 and in L^p. We have

$$|Im\ \langle \bar{H}_p u, sgn\ u \cdot |u|^{p-1}\rangle| \leq \mathcal{K}_1\|A^{\frac{1}{2}}u_p\|_2^2 + \Lambda_1\|u\|_p^p,$$

where $\mathcal{K}_1 = \varkappa(p)(\frac{|p-2|}{2\sqrt{p-1}}(1+\beta_1) + \beta_2)$, $\Lambda_1 = \varkappa(p)(\frac{|p-2|}{2\sqrt{p-1}}c_B(\beta_1) + c_C(\beta_2))$,

$$Re\ \langle \bar{H}_p u, sgn\ u \cdot |u|^{p-1}\rangle \geq \mathcal{K}\|A^{\frac{1}{2}}u_p\|_2^2 + \Lambda\|u\|_p^p,$$

where $\mathcal{K} = 4\frac{p-1}{p^2} - (\beta_1 + \frac{|p-2|}{2\sqrt{p-1}}\beta_2)\varkappa(p)$, $\Lambda = l - \varkappa(p)(c_B(\beta_1) + \frac{|p-2|}{2\sqrt{p-1}}c_C(\beta_2)) \equiv l - \Lambda_2$.
Setting $l = \Lambda_2 + \frac{\mathcal{K}}{\mathcal{K}_1}\Lambda_1$, we obtain

$$|Im\ \langle \bar{H}_p u, sgn\ u \cdot |u|^{p-1}\rangle| \leq \frac{\mathcal{K}_1}{\mathcal{K}}\ Re\ \langle \bar{H}_p u, sgn\ u \cdot |u|^{p-1}\rangle. \tag{4}$$

In order to complete the proof of iii) it remains only to pass to the limit in (4). Let $H_p(n,m) := A_p - B_{p,n} + iC_{p,m}$ and $u_{n,m} = \exp(-tH_p(n,m))f$, $u = \exp(-tH_p)f$, $f \in L^2 \cap L^p$. Using the general approximation theorem [K] it is not difficult to check that $u_{n,m} \underset{L^p}{\overset{s}{\to}} u$.
Furthermore by virtue of (4) the semigroups $\exp(-t\bar{H}_p(n,m))$ are holomorphic and uniformly bounded with respect to m, n in the sector $S_p = \{t; |\arg t| \leq \theta,\ \forall\theta \in (0,\theta_1)\}$. By (4) $u_{n,m}$ converges pointwise on the positive semiaxis. Consequently by the Vitali convergence theorem u is holomorphic in S_p and $u(z) = s - L^p - \lim_{n,m\to\infty} u_{n,m}(z)$, $z \in S_p$. In particular $u'(z) = s - L^p - \lim_{n,m\to\infty} u'_{n,m}(z)$, $z \in S_p$, and $\bar{H}_p(n,m)u_{n,m} \underset{L^p}{\overset{s}{\to}} \bar{H}_p u$. Moreover passing to a subsequence if required we have $g_{n,m} := sgn\ u_{n,m}|u_{n,m}|^{p-1} \longrightarrow g := sgn\ u|u|^{p-1}$ pointwise μ-a.e. and $\|g_{n,m}\|_{p'} = \|u_{n,m}\|_p^{p-1} \longrightarrow \|g\|_{p'} = \|u\|_p^{p-1}$ by virtue of (4). Therefore $g_{n,m} \underset{L^{p'}}{\overset{s}{\to}} g$. Thus, in (4) we can pass to the limit $(n,m) \longrightarrow \infty$. □

Theorem 3. Let A and B be submarkovian generators. Suppose that

1) $B \in PK_\beta(A)$ for some $\beta \in (0,1)$.

2) $\|f\|_{2j}^2 \leq c\, t_A[f]$ for some $j > 1$ and all $f \in Q(A)$.

Let $H = A \dot{-} B$ and $t(\beta), t'(\beta)$ be the corresponding roots of the equation $\beta \varkappa(q) = 4\frac{q-1}{q^2}$, $1 < q < \infty$. Then

$$(i) \qquad H_p^\sigma e^{-tH_p} \in \mathcal{L}(L^p, L^q), \quad \sigma = 0,1,2,\ldots, \quad t > 0$$

for arbitrary $p \in (t(\beta), t'(\beta))$ and $q \in [p, jt'(\beta))$.
If in addition $1/p - 1/q \leq 1/j'$ then

$$(ii) \qquad (l + H_p)^{-1} \in \mathcal{L}(L^p, L^q), \quad \forall l \in \rho(-H_p).$$

Moreover if $g \in L^2 \cap L^{t'(\beta)}$ then

$$(iii) \quad \mu(\{x : |(l+H)^{-1} g(x)| \geq s\}) \leq c \cdot s^{-jt'(\beta)} \cdot \ell n^j s, \quad \forall s > \exp \frac{1}{t'(\beta) - t(\beta)}.$$

The proof is based on Theorem 2(iv) and on Chebyshev's inequality.

Remarks.
1. Assumption 2) of Theorem 3 is an abstract version of Sobolev's imbedding.
2. Statements (i) and (ii) of Theorem 3 are sharp in the following sense: simple examples show that for arbitrary $B \in PK_\beta(A)$ the imbedding $(l+H)^{-1}[L^1 \cap L^\infty] \subset L_{loc}^{j \cdot t'(\beta)}$, $l \in \rho(-H)$ fails. On the other hand, one can prove that if $M = \mathbb{R}^d$, $d\mu = dx$, $A = -\Delta$ and $B = V^- \in PK_\beta(-\Delta) \cap L^{d/2,\infty}(\mathbb{R}^d)$ then

$$(l + H_p)^{-1} \in \mathcal{L}(L^p, L^{j \cdot t'(\beta), \infty}), \quad l \in \rho(-H_p), \quad p \in (t(\beta), t'(\beta)).$$

The last leads to the conjecture that the estimate for the distribution function of $(l + H)^{-1} g$ obtained in Theorem 3 can be improved (on $\ell n^j s$).

References

[CKS] Carlen, E.A., Kusuoka, S. and Stroock, D.W., Upper bounds for symmetric Markov transition functions, Ann. Inst. Henri Poincaré, **23**, (1987), 245–287.

[G] Goldstein, J., Semigroups of Linear Operators and Applications, Oxford University Press, New York, 1985.

[K] Kato, T., Perturbation theory for linear operators, Springer-Verlag, Berlin-Heidelberg, 1966.

[LSe] Liskevich, V.A. and Semenov, Yu. A., Some inequalities for submarkovian generators and their applications to the perturbation theory, Proc. AMS, to appear.

[St] Stein, E.M., Topics in harmonic analysis related to Littlewood–Paley theory, Princeton Univ. Press, Princeton, N.J. 1970.

[S] Stroock, D.W., An introduction to the theory of large deviations, Springer-Verlag, New York, 1984.

[V] Varopoulos, N. Th., Hardy-Littlewood theory for semigroups, J. Funct. Anal., **63**, 1985, 240–260.

V. A. Liskevich, Department of Mathematics, The Weizmann Institute of Science, Rehovot 76100, ISRAEL

Yu. A. Semenov, Department of Mathematics, Kiev Polytechnic Institute, Kiev 252056, UKRAINE

Operator Theory:
Advances and Applications, Vol. 70

Smoothness Estimates and Uniqueness for the Dirichlet Operator

V.A. Liskevich*

Abstract

We study the equation $du/dt + Hu = 0$, where H is the operator associated with Dirichlet form in $\mathbb{R}^d$. Estimates of the first and the second derivates of the solutions are obtained in L^p-spaces with weight. The results on strong uniqueness in L^p are also given.

The object of this note is the Dirichlet operator associated with a Dirichlet form on $\mathbb{R}^d$. The theory of Dirichlet forms is of increasing interest because of its relations to probability theory and quantum field theory (see. e.g. [ABR], [BK] and references therein). We are interested here in smoothness estimates of solutions of the parabolic equation $\frac{du}{dt} + Hu = 0$ in L^p, where H is a Dirichlet operator. In fact, we obtain the estimates for the first and the second order derivatives of solutions in L^p for the approximating equations with smooth coefficients. The main feature of these estimates is that they do not depend on the smoothness of the coefficients and on the dimension of $\mathbb{R}^d$. Moreover, as examples show, they correctly reflect the relationship between differential properties of solutions and L^p-properties of coefficients. We extend here the method and some of the results from [LS1]. The results on strong uniqueness in L^p are also given.

Let H be the operator in $L^2(\mathbb{R}^d, \rho dx)$ associated with the closure of the form

$$h[u,v] = \int_{\mathbb{R}^d} \sum_{i,j=1}^{d} \frac{\partial u}{\partial x_i} \frac{\partial v}{\partial x_j} \rho(x)dx, \quad \mathcal{D}(h) = C_0^\infty.$$

H is the generator of a Markov semigroup.

We use the following notations: $L^p \equiv L^p(\mathbb{R}, \rho dx)$, $\|\cdot\|_p$ is the norm in L^p, $\langle f,g \rangle \equiv \int_{\mathbb{R}^d} f(x)g(x)\rho(x)dx$, $\sum_i \equiv \sum_{i=1}^d$, $\nabla_i \equiv \frac{\partial}{\partial x_i}$, C_b is the set of all uniformly continuous bounded functions, C_b^∞ is the set of all infinitely differentiable bounded functions with all their derivatives. $\beta \equiv \nabla\rho/\rho$ is the logarithmic derivative of the measure ρdx. We assume below that $\rho > 0$ almost everywhere and $\int \rho dx = 1$, all functions are real-valued.

*Recipient of a Dov Biegun Postdoctoral Fellowship

Theorem 1. Let $|\beta| \in L^{2p}(\mathrm{IR}^d, \rho dx)$, $3/2 < p < \infty$, $\beta^n \in C_b^\infty$, $A_n = [-(\nabla + \beta^n)\nabla \upharpoonright C_b^\infty]^\sim_{C_b \to C_b}$, $u_n = e^{-tA_n} f$, $f \in C_b^\infty$, $t \geq 0$. Then

$$\int_0^t \|\|\nabla u_n\|\|_{2p}^{2p} ds \leq C_p t \|f\|_\infty^{2p} (\|\|\beta^n\|\|_{2p}^{2p} + \|\|\beta\|\|_{2p}^{2p}) + \tilde{C}_p \|f\|_\infty^2 \|\|\nabla f\|\|_{2p-2}^{2p-2},$$

$$\int_0^t \|(\sum_{i,j} (\nabla_i \nabla_j u_n)^2)^{1/2}\|_q^q ds \leq C_p t \|f\|_\infty^2 (\|\|\beta^n\|\|_{2p}^{2p} + \|\|\beta\|\|_{2p}^{2p} + 1)$$
$$+\tilde{C}_p (1 + \|f\|_\infty^2) \|\|\nabla f\|\|_{2p-2}^{2p-2}, cr$$

where $q = 2$ if $p \geq 2$ and $q = p$ if $3/2 < p < 2$.

Proof. Let us consider the problem

$$\frac{du_n}{dt} = (\nabla + \beta^n)\nabla u_n, \quad u_n(0) = f$$

(the equality in $L^2 \supset C$). Taking the inner product of both sides with $-(\nabla_j + \beta_j) w_j |w|_s^{\nu-1}$, where $w = \nabla u_n$, $|w| = (\sum_j w_j^2)^{1/2}$, $|w|_s = \max\{|w|, s\}$, $s > 0$, $\nu + 3 = 2p$ after summing we have

$$\begin{aligned} \frac{1}{2}\langle \frac{d|w|^2}{dt}, |w|_s^{\nu-1}\rangle + I_s + (\nu-1)J_s = -\sum_{i,j} \langle \beta_i \nabla_i w_j, w_j |w|_s^{\nu-1}\rangle \\ -\langle \beta^n \cdot w, |w|_s^{\nu-1} \frac{du_n}{dt}\rangle - (\nu-1)\sum_j \langle \beta^n \cdot w, w_j |w|_s^{\nu-2} \nabla_j |w|_s\rangle \\ +\langle \beta^n \cdot w, |w|_s^{\nu-1} (\beta^n - \beta) \cdot w\rangle, \end{aligned} \tag{1}$$

where $I_s = \sum_{i,j} \langle \nabla_i w_j, |w|_s^{\nu-1} \nabla_i w_j\rangle$, $J_s = \sum_j \langle \nabla_j |w|_s, |w|_s^{\nu-1} \nabla_j |w|_s\rangle$.

From the following equality

$$\begin{aligned} \|\|w|_s^{\frac{\nu-1}{2}} \frac{du_n}{dt}\|_2^2 = \langle (\beta^n - \beta) \cdot w, |w|_s^{\nu-1} \frac{du_n}{dt}\rangle - \frac{1}{2}\langle \frac{d|w|^2}{dt}, |w|_s^{\nu-1}\rangle \\ -(\nu-1)\sum_j \langle \frac{du_n}{dt} w_j |w|_s^{\nu-1} \nabla_j |w|_s\rangle \end{aligned}$$

we obtain

$$\frac{1}{4}\|\|w|_s^{\frac{\nu-1}{2}} \frac{du_n}{dt}\|_2^2 + \frac{1}{2}\langle \frac{d|w|^2}{dt}, |w|_s^{\nu-1}\rangle \leq \frac{1}{2}\|(\beta - \beta^n) \cdot w|w|_s^{\frac{\nu-1}{2}}\|_2^2 + (\nu-1)^2 J_s.$$

By virtue of the inequality $\|\|w|_s^{\frac{\nu-1}{2}}\|_{2p'}^{2p'} \leq \mathcal{T}_s \equiv \langle |w|^2, |w|_s^{\nu+1}\rangle$, $\frac{1}{p} + \frac{1}{p'} = 1$, $\forall \delta > 0$ we have

$$\|\|w|_s^{\frac{\nu-1}{2}} \frac{du_n}{dt}\|_2^2 + 2\langle \frac{d|w|^2}{dt}, |w|_s^{\nu-1}\rangle \leq 2\delta \mathcal{T}_s + 2C_p \delta^{1-p} \|\|\beta^n - \beta\|\|_{2p}^{2p} + 4(\nu-1)^2 J_s. \tag{2}$$

Estimating the terms from the right-hand side of (1) and using (2) we arrive at

$$\langle \frac{d|w|^2}{dt}, |w|_s^{\nu-1}\rangle + J_s \le \delta \mathcal{T}_s + C_p\delta^{1-p}(|||\beta|||_{2p}^{2p} + |||\beta^n|||_{2p}^{2p}) \quad \forall\delta > 0. \tag{3}$$

Now using integration by parts and the original equation we have

$$\mathcal{T}_s = -\langle u_n, |w|_s^{\nu+1}\frac{du_n}{dt}\rangle - (\nu+1)\langle u_n, |w|_s^{\nu} w\cdot\nabla|w|_s\rangle - \langle u_n, |w|_s^{\nu+1}(\beta-\beta^n)\cdot w\rangle. \tag{4}$$

Let us estimate the terms in the right-hand side of (4). From (2) it follows

$$|\langle u_n, |w|_s^{\nu-1}\frac{du_n}{dt}\rangle| \le \|f\|_\infty^2(2\hat{\delta}\mathcal{T}_s + 2C_p\delta^{1-p}|||\beta^n-\beta|||_{2p}^{2p} + 4(\nu-1)^2 J_s$$
$$-2\langle\frac{d|w|^2}{dt}, |w|_s^{\nu-1}\rangle) + \frac{1}{4}(\mathcal{T}_s + s^{\nu+3}), \quad \hat{\delta} > 0.$$

Choosing $\hat{\delta} = 1/16\|f\|_\infty^2$ we rewrite the last inequality in the form

$$|\langle u_n, |w|_s^{\nu-1}\frac{du_n}{dt}\rangle| \le \frac{3}{8}\mathcal{T}_s + 4(\nu-1)\|f\|_\infty^2 J_s + C_p\|f\|_\infty^{2p}|||\beta^n-\beta|||_{2p}^{2p}$$
$$+\frac{1}{4}s^{\nu+3} - 2\|f\|_\infty^2\langle\frac{d|w|^2}{dt}, |w|_s^{\nu-1}\rangle, \quad \forall s \ge 0.$$

Using Hölder's inequality in the second term of the right-hand side of (4) we arrive at

$$\mathcal{T}_s \le C_p\|f\|_\infty^2 J_s + \hat{C}_p\|f\|_\infty^{2p}|||\beta^n-\beta|||_{2p}^{2p} + \frac{5}{4}s^{\nu+3} - 8\|f\|_\infty^2\langle\frac{d|w|^2}{dt}, |w|_s^{\nu-1}\rangle. \tag{5}$$

Note that

$$\int_0^t \langle\frac{d|w|^2}{d\tau}, |w|_s^{\nu-1}\rangle d\tau = \langle |w|^2 - \frac{\nu-1}{\nu+1}|w|_s^2, |w|_s^{\nu-1}\rangle - \langle |\nabla f|^2 - \frac{\nu-1}{\nu+1}|\nabla f|_s^2, |\nabla f|_s^{\nu-1}\rangle$$

and $\langle |w|^2 - \frac{\nu-1}{\nu+1}|w|_s^2, |w|_s^{\nu-1}\rangle \ge 0$. From (3) and (5) after integration in t we obtain

$$\int_0^t \mathcal{T}_s d\tau \le C_p t\|f\|_\infty^{2p}(|||\beta|||_{2p}^{2p} + |||\beta^n|||_{2p}^{2p}) + \hat{C}ts^{\nu+3}$$
$$+\hat{C}\|f\|_\infty^2\langle |\nabla f|^2 - \frac{\nu-1}{\nu+1}|\nabla f|_s^2, |\nabla f|_s^{\nu-1}\rangle. \tag{6}$$

The last inequality holds $\forall s > 0$ and admits passing to the limit $s \downarrow 0$. This proves the first statement of the theorem. To prove the second statement note that for $0 < \nu < 1$ from Hölder's inequality it follows that

$$\|(\sum_{i,j}(\nabla_i w_j)^2)^{1/2}\|_p^p \le \frac{p}{2}I_s + \frac{2-p}{2}(\mathcal{T}_s + s^{\nu+1}).$$

Setting $s = \|f\|_\infty$ and using (3) and (6) after integration we obtain the desired inequality. For $\nu \geq 1$ note that from (3) it follows

$$\sum_{i,j} \langle \nabla_i w_j, \nabla_i w_j \rangle \leq I_1 \leq \mathcal{T}_1 + C_p(\||\beta|\|_{2p}^{2p} + \||\beta^n|\|_{2p}^{2p}) - \langle \frac{d|w|^2}{dt}, |w|_1^{\nu-1} \rangle.$$

It remains only to integrate in t and use (6) with $s = 1$. □

Remark. Passing to the limit $\beta^n \longrightarrow \beta$ in L^{2p} one can conclude that if $\beta \in L^{2p}$ then $\nabla u \in L^{2p}$ where u is the solution of the equation $\frac{du}{dt} + Hu = 0$.

Theorem 1 enables us to prove the following uniqueness result.

Theorem 2. Let $\beta \in L^{2p}$. Then the operator $(\nabla(\nabla + \beta) \restriction C_0^\infty(\mathbb{R}^d))^{\sim}_{L^p \to L^p}$ is the generator of a C_0-semigroup of contractions on L^p.

Modifying the arguments from [LS2] one can prove the following perturbation result.

Theorem 3. Let $\beta \in L^{2p}$. Let V be a measurable function satisfying $\langle V\varphi^2 \rangle \leq \gamma\|\nabla\varphi\|_2^2 + C_\gamma\|\varphi\|_2^2$, $\gamma \in (0,1)$, $C_\gamma \geq 0$, $\forall \varphi \in C_0^\infty(\mathbb{R}^d)$ and $V \in L^p$ with $p \in (\frac{2}{1+\sqrt{1-\gamma}}, \frac{2}{1-\sqrt{1-\gamma}})$. Then the operator $(\nabla(\nabla + \beta) - V \restriction C_0^\infty(\mathbb{R}^d))^{\sim}_{L^p \to L^p}$ is the generator of a C_0-semigroup of quasicontractions on L^p.

References

[ABR] Albeverio, S., Brasche, J. and Röckner, M., *Dirichlet forms and generalized Schrödinger operators*, Proc. Sonderborg Conf. "Schrödinger operators", (H.Holden and A.Jensen, eds.), Lect. Notes in Phys., Vol. 345, Springer-Verlag Berlin, 1988, pp. 1–45.

[BK] Berezanskii, Yu.M. and Kondratiev, Yu.G., *Spectral methods in infinite dimensional analysis*, Naukova Dumka, Kiev, 1988.

[LS1] Liskevich, V.A. and Semenov, Yu. A., *Dirichlet operators: a priori estimates and the uniqueness problem*, J. Funct. Anal., Vol. **109** (1992), 199–213.

[LS2] Liskevich, V.A. and Semenov, Yu. A., *Some problems on Markov semigroups*, to appear.

V.A. Liskevich, Department of Mathematics, The Weizmann Institute of Science, Rehovot 76100, ISRAEL

Operator Theory:
Advances and Applications, Vol. 70

Trace Ideal Properties of Perturbed Dirichlet Semigroups

Peter Stollmann

Introduction

We study semigroup differences of the form

$$e^{-t(H+\mu)} - e^{-t(H+\mu+\nu)},$$

where H is the generator of a regular Dirichlet form and μ and ν are suitable measures. In the first section the appropriate classes of measures are introduced. Moreover, we provide a list of examples which can be treated in the Dirichlet form framework.

In the second section the above semigroup is investigated in terms of Hilbert–Schmidt and trace class properties. It turns out that if the set on which ν "lives", the so–called *quasi–support*, has finite capacity then the above semigroup difference is Hilbert–Schmidt. A more restrictive condition on the quasi–support of ν is exhibited which implies that this difference is trace class.

1. Dirichlet forms plus measures

We consider trace ideal properties and their consequences in the framework of *regular Dirichlet forms*, see [10]. In [13] the more general setting of quasi–regular Dirichlet forms is described in detail; with minor changes, our results hold true in this setting. To begin we briefly recollect some basic material concerning the perturbation of Dirichlet forms by measures. Instead of reproducing all the definitions we illustrate the scope of this setting by some examples. In this respect, the monograph [6] is a good reference.

1.1. Examples. (1) On $L_2(\mathbb{R}^d)$ the form

$$\mathfrak{h}_0[u,v] := \int \nabla u \nabla \overline{v} dx, D(\mathfrak{h}_0) = W_0^{1,2}(\mathbb{R}^d)$$

is a regular Dirichlet form.

(2) More generally, forms of the type

$$\mathfrak{h}[u,v] = \sum_{i,j} \int a_{ij}(x)\partial_i u(x)\partial_j v(x)dx$$

with domain $W_0^{1,2}(\Omega)$ are regular Dirichlet forms for suitable a_{ij}; see [6], Section 1.2.

(3) In the last example also Neumann boundary conditions can be included, provided Ω has the extension property, see [1], p. 83f (in order that the form be regular, one needs that $C(\overline{\Omega})$ is dense in $W^{1,2}(\Omega)$).

(4) If $p : [0,\infty) \times E \times E \to \mathbb{R}$ satisfies the basic assumptions of stochastic spetral analysis, "BASSA", then the generator of the associated semigroup corresponds to a regular Dirichlet form on $L_2(E)$. Thus our setting incorporates all the examples listed in [7].

We consider measure perturbations of regular Dirichlet forms and refer the reader to [2, 3, 14, 19, 21] where he can find more details as well as more relevant literature. A recent investigation of spectral theoretic properties of $-\Delta + \mu$ can be found in [4]; see also the article of J. Brasche in the present volume.

Let us mention one point which makes measure perturbations interesting: They provide a convenient means of treating *potentials* and *Dirichlet boundary conditions* simultaneously.

We recall that a regular Dirichlet form $\mathfrak{h}$ with domain D defines a set–function, the *capacity*, by

$$\text{cap }(K) := \inf\{\mathfrak{h}[u,u] + \|u\|^2; u \in D, u \geq \chi_K\}$$

for compact K and regular extension to arbitrary sets. The right class for the positive part of our perturbations is

$$M_0 := \{\mu : \mathfrak{B} \to [0,\infty]; \mu \text{ countably additive}, \mu << \text{ cap}\},$$

the class of measures which do not carge sets of zero capacity. It is wellknown that

$$(\mathfrak{h} + \mu)[u,v] := \mathfrak{h}[u,v] + \int \tilde{u}\overline{\tilde{v}}d\mu$$

defines a closed form, where we write $\tilde{u}$ for the quasi–continuous representative of u.

1.2. Examples. (1) If $\mu = Vdx$ then $\mathfrak{h} + \mu = \mathfrak{h} + V$ is the usual form sum.
(2) (Cf. [3]) The measure
$\infty_A(B) := \infty \cdot \text{cap}(A \cap B)$ defines an element of M_0. It is clear that

$$D(\mathfrak{h} + \infty_A) = \{u \in D; \tilde{u} = 0 \text{ q.e. on } A\} =: D_0(A^c).$$

(3) For the classical Dirichlet form $\mathfrak{h}_0$ (see Example 1.1 (1)) and $A = \overline{A}$ it is not hard to check that

$$D(\mathfrak{h}_0 + \infty_A) = W_0^{1,2}(A^c),$$

so that $\mathfrak{h}_0 + \infty_A$ is the form corresponding to the Dirichlet Laplacian on A^c. For more details on that we refer to [12, 9].

As negative parts of perturbations measures which satisfy a Kato condition turn out to be especially well suited. This is due to the fact that the corresponding semigroups still act on different L_p–spaces. We fix a regular Dirichlet form $\mathfrak{h}$ with domain D on some $L_2(X, m)$ and denote the corresponding operator by H. Let us now briefly recall the *extended Kato class*, introduced in [19]. We denote

$$\hat{S}_K := \{\mu \in M_0; \langle \mu; (H+E)^{-1}\cdot \rangle \in L_1(X,m)' = L_\infty(X,m)\},$$

$$c_E(\mu) := \|\langle \mu, (H+E)^{-1}\cdot \rangle\|_\infty,$$

$$c(\mu) := \inf_{E>0} c_E(\mu),$$

which generalizes the Kato class K_d of potentials. More precisely: $V \in K_d$ implies that $Vdx \in \hat{S}_K$ with respect to the classical Dirichlet form $\mathfrak{h}_0$, and $c(Vdx) = 0$. We refer to [2, 19] for details. In [19] it is in particular shown that each $\mu^- \in \hat{S}_K$ is form bunded with respect to $\mathfrak{h}$ with bound $c(\mu)$ so that $\mu = \mu^+ - \mu^-$, $\mu^+ \in M_0$, $\mu^- \in \hat{S}_K, c(\mu^-) < 1$ gives rise to a closed form $\mathfrak{h} + \mu$. Due to the fact that μ^+ may be infinite this form is not necessarily densely defined, but we can associate a self adjoint operator $H + \mu$ in the Hilbert space $\overline{D(\mathfrak{h} + \mu)}$ which is of the form $L_2(Y)$ for a subset Y of X. We extend the semigroup $e^{-t(H+\mu)}$ by 0 to all of $L_2(X)$ and recall from [19] that it also acts on the spaces $L_p(X)$.

A simple way of defining a suitable notion of support for measures in M_0 is the following: A quasi–closed set $\Sigma(\nu)$ is called *quasi–support* of ν if

$$\mathfrak{h} + \infty\nu = \mathfrak{h} + \infty_{\Sigma(\nu)}.$$

This definition from [17] is based on the characterization of closed lattice ideals in [15] and extends the notion of quasi–support which was given for the smaller class of smooth measures in [11]. It is easy to see that, for Radon measures, $\Sigma(\nu)$ is smaller than the usual support and that these sets need not be equal; the examples in [21], Section 9 can be interpreted in this sense.

2. Trace ideal properties of semigroup differences

Throughout this section we assume that $\mathfrak{h}$ is a regular Dirichlet form with the additional property that $e^{-tH} : L_1 \to L_\infty$ is bounded. We write $\|\cdot\|_{HS}$ for the Hilbert–Schmidt norm.

2.1. Theorem. *Let $\mu^+, \nu \in M_0$ and $\mu^- \in \hat{S}_K$ with $c(\mu^-) < 1/2$. If* $\mathrm{cap}(\Sigma(\nu)) < \infty$ *then*

$$\|e^{-t(H+\mu)} - e^{-t(H+\mu+\nu)}\|_{HS} \leq 2\mathrm{cap}(\Sigma(\nu))^{1/2}\|e^{-t(H-2\mu^-)} : L_2 \to L_\infty\|$$

for all $t > 0$.

The proof goes essentially along the same lines as the proof of Theorem 1 in [16] and will be given in [5]. This Theorem can, in particular, be applied to $\mu = V \in L^1_{loc} - K_d$ and $\nu = \infty_\Gamma$ in the setting of "BASSA". As a consequence concerning large coupling limits we have (see [17], Corollary 2.4)

2.2. Corollary. *Let μ, ν be as in the Theorem. Then*

$$H + \mu + n \cdot \nu \stackrel{nrs}{\longrightarrow} H + \mu + \infty_{\Sigma(\nu)},$$

which means convergence in the norm resolvent sense.

Our next aim is an estimate of the trace norm $\|\cdot\|_{tr}$ of the semigroup difference if ν has sufficiently small support. A recent application of such an inequality can be found in [18], where $H = -\Delta, \mu = V \in L^1_{loc} - K_d$ and $\nu = \infty_S$ is considered. To formulate the appropriate condition on $\Sigma(\nu)$ we have to recall that there exists a Markov process which is associated with $\mathfrak{h}$ in the sense that

$$e^{-tH} f(x) = \mathbb{E}^x[f \circ X_t].$$

By τ_B we denote the first hitting time of B,

$$\tau_B(\omega) = \inf\{t > 0 : X_t(\omega) \in B\}$$

for $B \subset E$. We have:

2.3. Theorem. *Let If $c_{\Sigma(\nu)} := \int \left(\mathbb{P}^x\left[\tau_{\Sigma(\nu)} \leq t\right]\right)^{1/2} dx < \infty$ for some $t > 0$, where μ, ν are as in Theorem 2.1. Then*

$$\|e^{-2t(H+\mu)} - e^{-2t(H+\mu+\nu)}\|_{tr} \leq 2c_B \cdot \|e^{-t(H-2\mu^-)} : L_1 \to L_2\|^2.$$

Again, this may be applied to obstacle scattering, setting $\nu := \infty_B$. The slightly simpler case $\mu = 0$ had been studied in [16]; under the more restrictive regime of "BASSA", the corresponding results for $\mu = Vdx$, $\nu = \infty_\Gamma$ were obtained in [8]. Let us mention that the above Theorem can also be applied to scattering by non–closed obstacles. Of course, the Theorem implies trace class convergence

$$e^{-t(H+\mu+n\nu)} \longrightarrow e^{-t(H+\mu+\infty_{\Sigma(\nu)})}.$$

In the case of the classical Dirichlet form on $\mathbb{R}^d$, the assumption on $\Sigma(\nu)$ occuring in the Theorem is satisfied for bounded sets. Related estimates for $c_{\Sigma(\nu)}$ can be found in Section 2 of [18].

Acknowledgement. It is a pleasure to thank R. Lang for a fruitful discussion and for pointing out [20] to me.

References

[1] R. Adams: *Sobolev Spaces.* Academic Press, New York, 1975

[2] S. Albeverio and Z. Ma: Perturbation of Dirichlet forms–Lower semiboundedness, Closability, and Form Cores. *J. Funct. Anal.* **99** (1991), 332–356

[3] J. Baxter, G. DalMaso and U. Mosco: Stopping times and Γ–convergence. *Trans. Amer. Math. Soc.* **303**, 1–38 (1987)

[4] J. Brasche, P. Exner, Kuperin and P. Šeba: Schrödinger operators with singular interactions. Journ. Math. Anal. Appl., to appear.

[5] J. van Casteren, M. Demuth and P. Stollmann: Trace Class Properties for Integral Operators and Diffusion Semigroups. In preparation.

[6] E.B. Davies: "Heat kernels and spectral theory." Cambridge University Press, Cambridge, 1989

[7] M. Demuth: On topics in spectral and stochastic analysis for Schrödinger operators. Proceedings "Recent Developments in Quantum Mechanics", A. Boutet de Monvel et al. (eds.), Kluwer, 1991

[8] M. Demuth and J. van Casteren: Spectral Theory For Feller Generators. Book in preparation.

[9] D. Feyel and A. de la Pradelle: Espaces de Sobolev sur les ouvert fins. *C. R. Acad. Sci,* **280**, série A, 1125–1127(1975)

[10] M. Fukushima: *Dirichlet forms and Markov processes.* North Holland, Amsterdam, 1980

[11] M. Fukushima and Y. Le Jan: On quasi–supports of smooth measures and closability of pre–Dirichlet forms. *Osaka J. Math.* **28** (1991), 837–845

[12] L.I. HEDBERG: Spectral synthesis and stability in Sobolev spaces. In: Euclidean Harmonic Analysis, Proceedings Maryland, J.J. Benedetto (ed.), Lect. Notes in Math. **779**, 73–103(1980)

[13] Z. MA AND M. RÖCKNER: *An introduction to the theory of (non-symmetric) Dirichlet forms*. Springer, 1992

[14] P. STOLLMANN: Smooth perturbations of regular Dirichlet forms. *Proc. Amer. Math. Soc.* **117** (1992), 747–752

[15] P. STOLLMANN: Closed ideals in Dirichlet spaces. Potential Analysis, to appear

[16] P. STOLLMANN: Scattering by obstacles of finite capacity. J. Funct. Anal., to appear

[17] P. STOLLMANN: A convergence theorem for Dirichlet forms with applications to boundary value problems with varying domains. Preprint 1993

[18] P. STOLLMANN AND G. STOLZ: Singular continuous spectrum for multidimensional Schrödinger operators with potential barriers. Preprint 1993

[19] P. STOLLMANN AND J. VOIGT: Perturbation of Dirichlet forms by measures. Preprint, 1992

[20] P.G. STOLZMAN: Unpublished notes, 1993

[21] T. STURM: Measures charging no polar sets and additive functionals of Brownian motion. *Forum Math.* **4** (1992), 257–297

Peter Stollmann, Universität Frankfurt, Fachbereich Mathematik, D-60054 Frankfurt am Main.
E-mail: stollmann@mathematik.uni-frankfurt.dbp.de

Chapter 4

Many-body problems and statistical physics

Operator Theory:
Advances and Applications, Vol. 70

Quantum Dynamical Semigroups

Kalyan B. Sinha

1 Introduction

Feller [1] and Kato [2] proved the existence of a unique minimal semigroup associated with the classical Fokker-Planck equation:

$$\left.\begin{array}{l} \frac{dp_{ij}}{dt} = \sum\limits_k p_{ik}\Omega_{kj}, \quad t \geq 0 \\ \text{with initial condition } p_{ij}(0) = \delta_{ij} \end{array}\right\}, \tag{1.1}$$

and subject to the Markov condition:

$$\Omega_{kj} \geq 0 \text{ for } k \neq j \text{ and } \sum_{j \neq k} \Omega_{kj} = -\Omega_{kk} < \infty. \tag{1.2}$$

By analogy, one can consider the quantum mechanical Fokker-Planck equation in $V = \mathcal{B}_1(h)$ (the real Banach space of selfadjoint trace-class operators in a complex separable Hilbert space h):

$$\begin{aligned} \frac{d\rho(t)}{dt} &= Y\rho(t) + \rho(t)Y^* + \sum_{k \geq 1} L_k \rho(t) L_k^*, \\ \rho(0) &= \rho, \end{aligned} \tag{1.3}$$

where is the generator of a C_0 contraction semigroup $C(t)$ on $h, \{L_k\}$ are linear operators satisfying

$$\begin{array}{ll} \text{(i)} & D(L_k) \supseteq D(Y) \ \forall \ k \\ \text{(ii)} & (f, Yg) + (Yf, g) + \sum\limits_{k \geq 1} (L_k f, L_k g) = 0, \ \forall \ f, g \in D(Y). \end{array} \tag{1.4}$$

The condition (1.4) is the quantum mechanical equivalent of the Markov condition (1.2). We seek the solution of (1.3) as a strongly continuous semigroup σ_t on V such that $\rho(t) = \sigma_t(\rho)$. One can instead look at the dual picture i.e. look for a $w - *$ continuous semigroup τ_t on $\mathcal{B}(h)$ such that

$$\frac{d\tau_t(X)}{dt} = \tau_t(\mathcal{L}(X)), \tag{1.5}$$

for all X in a suitable subset of $\mathcal{B}(h)$ where the map $\mathcal{L}$ (Lindbladian) is defined as a form:

$$(f, \mathcal{L}(X)g) \equiv (f, XYg) + (Yf, Xg) + \sum_k (L_k f, X L_k g) \tag{1.6}$$

for all $f, g \in D(Y)$. Then formally the Markov condition (ii) reads as $\mathcal{L}(I) = 0$, but this should be rigorously interpreted only as a form.

Remark : If $Y \in \mathcal{B}(h)$ i.e. if $C(t)$ is a norm-continuous semigroup on h, then (1.4) implies that L_k's are all bounded and the sum $\sum_{k\geq 1} L_k^* L_k$ converges strongly. Then (1.6) actually defines a bounded linear map $\mathcal{L}$ on $\mathcal{B}(h)$. In such a case (1.5) obviously has a unique solution τ_t as a norm-continuous semigroup of completely positive maps on $\mathcal{B}(h)$ [3]. Lindblad [4] infact established the converse, viz. every norm continuous semigroup of completely positive maps on $\mathcal{B}(h)$ has its generator $\mathcal{L}$ given by (1.6) as operator on $\mathcal{B}(h)$ subject to (1.4). In general, however, the relation (1.4) implies that each L_k is Y-bounded with relative bound ≤ 1.

2 Construction of the minimal semigroup

We define two maps : $V \to V$ by $S_t(\rho) = C(t)\rho C(T)^*$, $\quad \pi(\rho) = (1-Y)^{-1}\rho(1-Y^*)^{-1}$, and set $\mathcal{D} = \pi(V)$. Then it is clear that S_t is a strongly continuous positive contraction semigroup on V with its generator Z given formally as :

$$Z(\rho) = Y\rho + \rho Y^*. \tag{2.1}$$

The following proposition sums up the results, omitting those parts of the proof given in Davies [5].

Proposition 2.1 : Let $\{L_k\}, Y$ be operators in h satisfying (1.4). Then

(i) $\mathcal{D}$ is a core for Z and (2.1) is valid on $\mathcal{D}$.

(ii) Define $\mathcal{J}$ on $\mathcal{D}$ by $\mathcal{J}(\rho) = \sum_k L_k \rho L_k^* \ \forall \ \rho \in \mathcal{D}$. Then $\mathcal{J}$ has a positive extension $\mathcal{J}'$ on $D(Z)$ such that for $\rho \in D(Z)$,

$$\mathrm{Tr}[Z(\rho) + \mathcal{J}'(\rho)] = 0. \tag{2.2}$$

(iii) For each fixed $\lambda > 0, \mathcal{J}'(\lambda - Z)^{-1}$ is a linear map from $\mathcal{D}$ into V and has a unique bounded positive extension A_λ in V with $\|A_\lambda\| \leq 1$ and $\mathcal{J}'(\rho) = A_1[\rho - Z(\rho)] \ \forall \ \rho \in D(Z)$.

(iv) For any fixed $r (0 \leq r < 1), W^{(r)} \equiv Z + r\mathcal{J}'$ is the generator of a strongly continuous positive contraction semigroup $\sigma_t^{(r)}$ on V whose resolvent is given by

$$R_\lambda^{(r)} \equiv (\lambda - W^{(r)})^{-1} = (\lambda - Z)^{-1} \sum_{k=0}^{\infty} r^k A_\lambda^k, \tag{2.3}$$

where the series converges in trace norm, and

$$\|R_\lambda^{(r)}\| \leq \lambda^{-1} \ \forall \ \lambda > 0 \quad \text{and} \quad 0 \leq r < 1.$$

(v) For each $\lambda > 0, R_\lambda^{(r)}$ and for each $\rho \in V_+,\ t \geq 0,\ \sigma_t^{(r)}(\rho)$ are increasing with r and converges strongly to $R_\lambda \equiv (\lambda - Z)^{-1} \sum_{k\geq 0} A_\lambda^k$ and $\sigma_t(\rho)$ respectively. The family σ_t so defined is a strongly continuous positive contraction semigroup and $R_\lambda = (\lambda - W)^{-1}$, the resolvent of W, the generator of σ_t. Also $W = Z + \mathcal{J}$ on $\mathcal{D}$.

(vi) The semigroup σ_t is minimal in the following sense: if there exists a positive strongly continuous semigroup σ_t' will generator W' extending $W|_\mathcal{D}$, then $\sigma_t' \geq \sigma_t \ \forall \, t \geq 0$.

Proof: For (i)-(iv) and part of (v), see Davies [5]. Set $R_{\lambda,n}^{(r)} \equiv (\lambda - Z)^{-1} \sum_{0\leq k\leq n} r^n A_\lambda^n$, $R_{\lambda,n} \equiv R_{\lambda,n}^{(1)}$ and $R_\lambda \equiv s-\lim\ R_\lambda^{(r)}$ as $r \uparrow 1$ (which exists since $R_\lambda^{(r)}$ is an increasing bounded family of positive maps). Then it is clear that

$$R_{\lambda,n}^{(r)} \leq R_\lambda^{(r)} \leq R_\lambda \text{ and hence } R_{\lambda,n} \leq R_\lambda.$$

But since $R_{\lambda,n}$ is also increasing with n, $\tilde{R}_\lambda \equiv s - \lim_{n\to\infty} R_{\lambda,n}$ exists and $\tilde{R}_\lambda \leq R_\lambda$. On the other hand, $R_{\lambda,n}^{(r)} \leq R_{\lambda,n} \leq \tilde{R}_\lambda$ so that

$$R_\lambda \equiv s - \lim_{r\uparrow 1} s - \lim_{n\to\infty} R_{\lambda,n}^{(r)} \leq \tilde{R}_\lambda.$$

For (vi), we note that W' is a closed extension of $W|_\mathcal{D}$ and since $\mathcal{D}$ is a core for $Z, D(W') \supseteq D(Z)$. The resolvent $R_\lambda' \equiv (\lambda - W')^{-1} = \int_0^\infty e^{-\lambda t} \sigma_t' dt$ exists for sufficiently large λ and is positive . Also by (ii), $R_\lambda' - R_\lambda^{(r)} = R_\lambda'(W' - W^{(r)})R_\lambda^{(r)} = (1-r)R_\lambda' \mathcal{J}' R_\lambda^{(r)}$ and therefore $R_\lambda' - R_\lambda$ is positive. Thus for $0 < t < \infty, \rho \in V_+ :$ $(1 - W't/n)^{-n}(\rho) \geq (1 - Wt/n)^{-n}$ for n large enough, leading to $\sigma_t' \geq \sigma_t$ for $0 < t < \infty$.

Having obtained the minimal semigroup σ_t in V associated with (1.3) and (1.4), the next natural question that arises is : Is σ_t *conservative* i.e. is $\text{Tr}\sigma_t(\rho) = \text{Tr}\ \rho \ \forall \ \rho \in V,\ t \geq 0$? The next theorem gives some necessary and sufficient conditions for this to happen.

Theorem 2.2 [6] : Let σ_t be the minimal semigroup constructed above, $W_0 = W|\mathcal{D}$ and let W_0^* be the adjoint of W_0. Then the following are equivalent:

(i) $\text{Tr}\sigma_t(\rho) = \text{Tr}\ \rho \, \forall \ \rho \in V,\ t \geq 0$;

(ii) for each fixed $\lambda > 0, A_\lambda^n \to 0$ strongly as $n \to \infty$;

(iii) for fixed $\lambda > 0, (\lambda - W_0)\mathcal{D}$ is dense in V;

(iv) for fixed $\lambda > 0$, the characteristic equation $W_0^* X = \lambda X$ has no non-zero solution in $\mathcal{B}(h)$;

(v) for any $\lambda > 0$,

$$\begin{aligned}\beta_\lambda &\equiv \{X \geq 0, X \in \mathcal{B}(h) | (f, \mathcal{L}(X)g) = \lambda(f, Xg), f, g \in D(Y)\}. \\ &= \{0\}.\end{aligned}$$

The proof of this theorem can be found in [6].

There is a dual version of the theorem 2.2. We define for $\lambda > 0$ a map $Q_\lambda : \mathcal{B}(h) \to \mathcal{B}(h)$ as :

$$(f, Q_\lambda(X)g) \equiv \int_0^\infty e^{-\lambda t} \sum_k (L_k C(t) f, X L_k C(t) g) dt. \tag{2.4}$$

Theorem 2.3 : Let $Q_\lambda(X)$ for $X \in \mathcal{B}(h)$ be defined as a form on $D(Y) \times D(Y)$ by (2.4). Then

(i) $Q_\lambda(X)$ is well-defined and extends to an operator in $\mathcal{B}(h)$;

(ii) $Q_\lambda : \mathcal{B}(h) \to \mathcal{B}(h)$ is contractive and completely positive;

(iii) $\beta_\lambda = \{X \in \mathcal{B}(h), 0 \leq X \leq I | Q_\lambda(X) = X\}$

(iv) $Q_\lambda = A_\lambda^*$

(v) $\bar{X}_\lambda \equiv s - \lim_{n\to\infty} Q_\lambda^n(I)$ exists and is the unique maximal element of β_λ;

(vi) σ_t is conservative iff $\overline{X}_\lambda = 0$ for some $\lambda > 0$.

The proof of this can be found in [7,8] and we give here a brief sketch only.

Proof : The condition (1.4) implies that $(f, Q_\lambda(X)g)$ is well defined and
$|(f, Q_\lambda(X)g)|^2 \leq \|X\|^2 (f, Q_\lambda(I)f)(g, Q_\lambda(I)g)$
$\leq \|X\|^2 (\int_0^\infty e^{-\lambda t} \frac{d}{dt} \|C(t)f\|^2)(\int_0^\infty e^{-\lambda t} \frac{d}{dt} \|C(t)g\|^2)$
$= \|X\|^2 \|f\|^2 \|g\|^2$ which proves (i) and (ii), complete positivity being obvious.
For (iv), we note that for $\rho \in \mathcal{D}$, Tr $[\rho Q_\lambda(X)]$
$= \int_0^\infty e^{-\lambda t} dt \sum_k \mathrm{Tr}[\rho C(t)^* L_k^* X L_k C(t)]$
$= \int_0^\infty e^{-\lambda t} dt\ \mathrm{Tr}[\mathcal{J}'(S_t(\rho))X] = \mathrm{Tr}[\mathcal{J}'(\lambda - Z^{-1})(\rho)X] = \mathrm{Tr}[A_\lambda(\rho)X]$.
Clearly Q_λ^n is a positive contraction for all $n = 1, 2, ...$ and hence for $f \in h$

$$\begin{aligned}(f, Q_\lambda^{n+1}(I)f) &= (f, Q_\lambda^n(Q_\lambda(I))f) \\ &\leq \|Q_\lambda(I)\| (f, Q_\lambda^n(I)f) \leq (f, Q_\lambda^n(I)f)\end{aligned}$$

showing that $Q_\lambda^n(I)$ is a decreasing family of positive contractions from which (v) follows. Thus $0 \leq \bar{X}_\lambda \leq I$ and $Q_\lambda(\bar{X}_\lambda) = \bar{X}_\lambda$ so that $\bar{X}_\lambda \in \beta_\lambda$. Now if $X \in \beta_\lambda$, then by the positivity of Q_λ one has $0 \leq Q_\lambda(X) = X \leq Q_\lambda(I)$ which on iteration gives $0 \leq X \leq \bar{X}_\lambda$.

Since $\bar{X}_\lambda$ is the unique maximal element of β_λ the 'if' part of (vi) follows easily. If on the other hand σ_t is conservative, then by Theorem 2.2 (ii) $\mathrm{Tr} A_\lambda^n(\rho) \to$

0 as $n \to \infty \ \forall \ \rho \in V_+$. But by (iv), Tr $[\rho Q_\lambda^n(I)] = \text{Tr } A_\lambda^n(\rho) \to 0$ and therefore $(f, Q_\lambda^n(I)f) \to 0 \ \forall \ f \in h$ which implies the strong convergence of $Q_\lambda^n(I)$ to 0 as $n \to \infty$. ■

Some examples of conservative and non-conservative minimal semigroups have been given in [7,9]. In particular, if L_k's are bounded or if L is normal (but possibly unbounded) for $k = 1$, then the corresponding minimal semigroups are conservative. On the other hand, it is shown in [7] that for a very standard closed symmetric operator L (for $k = 1$) with $Y = -\frac{1}{2}L^*L$, the semigroup is not conservative. Thus a natural question arises, viz. if the minimal semigroup σ_t is *not* conservative, then what are its possible conservative extensions ? There does not seem to exist too much literature dealing with this, but the original article of Feller [10] gives some hints, and recently a fresh attempt has been made in [11,12]. Next, we discuss briefly some of these results.

3 Perturbations of the minimal semigroup

In this section, we study some perturbations of the minimal semigroup σ_t when σ_t is not conservative. If we denote by τ_t the semigroup on $\mathcal{B}(h)$, dual to σ_t, then we note that if σ_t is conservative i.e. if $\tau_t(I) = I$ then there is no other contractive positive semigroup on $\mathcal{B}(h)$ satisfying (1.5). For if τ_t' is one such, then for $0 \leq X \leq I$,

$$\begin{aligned} \tau_t'(X) &= \tau_t'(I) - \tau_t'(1 - X) \leq I - \tau_t(I - X) \\ &= \tau_T(I) - \tau_t(I - X) = \tau_t(X) \leq \tau_t'(X). \end{aligned}$$

As in section 2, let $R_\lambda, \bar{X}_\lambda$ be the resolvent of the minimal semigroup σ_t and the maximal element of β_λ (assumed to be non-zero in most of this section) respectively. Also we choose and fix a real number m and a state ω, i.e. $\omega \in V_+$ s.t. $Tr\ \omega = 1$, and set for $\lambda > 0$

$$\begin{aligned} \tilde{R}_\lambda &\equiv R_\lambda(I + (m + 1 - \alpha_\lambda)^{-1}|\omega\rangle\langle\bar{X}_\lambda|) \\ &\equiv R_\lambda(I + H_\lambda), \end{aligned} \tag{3.1}$$

where we have used the Dirac notation

$$|\omega\rangle\langle X| : V \to V \text{ by } |\omega\rangle\langle X|\rho \equiv \text{Tr}(\rho X)\omega,$$

with $X \in \mathcal{B}(h) \ \forall \ \rho \in V$ and $\alpha_\lambda \equiv \text{Tr}(\omega\bar{X}_\lambda)$. It is also convenient sometime to write the relation dual to (3.1) :

$$\begin{aligned} \tilde{P}_\lambda &\equiv (1 + (m + 1 - \alpha_\lambda)^{-1}|\bar{X}_\lambda\rangle\langle\omega|)P_\lambda \\ &= (I + K_\lambda)P_\lambda, \end{aligned} \tag{3.2}$$

where P_λ and $\tilde{P}_\lambda$ are dual to R_λ and $\tilde{R}_\lambda$ respectively. Then we have

Theorem 3.1 : (i) $\bar{X}_\lambda = I - \lambda P_\lambda(I)$ so that $\tau_t(I) = I$ iff $P_\lambda(I) = \lambda^{-1}$.

(ii) For $\mu > 0, \mu \neq \lambda$

$$P_\lambda(\bar{X}_\mu) = (\mu - \lambda)^{-1}(\bar{X}_\lambda - \bar{X}_\mu)$$

(iii) $\tilde{P}_\lambda$ satisfy resolvent equation :

$$\tilde{P}_\lambda - \tilde{P}_\mu = (\mu - \lambda)\tilde{P}_\lambda \tilde{P}_\mu \tag{3.3}$$

and hence by duality so does $\tilde{R}_\lambda$ i.e.

$$\tilde{R}_\lambda - \tilde{R}_\mu = (\mu - \lambda)\tilde{R}_\lambda \tilde{R}_\mu. \tag{3.4}$$

(iv) The null space of $\tilde{R}_\lambda$ is trivial for each $\lambda > 0$ and $\tilde{R}_\lambda$ is the resolvent of a closed operator $\tilde{W}$, called the perturbation of W.

(v) $\|\tilde{R}_\lambda\| \leq \lambda^{-1}$ if $m \geq 0$ and hence $\tilde{W}$ is the generator of a positive strongly continuous semigroup $\tilde{\sigma}_t$, the perturbed semigroup, which is conservative iff $m = 0$ in (3.1).

(vi) $D(\tilde{W}) = D(W)$ and for $\rho \in D(W)$

$$\begin{aligned} \tilde{W}\rho &= W\rho - (m+1)^{-1}\omega \mathrm{Tr}[\bar{X}_\lambda(W - \lambda)\rho] \\ &= W\rho - (m+1)^{-1}\omega \mathrm{Tr}(W\rho). \end{aligned} \tag{3.5}$$

If $f, g \in D(Y)$, then $\rho \equiv |f\rangle\langle g| \in \mathcal{D} \subseteq D(W)$ and $\mathrm{Tr}(X\tilde{W}\rho) = \mathrm{Tr}(XW\rho) = (g, \mathcal{L}(X)f)\ \forall\, X \in \mathcal{B}(h)$, where $\mathcal{L}(X)$ is defined in (1.6).

Proof (i) By the notation of the proof of Proposition 2.1, $(\mathcal{J}' + Z)R_{\lambda,n}\rho = -\rho + \lambda R_{\lambda,n}\rho + A_\lambda^{n+1}(\rho)$ and taking the trace, using (2.2) we have for $\rho \in V$,

$$\mathrm{Tr}[Q_\lambda^{n+1}(I)\rho] = \mathrm{Tr}\ A_\lambda^{n+1}(\rho) = \mathrm{Tr}([I - \lambda R_{\lambda,n}]\rho).$$

Taking limit $n \to \infty$, using Theorem 2.3 (v) and Proposition 2.1 (v), this leads to the relation $\bar{X}_\lambda = I - \lambda P_\lambda(I)$. The last part follows easily from Theorem 2.3 (vi).

(ii) Let $\lambda \neq \mu$. Then by (i) and the fact that P_λ satisfies resolvent idenity: $P_\lambda - P_\mu = (\mu - \lambda)P_\lambda P_\mu$, we have

$$\begin{aligned} P_\lambda(\bar{X}_\mu) &= P_\lambda(I) - \mu P_\lambda P_\mu(I) = P_\lambda(I) - \frac{\mu}{\mu - \lambda}[P_\lambda(I) - P_\mu(I)] \\ &= \frac{\mu P_\mu(I) - \lambda P_\lambda(I)}{\mu - \lambda} = \frac{\bar{X}_\lambda - \bar{X}_\mu}{\mu - \lambda}. \end{aligned}$$

(iii) Note that by (ii), $P_\lambda K_\mu = \frac{1}{m+1-\alpha_\lambda} \frac{|\bar{X}_\lambda - X_\mu\rangle\langle\omega|}{\mu - \lambda}$ and hence

$$\begin{aligned} \tilde{P}_\lambda \tilde{P}_\mu &= (I + K_\lambda)(P_\lambda P_\mu + P_\lambda K_\lambda P_\mu) \quad \text{or} \\ (\mu - \lambda)\tilde{P}_\lambda \tilde{P}_\mu &= P_\lambda - P_\mu + K_\lambda P_\lambda - K_\lambda P_\mu + \\ &\quad [(m+1-\alpha_\lambda)(m+1-\alpha_\mu)^{-1}K_\lambda - K_\mu]P_\mu \\ &\quad +((m+1-\alpha_\lambda)(m+1-\alpha_\mu)^{-1}K_\lambda^2 - K_\lambda K_\mu)P_\mu. \end{aligned}$$

Noting that by the definition of K_λ and α_λ in (3.2), $K_\lambda K_\mu = \alpha_\mu(m+1-\alpha_\mu)^{-1}K_\lambda$ for all μ and $\lambda > 0$ we have
$(\mu - \lambda)\tilde{P}_\lambda \tilde{P}_\mu = (1+K_\lambda)P_\lambda - (I+K_\mu)P_\mu$

$+K_\lambda P_\mu[-1+(m+1-\alpha_\lambda)(m+1-\alpha_\mu)^{-1}+\alpha_\lambda(m+1-\alpha_\mu)^{-1}-\alpha_\mu(m+1-\alpha_\mu)^{-1}]$

$= \tilde{P}_\lambda - \tilde{P}_\mu.$

(iv) From the definition of α_λ and H_λ, it is clear that $(I+H_\lambda)^{-1} = (I - (m+1)^{-1}|\omega\rangle\langle\bar{X}_\lambda|)$ which is a bounded linear map in V. Since R_λ is one-to-one, it follows that $\tilde{R}_\lambda$ is one-to-one. The rest of the conclusion of (iv) follows from a standard argument (see page 428 of [13]).

(v) Since H_λ is a positive map, it is clear that $\tilde{R}_\lambda$ is also positive. Let $\rho \in V_+$, then

$$\begin{aligned} \mathrm{Tr}\tilde{R}_\lambda(\rho) &= \mathrm{Tr}\, R_\lambda(\rho) + (m+1-\alpha_\lambda)^{-1}(\mathrm{Tr}\, R_\lambda\omega)(\mathrm{Tr}\, \bar{X}_\lambda\rho) \\ &= \mathrm{Tr}\, R_\lambda(\rho) + (m+1-\alpha_\lambda)^{-1}\lambda^{-1}(1-\alpha_\lambda)[\mathrm{Tr}\,\rho - \lambda\,\mathrm{Tr}\, R_\lambda\rho], \end{aligned}$$

where we have used (i) to conclude that

$$\lambda\,\mathrm{Tr}\, R_\lambda\omega = \mathrm{Tr}[\omega(\lambda P_\lambda(I)] = \mathrm{Tr}(\omega(1-\bar{X}_\lambda)) = 1-\alpha_\lambda.$$

Thus

$$\mathrm{Tr}\,\tilde{R}_\lambda(\rho) = \frac{m}{m+1-\alpha_\lambda}\,\mathrm{Tr}\, R_\lambda(\rho) + \frac{1-\alpha_\lambda}{m+1-\alpha_\lambda}\lambda^{-1}\,\mathrm{Tr}\,\rho. \tag{3.6}$$

Since $0 \le \bar{X}_\lambda \le I$ and since R_λ is positive with $\|R_\lambda\| \le \lambda^{-1}$, it follows that $1-\alpha_\lambda \ge 0$ and hence for $m \ge 0, \rho \in V_+,$ $\mathrm{Tr}\,\tilde{R}_\lambda(\rho) \le \lambda^{-1}\,\mathrm{Tr}\,\rho$ which implies the first part of (v). Therefore by Hille-Yoshida theorem [3], the associated semigroup is a positive contraction semigroup with generator $\tilde{W}$, obtained in (iv).

¿From the formulae $\tilde{\tau}_t$ (the dual semigroup of $\tilde{\sigma}_t$) $= s - \lim_{n\to\infty}(\frac{n}{t}\,\tilde{P}_{n/t})^n$ and $\tilde{P}_\lambda = \int_0^\infty e^{-\lambda t}\tilde{\tau}_t\,dt$, it is clear that $\tilde{\tau}_t(I) = I$ iff $\tilde{P}_\lambda(I) = \lambda^{-1}$. On the other hand, it follows from (3.2) and Theorem 3.1 (i) that $\lambda\tilde{P}_\lambda(I) = I + m(m+1-\alpha_\lambda)^{-1}\bar{X}_\lambda$ which leads to the last part of (v).

(vi) It follows from the resolvent equation (3.4), the fact that $(I - H_\lambda)$ is bounded invertible and standard arguments that Range $\tilde{R}_\lambda =$ Range R_λ, which is independent of λ. This implies that $D(\tilde{W}) = D(W) =$ Range $\tilde{R}_\lambda$. Since by (iv) $\tilde{R}_\lambda$ is invertible, for $\rho \in D(W)$

$$\begin{aligned} \tilde{R}_\lambda^{-1}\rho &\equiv (\lambda - \tilde{W})\rho = (I+H_\lambda)^{-1}R_\lambda^{-1}\rho \\ &= (I-(m+1)^{-1}|\omega\rangle\langle\bar{X}_\lambda|)(\lambda - W)\rho \\ &= (\lambda - W)\rho + (m+1)^{-1}\omega\,\mathrm{Tr}\,[\bar{X}_\lambda(W-\lambda)\rho] \end{aligned}$$

which is the first part of (3.5). For the second part, we need only to observe that

$$\begin{aligned} \mathrm{Tr}[\bar{X}_\lambda(W-\lambda)\rho] &= \mathrm{Tr}[(I-\lambda P_\lambda(I))(W-\lambda)\rho] \\ &= \mathrm{Tr}(W-\lambda)\rho - \lambda\,\mathrm{Tr}\, R_\lambda(W-\lambda)\rho = \mathrm{Tr}(W\rho). \end{aligned}$$

Finally, if $\rho \equiv |f\rangle\langle g|$ with $f, g \in D(Y)$ then clearly $\rho \in \mathcal{D}$ as defined in section 1 and since by Proposition 2.1 (v) $W = Z + \mathcal{J}'$ on $\mathcal{D}$, we have

$$\begin{aligned} \mathrm{Tr}(X\tilde{W}\rho) &= \mathrm{Tr}(XW\rho) - (m+1)^{-1}\mathrm{Tr}(\omega X)(g, [\mathcal{L}(\bar{X}_\lambda) - \lambda\bar{X}_\lambda]f) \\ &= \mathrm{Tr}(XW\rho) = (g, \mathcal{L}(X)f). \end{aligned}$$

Here we have used the fact that $\bar{X}_\lambda \in \beta_\lambda$ (Theorem 2.3 (v)). ∎

The above thoerem says that $\tilde{W}$ is a sort of rank-one perturbation of W, though both are associated with the same Lindblad-form $\mathcal{L}(X)$. It is interesting to note that $\tilde{W}$ is essentially the same as the "extension" considered by Davies in [5] and that the maximal element $\bar{X}_\lambda$ in the *Feller set* β_λ (defined in Theorem 2.2 and 2.3) determines the perturbations written down in (3.1) and (3.2). This has some formal similarity to the theory of extensions of a symmetric operator in a Hilbert space. If we formally write $Y = -\frac{1}{2}L^*L + iH$ (with $k = 1$), then

$$\mathcal{L}(X) = \frac{1}{2}\{[L^*, X]L - L^*[L, X]\} - i[H, X]$$

in which we can think of the first term in the R.H.S. as the abstract generalization of $\frac{1}{2}$ Laplacian while the second term is a derivation.

References

1. W. Feller, An introduction to probability theory and its applications, Vol. 2, John Wiley, New York 1966.

2. T. Kato, On the semigroup generated by Kolmogoroff's differential equation, Jour. Math. Soc. Japan **6**, 1(1954).

3. E.B. Davies, One parameter semigroups, Academic Press, London, 1980.

4. G. Lindblad, On the generators of quantum dynamical semigroups, Commun. Math. Phys. **48**, 119 (1976).

5. E.B. Davies, Quantum dynamical semigroups and neutron diffusion equation, Rep. Math. Phys. **11**, 169 (1977).

6. A. Mohari and K.B. Sinha, Stochastic dilation of minimal quantum dynamical semigroup, Proc. Indian Acad. Sciences (Math.) **102** (3), 159 (1993).

7. B.V.R. Bhat and K.B. Sinha, Examples of unbounded generators leading to non-conservative semigroups, Indian Statistical Institute, New Delhi, preprint, February 1993.

8. F. Fagnola, Unitarity of solutions to q.s.d.e. and conservativity of the associated semigroups, Quantum Probability and Related topics VII, 139 (1992), World Scientific.

9. B.V.R. Bhat and K.B. Sinha, A stochastic differential equation with time-dependent unbounded operator coefficients, Jour. Funct. Anal. **114**, 12 (1993).

10. W. Feller, On boundaries and lateral conditions for the Kolmogoroff differential equations, Ann. Math., **65**, 527 (1957).

11. B.V.R. Bhatt, Markov dilations of non-conservative quantum dynamical semigroups and a quantum boundary theory, Ph. D. Thesis, Indian Statistical Institute, New Delhi, 1993.

12. B.V.R. Bhatt and K.R. Parthasarathy, Markov dilations of non-conservative dynamical semigroups and a quantum boundary thoery, Indian Statistical Institute, New Delhi, preprint, July 1993.

13. T. Kato, Perturbation theory for linear operators, Springer-Verlag, New York 1966.

Kalyan B. Sinha, Indian Statistical Institute, Delhi Centre, 7, S.J.S. Sansanwal Marg, New Delhi – 110016, India

Operator Theory:
Advances and Applications, Vol. 70

Limits of Infinite Order, Dimensionality or Number of Components for Random Finite-Difference Operators

A.M.Khorunzhy and L.A.Pastur

Abstract

We consider random operators that are analogs of the statistical mechanics Hamiltonians with a varying interaction radius R , the dimensionality of space d and the number of the field components (orbitals) n . We prove that all the moments of the Green functions for nonreal energies of these operators converge as $R,\ d,\ n \to \infty$ to the products of the average Green functions, just as in the mean field approximation of statistical mechanics. We find in particular the selfconsistent equation for the limiting integrated density of states and the limiting form of the conductivity, which is nonzero on the whole support of the integrated density of states.

1. INTRODUCTION

The spectral and related properties of random operators have attracted considerable interest in both physical and mathematical literature. It is believed, in particular, that under suitable conditions the spectrum of these operators is pure point and dense. This have been proven under various circumstances (in the one-dimensional case or in any dimension near the edges of the spectrum or for a sufficiently large random potential). Therefore, although many important problems still remain open here (two-dimensional localization, calculation of the low frequency conductivity and other physical quantities, etc.), the strong disorder (or low energy) regime in the spectral theory of random operators can be regarded as rather well understood rigorously, especially in comparison with the weak disorder (or high energy) regime. This regime is almost unexplored rigorously despite extensive numerical and theoretical physics studies. In particular, the weak localization theory (see e.g. review [15]) allows us to calculate the so-called quantum corrections for many important physical quantities and, being supplemented by some renormalization group ideas, predicts complete localization in one and two dimensions, a mixed spectrum in higher dimensions and, as a result, the metal-insulator transition. The latter is largely similar to phase transition in statistical mechanics. Thus, based on the statistical mechanics experience it is natural to try to develop some versions of the selfconsistent approaches that are widely accepted tools of study of difficult phase transition problems. In statistical mechanics the most widely accepted selfconsistent schemes, such as numerous versions of the molecular field approximation or the spherical model, can be obtained as the limits of the infinite interaction radius, dimensionality of space or the number of field components (dimensionality of the spin space). An important feature of these limits is that they are nonperturbative in the sense that the interaction, responsible for the phase transition, is not assumed to be small in corresponding models. Small (or at least suppressed) in these limits are the fluctuations of the order parameter and other important physical quantities.

In this paper we study some random operators that can be regarded as analogs of statistical mechanics models with a large interaction radius R, a large dimensionality of space d or a large number of components n . We calculate the integrated density of states (IDS) of these operators and the conductivity in the limits of infinite R, d and n . In fact, the latter model was introduced and studied at the physical level of rigour by Wegner [25] (for some rigorous results on the IDS of this model see also [6]).

We use the method which is analogous to the method of correlation equations (or cluster expansion) of statistical mechanics and allows us to calculate the IDS and the conductivity (more exactly, the measure which is naturally associated with the conductivity) in all the three limits.

In principle, our method can also be used to construct the respective R^{-1}-, d^{-1}- and n^{-1} -expansions for which our limit expressions are the leading terms. We hope to discuss these expansions in subsequent publications (see however [27] where the physical n^{-1} -expansion scheme was developed for the conductivity and [6] for rigorous n^{-1} expansion for the IDS).

The matherial is organized as follows. In Section 2 we introduce the models and formulate the main results, according to which the IDS and the conductivity are practically the same for all the three models and can be calculated from some selfconsistent equation. The former fact should be contrasted with statistical mechanics, where the limits $R = \infty$ and $d = \infty$ coincide with the mean field approximation, while the limit $n = \infty$ coincides with the spherical model. In Section 3 we derive infinite systems of equations for the moments of the Green functions of the respective operators, that are our main technical tools. In Section 4 we solve these equations in the limits $R,\ d,\ n = \infty$, derive a selfconsistent equation for the limiting IDS (see equation (2.9) below) and in Section 5 we study some properties of the IDS (existence of the bounded density, location of the support, the form of singularities at the edges of the support). In Section 6 we calculate the conductivity of the respective disordered system in the same limits. Section 7 is devoted to discussion of our results, in particular their relation to the random matrix theory and their possible interpretations.

2. MODELS AND RESULTS

We start from the random operator of varying order. Let H_R be the selfadjoint operator, acting in $\ell^2(\mathbf{Z}^d)$ and defined by the matrix

$$H_R(x,y) = h(x-y) + R^{-d/2}\,\phi((x-y)/R)\,W(x,y). \qquad (2.1a)$$

Here $x, y \in \mathbf{Z}^d$, $h(-x) = h^*(x)$, $\sum \mid h(x) \mid < \infty$, $R > 0$, $\phi(t), t \in \mathbf{R}^d$, is a piece-wise continuous, nonnegative and such that $\mid \phi(t) \mid \leq \phi_0 < \infty$, $\phi(t) = 0$, $\mid t \mid > 1$, $\int \phi^2(t)dt = 1$ and $W(x,y) = W(y,x)$ are identically distributed and independent (modulo the above symmetry condition) random variables with zero mean value and such that for all $x, y \in \mathbf{Z}^d$

$$\mathbf{E}\{W(x_1,y_1)W(x_2,y_2)\} = w^2\{\delta(x_1-x_2)\delta(y_1-y_2)+\delta(x_1-y_2)\delta(y_1-x_2)\}, \qquad (2.1b)$$

where $\delta(x)$ is the Kronecker symbol in $\mathbf{Z}^d$.

Our second random operator H_d contains explicitly the dimensionality d of the space $\mathbf{Z}^d$. It acts also in $\ell^2(\mathbf{Z}^d)$ and is defined by the matrix

$$H_d(x,y) = h_d(x-y) + (2d)^{-1/2}\,W_1(x,y), \qquad (2.2a)$$

where

$$h_d(x) = d^{-1/2}\sum_{j=1}^{d} h_1(x_j)\prod_{k\neq j}\delta(x_k), \quad h_1(0) = 0, \qquad (2.2b)$$

$h_1(x)$, $x \in \mathbf{Z}^1$, satisfies (2.1b), $W_1(x,y) = W(x,y)\;\delta(|x-y|-1)$ and $W(x,y)$ are as in (2.1b). The simplest and quite important example of the operator h_d is the discrete Laplacian for which $h_1(x) = 0$, $\mid x \mid \neq 1$.

The third operator H_n acts in $\ell^2(\mathbf{Z}^d) \otimes \mathbf{C}^n$ and is defined by the matrix

$$H(\alpha,x;\beta,y) = h(x-y)\;\delta_{\alpha\beta} + n^{-1/2}\;\delta(x-y)\;W_{\alpha\beta}(x) \qquad (2.3a)$$

where $x, y \in \mathbf{Z}^d$, $\alpha, \beta = 1,...,n$, $h(x)$ is the same as in (2.1), $\delta_{\alpha\beta}$ is the Kronecker symbol, $W_{\alpha\beta}(x) = W_{\beta\alpha}(x)$ and $W_{\alpha\beta}(x)$ are identically distributed and independent for $1 \leq \alpha \leq \beta \leq n$ random variables with zero mean value and such that (cf. (2.1b))

$$\mathbf{E}\{W_{\alpha_1\beta_1}(x_1)\,W_{\alpha_2\beta_2}(x_2)\} = w^2\;\delta(x_1-x_2)\;\{\delta_{\alpha_1\alpha_2}\,\delta_{\beta_1\beta_2}+\delta_{\alpha_1\beta_2}\,\delta_{\alpha_2\beta_1}\}. \qquad (2.3b)$$

The random operator (2.3) is a special case of the operator introduced by Wegner [25] (the case of the site-diagonal disorder, according to Wegner's terminology). It can be regarded as the n-component analog of the discrete Schrodinger operator (the Anderson model) or as the Hamiltonian of a disordered system in the dimensions $d + d_1$, $d_1 = n$, in which the random potential in n "transverse" dimensions is considered in the "mean field" approximation.

All the three families of random matrices (2.1)-(2.3) define the essentially selfadjoint metrically transitive operators in the sense of book [21]. Our intention is to study, first of all, the simplest though rather important in several respects spectral characteristic of random operators known as the integrated density of states (IDS). It is defined as the measure

$$N_a(\mathrm{d}\lambda) = \mathbf{E}\{\mathrm{E}_a(0,0;d\lambda)\}, \tag{2.4}$$

where a is R or d and $\mathrm{E}_a(\mathrm{d}\lambda)$ is the resolution of identity of the operators H_a, $a = R,\ d$, and $\mathrm{E}_a(x,y;\mathrm{d}\lambda)$, $x,y \in \mathbf{Z}^d$ are the respective matrices. For the model (2.3),

$$N_n(\mathrm{d}\lambda) = \mathbf{E}\{n^{-1}\sum_{\alpha=1}^{n} \mathrm{E}_n(\alpha,0;\alpha,0;\mathrm{d}\lambda)\}, \tag{2.5}$$

where $\mathrm{E}_n(\alpha,x;\beta,y;\mathrm{d}\lambda)$ is the matrix of the resolution of identity of the operator H_n. For another definition of the IDS, which is based on a kind of the thermodynamic limiting transition, and for equivalence of this definitions, see [21].

For an arbitrary nonnegative measure $\mu(\mathrm{d}t)$ on $\mathbf{R}$, such that $\int \mu(\mathrm{d}t) = 1$, we define its Stieltjes transform $f(z) = \int (t-z)^{-1}\,\mu(\mathrm{d}t)$, $\mathrm{Im}\, z \neq 0$. It is an analytic function for $\mathrm{Im}\, z \neq 0$ such that

$$\mathrm{Im}\, f(z)\cdot \mathrm{Im}\, z > 0; \quad f(z) = -\, z^{-1} + o(z^{-1}), \quad z \to \infty, \tag{2.6}$$

The Stieltjes transform uniquely determines the respective measure, since due to the Stieltjes-Perron inversion formula for any interval $\Delta = (\lambda_1, \lambda_2)$ whose end-points are continuity points of $\mu(\mathrm{d}t)$ [1]

$$\mu(\Delta) = \lim_{\eta\downarrow 0}\, (2\pi i)^{-1} \int_{\lambda_1}^{\lambda_2} [f(\lambda + i\eta) - f(\lambda - i\eta)]\ \mathrm{d}\lambda. \tag{2.7}$$

Denote by $N_0(\mathrm{d}\lambda)$ the IDS of the unperturbed operators in (2.1)-(2.3). It is easy to show that for this Toeplitz operators,

$$N_0(d\lambda) = \ \mathrm{meas}\{\tilde{h}(k) \in d\lambda,\ k \in \mathbf{T}^d\}, \tag{2.8}$$

where $\mathbf{T}^d = [0,1]^d$ is d-dimensional torus and $\tilde{h}(k) = \sum h(x)\, \exp\{2\pi i k x\}$ is the symbol of these operators.

Our result for the IDS states (see Theorem 5.1 below) that the measures $N_a(\mathrm{d}\lambda), a = R,\ d,\ n$ converge weakly as $a \to \infty$ to the limit $N(\mathrm{d}\lambda)$ whose Stieltjes transform $r(z) = \int (\lambda - z)^{-1}\, N(\mathrm{d}\lambda)$ can be found as a unique solution in the class (2.6) of the equation

$$r(z) = r_0(z + w^2 r(z)) \tag{2.9}$$

in which $r_0(z)$ is the Stieltjes transform of the IDS of the unperturbed (nonrandom) operators in (2.1)-(2.3).

Note that for the operator H_d of (2.2) the unperturbed operator and its IDS depend also on the parameter d . Therefore, unlike H_R and H_n, in the case of

H_d the limiting transition $d \to \infty$ affects $N_0(\mathrm{d}\lambda)$ also. More precisely, in this case $N_0(\mathrm{d}\lambda)$ is

$$N_0(\mathrm{d}\lambda) = (2\pi h_2)^{-1/2} \exp\{-\lambda^2/2h_2\} \ \mathrm{d}\lambda, \quad h_2 = \sum h_1^2(x). \tag{2.10}$$

Equation (2.9) was found by Wegner for the ensemble (2.3) in the limit $n = \infty$ (infinite number of the orbitals in the terminology of Wegner). In fact, this equation had appeared before the Wegner's paper as the equation for the limit eigenvalue distribution of some $n \times n$ random matrices in the limit $n = \infty$ [20]. This distribution is known as the deformed semicircle law. The situation here is similar to that in statistical mechanics, where selfconsistent equations of the Curie-Weiss model (the mean field approximation) and the spherical model had been proposed long before than it was understood that these are the limits of an infinite interaction radius [16] and an infinite number of components [11, 22] of the classical Heisenberg (n-vector) model.

Now we shall outline the strategy of derivation of (2.9). Consider the family $\mathbf{E}\{\prod_{i=1}^k G(x_i, y_i)\}$, $k \geq 1$, of the moments of the Green function of the random operators (2.1)-(2.3). By using the resolvent identity, we derive for this family an infinite system of linear relations (that can be regarded as analogs of the BBGY or the Kirkwood-Salzburg equations in statistical mechanics). Some terms in these relations contain the small parameter a^{-1} in front of them. Treating these relations as an infinite system of equations for the moments and neglecting the small terms, we observe that the truncated system admits the factorized solution $\prod_{i=1}^k \Gamma(x_i - y_i)$, $k \geq 1$, where $\Gamma(x-y)$ is the Green function of the Toeplitz operator which is the sum of the nonrandom part of H_a and the effective coordinate-independent potential $w^2 \, \delta(x-y) \, \Gamma(0)$. This yields equation (2.9) which is in this scheme the solvability condition for the truncated system.

To justify this procedure, we act again as in statistical mechanics. Namely, we consider our infinite system as a linear equation in some Banach space containing, in particular, our family of moments. We prove that if the imaginary part of the energy is large enough, then the nonsmall part of the equation defines the contracting linear operator and that the norm of the remainder is small.

Thus, our central technical result (see Theorem 4.1) says that in the limit $a = \infty$ the moments of the Green function of our random operators (2.1)-(2.3) are factorized into the products of the first moments, and these first moments are to be found selfconsistently, by solving the nonlinear functional equation (2.9). This result is fairly similar to the main technical result of the mean field approximation (R, $d = \infty$ limit) and the spherical limit ($n = \infty$ limit) in statistical mechanics, where the correlation functions of all orders are factorized into products of the correlation function of order one (mean field approximation) or of orders one and two (spherical model). We refer to the papers [2, 10, 14] for some form of the latter results and for references.

Consider now the conductivity of a disordered system, described by the Hamiltonians (2.1)-(2.3). According to the Kubo formula, the conductivity of d-

dimensional ideal Fermi gas at a temperature T, described by a one-body Hamiltonian H and subjected to an external alternating electric field of the frequency ν, is [17]

$$\sigma_{ac}(\nu, T) = 2e^2\ \pi^{-1} \int_{-\infty}^{\infty} \nu^{-1}\ [\ n_F(E+\nu) - n_F(E)\ ]\ \sigma(\nu, E)\ \mathrm{d}E \quad . \tag{2.11}$$

Here e is the electron charge, $n_F(E) = (\exp\{(E-E_F)/T\}+1)^{-1}$ is the Fermi distribution function, E_F is the Fermi energy, $\sigma(\nu, E) = m(E+\nu, E)$, and $m(\lambda_1, \lambda_2)$ is the density of the measure

$$M(\mathrm{d}\lambda_1, \mathrm{d}\lambda_2) = \sum_{k=1}^{d} \mathbf{E}\{[\hat{v}_k \mathrm{E}(\mathrm{d}\lambda_1)\hat{v}_k \mathrm{E}(\mathrm{d}\lambda_2)](0,0)\}$$

on $\mathbf{R}^2$ in which $\hat{v} = i[H, \hat{x}]$ is the velocity operator, $\hat{x} = (x_1, ..., x_d)$ is the coordinate operator and $\mathrm{E}(\mathrm{d}\lambda)$ is the resolution of identity of the Hamiltonian H . For $T = 0$, (2.11) has the form

$$\sigma_{ac}(\nu, 0) = \nu^{-1} \int_{E_F}^{E_F+\nu} \sigma(\nu, E)\ \mathrm{d}E \tag{2.13}$$

and for low frequencies $\nu \ll E_F$,

$$\sigma_{ac}(\nu, 0) = 2e^2\ \pi^{-1}\ m(E_F+\nu, E_F)\ (1+o(1)), \quad \nu \to 0 \tag{2.14}$$

i.e. the low frequency conductivity (*dc* conductivity in particular) can be expressed through the density $m(\lambda_1, \lambda_2)$ itself.

Consider first the case of $a = R$. According to (2.10) and (2.1),

$$M(\mathbf{R}^2) = \sum_x x^2\ (\ |\ h(x)\ |^2 + w^2\ R^{-d}\ \phi^2(x/R)\ ).$$

Thus, to obtain a finite and nontrivial answer in the limit $R = \infty$, we have to consider the normalized measure $M^{(R)}(\mathrm{d}\lambda_1, \mathrm{d}\lambda_2) = R^{-2}\ M(\mathrm{d}\lambda_1, \mathrm{d}\lambda_2)$ and with this normalization we can set without loss of generality the nonrandom part of (2.1) to be zero $(h(x) \equiv 0)$. Similar arguments show that for $a = n$, d the properly normalized measures are $M^{(d)}(\mathrm{d}\lambda_1, \mathrm{d}\lambda_2) = M(\mathrm{d}\lambda_1, \mathrm{d}\lambda_2)$ and

$$M^{(n)}(\mathrm{d}\lambda_1, \mathrm{d}\lambda_2) = n^{-1} \sum_{\alpha=1}^{n} \sum_{j=1}^{d} \mathbf{E}\{\ [\hat{v}_j\ \mathrm{E}^{(n)}(\mathrm{d}\lambda_1)\ \hat{v}_j\ \mathrm{E}^{(n)}(\mathrm{d}\lambda_2)](\alpha, 0; \alpha, 0)\ \},$$

respectively.

3. EQUATIONS FOR THE GREEN FUNCTION MOMENTS

Since derivation of the infinite system of linear relations for the moments of the Green functions is rather cumbersome, we divide the problem into two parts. First, we consider these relations for the operators (2.1)-(2.3) with the Gaussian-distributed random variables W's. After that we consider arbitrarily distributed W's with finite third moment. However, for $a = d$ we restrict ourselves to the case of $h(x) \equiv 0$ in (2.3a).

Definition. Let $f(X_k; Y_k)$ be a complex-valued function of the arguments $X_k = (x_1, \ldots, x_k)$, $Y_k = (y_1, \ldots, y_k)$, $x_i, y_i \in \mathbf{Z}^d$, $i = 1, \ldots, k$, $k \geq 1$. Then,

$$\|f\|_k^{(a)} = \begin{cases} \sup_{X_k, Y_k} |f(X_k, Y_k)|, & a = R, \ n; \\ [\sup_{y_1} \sum_{x_1} \cdots \sup_{y_k} \sum_{x_k} |f(X_k, Y_x)|^2 \ ]^{1/2}, & a = d. \end{cases} \tag{3.1}$$

Proposition 3.1. *Let H_a, $a = R, \ d, \ n$ be the random operators, defined by (2.1)-(2.3) with the Gaussian-distributed random variables W's ,*

$$G(x, y; z) = \begin{cases} (H_a - z)^{-1}(x, y) \quad , & \text{if } a = R, \ d; \\ n^{-1} \sum_{\alpha=1}^{n} (H_n - z)^{-1}(\alpha, x; \alpha, y), & \text{if } a = n, \end{cases} \tag{3.2}$$

$\eta = |\operatorname{Im} z| \neq 0$, and $g(x - y) = (h - z)^{-1}(x, y)$ *be the Green function of the Toeplitz operator h , defined by the nonrandom part in (2.1)-(2.3). Introduce the moments*

$$F_k(X_k; Y_k) = \mathbf{E}\{ \prod_{i=1}^{k} G(x_i, y_i) \ \}. \tag{3.3}$$

Then,

$$F_k(X_k; Y_k) = g(x_1 - y_1)\, \delta(k-1) + (1 - \delta(k-1)) F_{k-1}(X_{k-1}; Y_{k-1}) \ g(x_k - y_k) +$$

$$w^2 \sum_{s,t} \chi_a(s - t) \ F_{k+1}(X_k, s; Y_{k-1}, t, s) \ g(t - y_k) + S^{(a)}(X_k; Y_k), \tag{3.4a}$$

$$\chi_a(x) = \begin{cases} R^{-d} \ \phi^2(x/R), & \text{if } a = R; \\ \delta(x), & \text{if } a = n; \\ (2d)^{-1} \ \delta(x - 1), & \text{if } a = d, \end{cases} \tag{3.4b}$$

$\delta(x)$ is the Kronecker symbol, and for some $\eta_0 > 0$ which is independent of a and $\eta = |\operatorname{Im} z| \geq \eta_0$,

$$\|S^a\|_k^{(a)} \leq C \ k \ \eta^{-k} \ \begin{cases} a^{-1}, & \text{if } a = R, n; \\ a^{-1/2}, & \text{if } a = d. \end{cases} \tag{3.5}$$

Remark. We use here and below the common symbol C for quantities, independent of a, k, η, but dependent, generally speaking, on the moments of $W(x, y)$, the function $\phi(x)$ and η_0.

Proposition 3.2. *Let H_a, $a = R,\ n,\ d$ be defined in (2.1)-(2.3), where the arbitrarily distributed i.i.d. W's satisfy (2.1b) and (2.3b) and, in addition,*

$$\mathbf{E}\{|\ W(x,y)\ |^3\}\ < \infty, \quad \mathbf{E}\{|\ W_{\alpha\beta}(x)\ |^3\}\ < \infty.$$

Assume also that $h(x) \equiv 0$ in (2.3a) . Then, the moments (3.3) satisfy (3.4) , where

$$\sup_{X_k, Y_k}\ |\ S_k(X_k; Y_k)\ | \le C\ k\ a^{-1/2}\ \eta^{-k} \tag{3.6}$$

with $\eta \geq \eta_0$, some η_0 and C being independent of k, η and a.

4. ASYMPTOTIC SOLUTION OF THE MOMENT EQUATIONS

In the previous section we have seen that the moments (3.3) of the Green functions of the operators (2.1)-(2.3) satisfy the infinite system of relations (3.4a). In this section we treat this system as a linear equation in some Banach space and show that for $a \to \infty$ this equation admits a rather simple solution. As a result, we obtain (2.9).

Theorem 4.1. *Let $G_a(x, y; z)$ be defined in (3.2), the respective operators H_a satisfy the conditions of Proposition 3.1 in the case of Gaussian W's and the conditions of Proposition 3.2 in the general case and*

$$\eta_1 = \max\{4w, 2\bar{h}\}, \quad \xi = \begin{cases} 3w/2, & \text{if } a = R,\ n\ ; \\ w, & \text{if } a = d\ , \end{cases}$$

where $\bar{h} = \sum |h(x)|$. If a is large enough, then

$$\sup_{|\mathrm{Im}\, z| \geq \eta_1}\ \| \ \mathbf{E}\{\prod_{i=1}^{k} G_a(x_i, y_i; z)\} - \prod_{i=1}^{k} \Gamma_a(x_i - y_i; z)\ \|_k^{(a)}\ \le C\xi^{-k} a^{-1/2}. \tag{4.1}$$

Here $k \geq 1$, C is independent of k and a,

$$\Gamma_a(x; z) = \int_{\mathbf{T}^d} (\tilde{h}(k) - z - w^2 \chi_a r_a(z)\ )^{-1}\ \exp\{2\pi i k x\}\ \mathrm{d}k,$$

where $\tilde{h}(k)$ is given by (2.8),

$$\chi_a \equiv \sum_x \chi_a(x) = \begin{cases} \int x^2 \phi^2(x) \mathrm{d}x, & \text{if } a = R\ ; \\ 1, & \text{if } a = n,\ d, \end{cases}$$

and $r_a(z)$ is a unique solution of the equation

$$r_a(z) = \int (\mu - z - w^2\ \chi_a\ r_a(z))^{-1}\ N_0(\mathrm{d}\mu), \tag{4.2}$$

in the class of functions analytic in z for nonreal z and such that

$$\mathrm{Im}\, r(z) \cdot \mathrm{Im}\, z\ \geq 0,\ \mathrm{Im}\, z \neq 0, \quad \sup_{\eta > 0}\ \eta\, |r(i\eta)| = 1 \tag{4.3}$$

and $N_0(\mathrm{d}\mu)$ is specified by (2.8) for $a = R$, n and (2.10) for $a = d$.

Proof. Let us consider the Banach space $\mathbf{B}_a$, $a = R, n, d$ whose elements are sequences $f = \{f_k(X_k; Y_k; z)\}_{k\geq 1}$, $X_k = (x_1, ..., x_k)$, $Y_k = (y_1, ..., y_k)$, $x_i, y_i \in \mathbf{Z}^d$, and each component f_k for fixed k, X_k, Y_k is an analytic function of z for nonreal z . The norm in $\mathbf{B}_a$ is defined as follows:

$$\|f\|^{(a)} = \sup_{|\mathrm{Im}\, z|\geq\eta_1} \sup_{k\geq 1} \xi^k \|f_k\|_k^{(a)}. \tag{4.4}$$

According to their definition, the moments (3.3) satisfy the inequalities $\|F_k\|_k \leq \eta^{-k}, \quad k \geq 1$. Since $\xi < \eta_1$, the sequence $F \equiv \{F_k\}_{k\geq 1}$ of the moments (3.3) belongs to $\mathbf{B}_a$. Now, according to (3.1), the sequence $S \equiv \{S_k\}_{k\geq 1}$ for $\eta \geq \eta_0$ satisfies the inequality

$$\|S\|^{(a)} \leq C\, a^{-1/2} \sup_{k\geq 1} \xi^{-k}\, \eta_0^{-k}. \tag{4.5}$$

Since our bounds in Proposition 3.1 are monotone in η_0, we can assume without loss of generality that $\eta_0 = \eta_1$, where η_1 is given by (4.1). Then, $\xi \cdot \eta_0^{-1} < 1$, and (4.5) implies that

$$\|S\|^{(a)} \leq C_1\, a^{-1/2} \tag{4.6}$$

where C_1 is independent of a .

Consider now the linear operator $\mathbf{A}$, defined by the second and the third terms of the r.h.s. of (3.4a):

$$(\mathbf{A}f)_k(X_k, Y_k) = \delta(k-1) \sum_{s,t} \chi_a(s-t)\, f_2(x_1, s; t, s)\, g(t-y_k) +$$

$$(\, 1 - \delta(k-1)\,)\{f_{k-1}(X_{k-1}; Y_{k-1})\, g(x_k - y_k) +$$

$$w^2 \sum_{s,t} \chi_a(s-t)\, f_{k+1}(X_k, s; Y_{k-1}, t, s)\, g(t-y_k).$$

It is easy to prove that $\|\mathbf{A}\| \leq 3/4$ if a is large enough. Therefore we can regard relation (3.4) as a linear relation in $\mathbf{B}_a$:

$$F = \mathbf{A}F + U + S \tag{4.7}$$

where $U = \{g(x_1, y_1), 0, ...\}$ belongs obviously to $\mathbf{B}_a$. Since, according to (4.5), S is small for $a \to \infty$, it is natural to consider the following linear equation in $\mathbf{B}_a$:

$$J = \mathbf{A}J + U. \tag{4.8}$$

This equation has a unique solution in $\mathbf{B}_a$. The ansatz

$$J_k(X_k; Y_k) = \prod_{i=1}^{k} \Gamma(x_i - y_i) \tag{4.9}$$

reduces the infinite sustem (4.8) to the single equation

$$\Gamma(x-y) = g(x-y) + w^2 \sum_{s,t} \chi_a(s-t)\,\Gamma(0)\,\Gamma(x-t)\,g(t-y).$$

The formal solution of this equation is $\Gamma(x-y) = g(x-y; z + w^2\chi_a\Gamma(0;z))$, where χ_a is specified by (3.4b) and the dependence on the complex energy z is indicated explicitly. Respective compatibility condition

$$\Gamma(0;z) = g(0; z + w^2\chi_a\Gamma(0;z))$$

can be rewritten in the form (4.5).

We prove that equation (4.2) has a unique solution in the class of functions that are analytic for $\operatorname{Im} z \neq 0$ and satisfy (4.3). Therefore formula (4.9) give the unique solution of the infinite system (4.8) in our spaces $\mathbf{B}_a$. Subtracting (4.8) from (4.7) and iterating the resulting relation, we find that $F - J = (I - \mathbf{A})^{-1} S$. Thus, in view of (4.6),

$$\|F - J\|^{(a)} \leq 4C_1\, a^{-1/2}. \tag{4.10}$$

According to (4.4) and (4.9), the k-th component of (4.10) is (4.1). The theorem is proved.

Corollary 4.1 . *For every fixed* $k \geq 1$, $X_k = (x_1, ..., x_k)$, $Y_k = (y_1, ..., y_k)$ *and uniformly in* z *belonging to a fixed compact set such that* $\operatorname{Im} z \neq 0$, *we have*

$$\lim_{a\to\infty} \left(F_k(X_k; Y_k) - \prod_{i=1}^{k} \Gamma(x_i - y_i)\right) = 0.$$

5. THE DEFORMED SEMICIRCLE LAW

Theorem 5.1 . *Let* H_a, $a = R, n, d$ *be the nonrandom operators defined in (2.1)-(2.3),* $N_0(\mathrm{d}\lambda)$ *be defined by (2.8) for* $a = R$, n *and by (2.10) for* $a = d$, $N_a(\mathrm{d}\lambda)$ *be the IDS of* H_a *given by (2.4) and (2.5) and* $N_a(\lambda) \equiv N_a((-\infty, \lambda])$. *Then, for each* $\lambda \in \mathbf{R}$

$$\lim_{a\to\infty} N_a(\lambda) = N(\lambda) \tag{5.1}$$

where the Stieltjes transform of $N(\mathrm{d}\lambda)$ *can be found as a unique solution of (2.9) in the class (2.6).*

Proof. According to the spectral theorem, (2.4), (2.5) and (3.2), the Stieltjes transform of $N_a(\mathrm{d}\lambda)$ is $\mathbf{E}\{G_a(0,0;z)\}$ and, according to Corollary 4.1, for $k = 1$

$$\lim_{a\to\infty} [\mathbf{E}\{G_a(0,0;z)\} - r_a(z)] = 0$$

where $\operatorname{Im} z \neq 0$ and $r_a(z)$ is a unique solution of (4.2). By Lemma 4.2, we can perform the limiting transition $a \to \infty$. Thus, we have proved that the Stieltjes transform of $N_a(\mathrm{d}\lambda)$ converges for $\operatorname{Im} z \neq 0$ to the solution of (2.9). Since this convergence implies weak convergence of the respective measures, we have thus proved weak convergence of $N_a(\mathrm{d}\lambda)$ to $N(\mathrm{d}\lambda)$. According to the property (5.4) below, $N(\mathrm{d}\lambda)$ posesses a bounded density for any $N_0(\mathrm{d}\lambda)$. This proves pointwise convergence in (5.1). The theorem is proved.

The simplest case of equation (2.9) corresponds to $N(\mathrm{d}\lambda) = \delta(\lambda)$, when the unperturbed (nonrandom) part of (2.1)-(2.3) is zero. Then, $r = -(z + w^2 r)^{-1}$ and

$$r = (2w^2)^{-1} \left((z^2 - 4w^2)^{1/2} - z\right), \tag{5.2}$$

where we use the branch of the radical that has a positive imaginary part on the upper edge of the cut $\mathrm{Im}\, z = 0$, $|\mathrm{Re}\, z| \leq 2w$. (5.2) and (2.7) yield $N(\mathrm{d}\lambda) = \rho(\lambda)\, \mathrm{d}\lambda$, where

$$\rho(\lambda) = (2\pi\, w^2)^{-1} \begin{cases} (4\, w^2 - \lambda^2)^{1/2}, & \text{if } |\, \lambda \,| \leq 2w; \\ 0, & \text{if } |\, \lambda \,| > 2w. \end{cases} \tag{5.3}$$

This is the well known Wigner or semcircle law which is the $n = \infty$ limiting eigenvalue distribution for the ensemble of $n \times n$ symmetric random matrices whose entries are independent identically distributed random variables with zero mean value and variance w^2/n (this ensemble is called the Wigner ensemble of random matrices, see [18,19,26] for the references, the history and numerous related results). The limiting eigenvalue distribution defined by equation (2.9) is known as the deformed semicircle law. This distribution was found in [20] (see also [12,14]).

Now we will list some useful properties of the deformed semicircle law, defined by equation (2.9).

(i) *For any nonperturbed* $N_0(\mathrm{d}\lambda)$,

$$N(\mathrm{d}\lambda) = \rho(\lambda)\, \mathrm{d}\lambda, \quad 0 \leq \rho(\lambda) \leq (\pi w)^{-1}. \tag{5.4}$$

(ii) *Let us call* supp$N$ *and* suppN_0 *the spectrum* σ *and the unperturbed spectrum* σ_0 . *Then,* σ *is contained in the* $2w$*-neighbourhood of* σ_0 .

(iii) *If* a *and* b *are the left and right endpoints of the interval containing* σ *and* a_0 *and* b_0 *are the same points for* σ_0 , *then* $a < a_0$ *and* $b > b_0$.

(iv) *If* $\sigma_0 = (a_0, b_0)$, *then* $\sigma = (a, b)$ *(and according to (ii) and (iii),* $(a_0, b_0) \subset (a, b) \subset (a_0 - 2w, b_0 + 2w)$*).*

(v) *Consider the intervals that comprise the complement of* σ_0 , *find the inverse* $\lambda_0(r)$ *to* $r_0(\lambda)$ *for these intervals, locate the intervals on which the function* $\lambda_0(r) - w^2 r$ *is monotonically increases and then determine the set of values of this function on these intervals. The spectrum* σ *is the complement of this set.*

If α_0 is an endpoint of one of the above mentioned intervals, then we get that $a = \lambda_0(\alpha) - w^2\alpha$ is the endpoint of one of the components of σ . Suppose that in the neighbourhood of α, $\lambda_0(r) - w^2 r$ is analytic. Then simple arguments show that the density of states in the neighbourhood of α behaves as follows:

$$\rho(\lambda) = \text{const } |\, \lambda - a \,|^{1/2k} \; (1 + o(1)), \quad |\, \lambda - a \,| \to 0$$

for some $k = 1, 2, \ldots$. Generically, $k = 1$, and this corresponds to the nondegenerated extremum of $\lambda_0(r) - w^2 r$ (a maximum if α is the right endpoint, or a minimum if α is the left endpoint of such an interval).

6. THE CONDUCTIVITY

As was explained in Section 2, we are going to prove weak convergence of the measures $M^{(a)}$, $a = R, d, n$ for $a \to \infty$ and calculate the respective limits. Therefore it suffices to consider their Stieltjes transforms

$$C^{(a)}(z_1, z_2) = \int \frac{M^{(a)}(d\lambda_1, d\lambda_2)}{(\lambda_1 - z_1)(\lambda_2 - z_2)}, \quad \operatorname{Im} z_{1,2} \neq 0 \tag{6.1}$$

and find their limits for $a \to \infty$ for $\operatorname{Im} z_{1,2} \neq 0$. After that we can apply the inversion formula (2.7) to each of the two variables and find the limiting measure. We will consider here the simplest case of the Gaussian-distributed randomness in (2.1)-(2.3), postponing the proof for an arbitrarily distributed randomness for a subsequent publication.

Theorem 6.1 . *Let* H_a, $a = R, d, n$ *be given by (2.1)-(2.3) in which*

$$h(x) = h(-x), \quad \sum_{x \in \mathbf{Z}^d} |x| \, |h(x)| < \infty \tag{6.2}$$

and W's be the Gaussian-distributed random variables. Then for $\operatorname{Im} z_{1,2} \neq 0$

$$\lim_{a \to \infty} C^{(a)}(z_1, z_2) =$$

$$\begin{cases} w^2 \, r(z_1) \, r(z_2) \int x^2 \, \phi^2(x) \, dx, & \text{if } a = R \\ w^2 \, r(z_1) \, r(z_2), & \text{if } a = d \\ w^4 \int \dfrac{|\nabla h(k)|^2 \, dk}{(h(k) - z_1 - w^2 r(z_1)) \, (h(k) - z_2 - w^2 r(z_2))}, & \text{if } a = n, \end{cases} \tag{6.3}$$

where $r(z)$ *is given by (2.9) and (2.8) for* $a = R, n$ *and by (2.9) and (2.10) for* $a = d$.

Proof. Since the proof of (6.3) is somewhat tedious even for the Gaussian W's in (2.1)-(2.3), we describe in detail the case of $a = n$ that turns out to be simplest. Note also that the case $a = n$ was considered for the first time in [25] where respective results on the density of states and the conductivity were obtained by the perturbation theory arguments.

The velocity operator in this case is identity with respect to the orbital indices α, β and its coordinate-dependent part is defined by the matrix

$$\hat{v}^{(n)}(x - y) = i \, (x - y) \, h(x - y) \tag{6.4}$$

and is bounded in view of (6.2). Therefore, according to (6.1) and (6.4),

$$C^{(n)}(z_1, z_2) = \sum_{y,t} \sum_{i=1}^{d} \hat{v}_i(y) \, \hat{v}_i(t) \, T_0(0, y, t; z_1, z_2), \tag{6.5}$$

where

$$T_0(x,y,t;z_1;z_2) = \mathbf{E}\{n^{-1}\sum_{\alpha=1}^{n} Q(x,y,t,z_1,z_2,\alpha,\alpha)\} \tag{6.6}$$

and

$$Q(x,y,t,z_1,z_2,\alpha,\gamma) = \sum_{\beta,u} G(\alpha,x;\beta,u+t;z_1)\, G(\beta,u;\gamma,y;z_2).$$

Consider now the infinite sequence of moments containing (6.6)

$$T_k(X_k;Y_k;Z_k;x,y,t,z_1,z_2) =$$
$$\mathbf{E}\{\prod_{j=1}^{k} G(x_j,y_j;\zeta_j)\; n^{-1}\sum_{\alpha=1}^{n} Q(x,y,t,z_1,z_2,\alpha,\alpha)\}, \quad k \geq 0$$

where $G(x,y;\zeta)$ is specified by (3.2) for $a = n$ and $z = \zeta$, $|\operatorname{Im}\zeta_i| \geq \eta$, $|\operatorname{Im} z_{1,2}| \geq \eta$. By using the resolvent identity for $G(\beta,u;\alpha,y;z_2)$ and the arguments similar to those in the derivation of (3.4), we obtain the system of relations

$$\begin{aligned}
&T_k(X_k;Y_k;Z_k;x,y,t,z_1,z_2) &&= \\
&\textstyle\sum_u F_{k+1}(X_k,x;Y_k,(u+t);Z_k,z_1)\, g_2(u-y) &&+ \\
&w^2 \textstyle\sum_s \left[T_{k+1}(X_k,x;Y_k,s;Z_k,z_1;s,s,t,z_1,z_2) \right. &&+ \\
&\left. T_{k+1}(X_k,s;Y_k,s;Z_k,z_2;x,s,t,z_1,z_2) \right] g_2(s-y) &&+ \quad R_k
\end{aligned} \tag{6.7}$$

where $g_2(x-y) = (h-z_2)^{-1}(x-y)$ and

$$|\, R_k \,| \leq 4n^{-1}\, w^2\, (k+1)\, \eta^{-k-4}. \tag{6.8}$$

Consider now the Banach space of the sequences

$$T = \{T_k(X_k;Y_k;Z_k;x,y,t,z_1,z_2)\}_{k=0}^{\infty} \tag{6.9}$$

with the norm

$$\|T\| = \sup_{k\geq 0} \xi^k \sup_{|\operatorname{Im}\zeta_k|,|\operatorname{Im} z_{1,2}|\geq\eta} \; \sup_{X_k,Y_k,x,y,t} |\, T_k \,| \,.$$

If $\mathbf{A}_1$ is the linear operator defined by the sum over s in (6.7), then it is easy to show that $\|\mathbf{A}_1\| \leq 4w^2\, \xi^{-1}\, \eta^{-1}$ and if R is the sequence specified by R_k, then in view of (6.8)

$$\|R\| \leq 4n^{-1}\, w^2\, \eta^{-4} \sup_{k\geq 0}\; k\, (\xi\, \eta^{-1})^k.$$

Thus, to guarantee a finite norm of T in (6.9), the contractivity property of $\mathbf{A}_1$ and the finiteness of $\|R\|$, it suffices to take

$$\xi = \eta/2, \quad \eta = \max\{3w,\; 2\bar{h}\}. \tag{6.10}$$

Then, $\|\mathbf{A}_1\| \le \frac{8}{9} < 1$, $\|R\| \le Cn^{-1}$, where C is independent of n. If, in addition, $\bar{F} = \{\bar{F}_k\}_{k\ge 0}$ is the sequence defined by the first sum in the r.h.s. of (6.7), then $\|\bar{F}\| \le \eta^{-2} \sup_{k\ge 1} (\xi\, \eta^{-1})^k < \infty$. Theorem 4.1 and summability of $g_2(x)$ imply that $\bar{F} = \bar{\Gamma} + \epsilon_n$, where $\lim_{n\to\infty} \|\epsilon_n\| = 0$ and $\bar{\Gamma} = \{\Gamma_k\}_{k=0}^{\infty}$ with

$$\Gamma_k(X_k; Y_k; Z_k; x, y, t, z_1, z_2) = \prod_{j=1}^{k} \Gamma(x_j - y_j; \zeta_j) \sum_u \Gamma(x - u - t; z_1)\, g(u - y; z_2)$$

where $\Gamma(x; z)$ is defined in Theorem 4.1.

Now, if $\bar{T} = \{\bar{T}_k\}_{k=0}^{\infty}$ is a unique solution of the equation $\bar{T} = \bar{\Gamma} + \mathbf{A}_1 \bar{T}$, then $\lim_{n\to\infty} \|T - \bar{T}\| = 0$ and $\bar{T}_k(X_k; Y_k; Z_k; x, y, t, z_1, z_2) = \prod_{j=1}^{k} \Gamma(x_j - y_j; \zeta_j)\, \bar{\Gamma}(x - y, t; z_1, z_2)$, where $\bar{\Gamma}(x, t; z_1, z_2)$ is a solution of the equation

$$\begin{aligned} \bar{\Gamma}(x-y,t) &= \textstyle\sum_u \bar{\Gamma}(x-t-u; z_1)\, g(u-y; z_2) + \\ & w^2 \textstyle\sum_s [\Gamma(0; z_2)\, \bar{\Gamma}(x-s,t) + \Gamma(x-s; z_2)\, \bar{\Gamma}(0,t)\,], \end{aligned} \tag{6.11}$$

provided that this equation is uniquely solvable. The latter fact can readily be proved in the space with the norm

$$\|\bar{\Gamma}\| = \sup_{|\operatorname{Im} z_{1,2}| \ge \eta} \; \sup_{x,t} \; |\, \bar{\Gamma}(x, t; z_1, z_2)\, |$$

with η_1 specified by (6.10). Besides, it is easy to check by direct calculation that

$$\bar{\Gamma}(x, t; z_1, z_2) = \Gamma_{12}(x+t) + w^2\, \Gamma_{12}(x)\, \Gamma_{12}(t)\, (1 - \Gamma_{12}(0))^{-1} \tag{6.12}$$

where $\Gamma_{12}(x-t)$ is the product of the Toeplitz operators $\Gamma|z = z_{1,2}$ whose kernel is specified by (4.3), i.e.

$$\Gamma_{12}(x) = \int_{\mathbf{T}^d} \frac{exp\{2\pi i k x\} dk}{(\tilde{h}(k) - z_1 - w^2 r(z_1))\, (\tilde{h}(k) - z_2 - w^2 r(z_2))}.$$

According to (6.2), $\tilde{h}(k)$ is even. Thus, $\Gamma_{12}(x)$ has the same property and since $\hat{v}(x)$ specified by (6.4) is odd, the second term of (6.12) gives no contribution to (6.5). Besides, since the Fourier transform of $\hat{v}(x)$ is $\nabla \tilde{h}(k)$, the substitution of (6.11) into (6.5) yields (6.3) for $a = n$ and $|\operatorname{Im} z_{1,2}| \ge \eta$. By using the analytic continuation arguments it is easy to show that the same limiting relation is true for all $\operatorname{Im} z_{1,2} \neq 0$. Thus, we have proved the proposition for $a = n$.

7. DISCUSSION

As we mentioned in Section 5, the deformed semicircle law appeared for the first time [20] in a somewhat different problem on the limiting eigenvalue distribution of the random matrices, known as the deformed Wigner ensemble. This ensemble arose in nuclear physics (see e.g. [18,26]), where it was proposed in order to describe the statistics of low-lying levels of heavy nuclei. Later similar ensembles appeared in quantum field theory [4], quantum chaology [9] and statistical mechanics [8].

The main difference of the respective random matrices from those modelling elementary excitations in disordered condensed matter (in particular, the random operators (2.1)-(2.3)) is that the former have all the entries of the same order of magnitude (e.g. identically distributed), while the latter have nonzero entries only on the finite number b of diagonals adjacent to the principal diagonal (for instance in (2.1) $b = (2\rho R)^d + 1$, where ρ is the radius of the support of $\phi(t)$). In other words the former random matrices correspond to nonlocally interacting disordered systems (the range of the interaction is of the order of the size of the system), while the latter correspond to the locally interacting (short range) systems. Therefore it is rather natural that we have obtained the deformed semicircle law as the result of the limiting transition $R,\ d,\ n \to \infty$.

Let us consider the finite volume version of (2.1), i.e. the restriction $H_\Lambda^{(R)}$ of (2.1) to a finite cube $\Lambda \subset \mathbf{Z}^d$ centred at the origin and having the side length L . Then the Wigner ensemble of random matrices corresponds to (2.1) with $d = 1,\quad L = l = 2m+1,\quad R = 2m$ and $\phi(t) = \chi_1(t)$, where $\chi_1(t)$ is the indicator of the interval $[-1/2, 1/2]$. Therefore Theorem 5.1 of this paper and the results of [20] (see also [12]) show that the deformed semicircle law is the limiting eigenvalue distribution in the two extreme cases of $H_\Lambda^{(R)}$: the first one corresponds to the two successive limiting transitions $L \to \infty$ and then $R \to \infty$ (Theorem 5.1), while the second one corresponds to the simultaneous limits $L \to \infty,\quad R \to \infty,\quad LR^{-1} \to 1$. In view of these results it is natural to analyse the intermediate cases when $L \to \infty$ and $R \to \infty$ but $0 \le \nu \equiv \lim\, L\,R^{-1} \le 1$. It is the so-called band random matrices, which appear for instance in the studies of quantum chaos [5]. In paper [13] it was shown that under fairly general conditions (condition (2.1b) in essence) the limiting eigenvalue distribution of these random matrices with $H_0 = 0$ is again the semicircle law if $\nu = 0$. The case of $0 < \nu < 1$ is more complicated (for details see [3,13]).

We have mentioned in Introduction that there exists an analogy between the spectral problems which we are studying in this paper and the mean field theories in statistical mechanics and solid state theory. Therefore it is natural to compare our results with results of mean field type approximations developed in the theory of disordered systems (see reviews [4,17,28]). These approximations are known also as single-site approximations and are applied mainly to the averaged Green function of the respective random operator.

The most widely accepted approximation of this type is the coherent potential approximation (CPA). It was proposed and applied to random operators with a diagonal disorder, i.e.the discrete Schrodinger operator

$$-h\,\Delta_{disc} + q(x), \quad x \in \mathbf{Z}^d \tag{7.1}$$

with an i.i.d. random potential first of all. Therefore one cannot expect too much similarity between our results obtained for the "opposite" case of the "maximum" off-diagonal disorder and results of the CPA. In particular for the Cauchy-distributed random potential (the Lloyd model), when

$$P\{ q(x) \in \mathrm{d}q\} = \gamma\, \pi^{-1}\, (q^2+\gamma^2)^{-1}\, \mathrm{d}q, \quad \gamma > 0, \tag{7.2}$$

the basic relations of the CPA [8,21,34]

$$\mathbf{E}\{ G(x,x;z) \} = G_0(0, z+\Delta(z)), \tag{7.3}$$

$$\mathbf{E}\{ \frac{q(x)+\Delta(z)}{1+(q(x)+\Delta(z))\, G_0(0,z+\Delta(z))} \} = 0 \tag{7.4}$$

yield the following for the selfenergy $\Delta(z)$

$$\Delta(\lambda+i\varepsilon) = i\gamma\, \mathrm{sign}\, \varepsilon. \tag{7.5}$$

Since according to our notation the l.h.s. of (7.3) is $r(z)$, we can write (7.3) and (7.5) for $\mathrm{Im}\, z \neq 0$ as

$$r(z) = r_0(z+i\gamma). \tag{7.6}$$

This formula is obviously different from our basic equation (2.9) which determines the selfenergy

$$\Delta(z) = w^2\, r(z) \tag{7.7}$$

as the solution of this equation. Nevertheless, as was shown in [25], the equations (7.3) and (7.4) yield (2.9) if the probability distribution of $q(x)$ is the semicircle law (5.3).

According to many suggestions and numerical results, the accuracy of the CPA increases with increase of the coordination number of a lattice, i.e. its dimensionality in particular (note that no general quantitative criteria for the validity of the CPA seem to be available). Consider in this connection the Lloyd model (7.1), (7.2) again. It is well known that for this model [17]

$$\mathbf{E}\{G(x,y;\lambda+i\varepsilon)\} = G_0(x-y;\lambda+i\varepsilon+i\gamma\cdot \mathrm{sign}\,\varepsilon). \tag{7.8}$$

Comparing this relation with (7.3) and (7.5) we conclude that the CPA is exact for the Lloyd model. Now, rescaling the translation invariant part of (7.1) in accordance with (2.2b) $h = h_d\, d^{-1/2}$, we find from (7.8) with $x=y$ and $\mathrm{Im}\, z > 0$ that

$$r(z) \equiv \lim_{d\to\infty} \mathbf{E}\{G(x,x;z)\} = \int (\, \lambda \; - \; z \; - \; \Delta(z)\,)^{-1} N_0(\mathrm{d}\lambda),$$

where $N_0(\mathrm{d}\lambda)$ is the Gaussian distribution given by (2.10) and $\Delta(z)$ is given by (7.5). This formula as well as (7.6) is different from equation (2.9) (with the same $N_0(\mathrm{d}\lambda)$) according to which the selfenergy is given by (7.7).

Thus, at least in the case of the Lloyd model, the $d = \infty$ limits of the exact averaged diagonal element of the Green function and of its CPA form are different from the same limit of $\mathbf{E}\{G(x,x;z)\}$ for the random operator (2.2) with an off-diagonal disorder satisfying (2.1). This conclusion is supported by recent diagrammatic analysis [24], according to which the $d = \infty$ limit of $\mathbf{E}\{G(x,x;z)\}$ and of its CPA form coincide for an arbitrary random i.i.d. potential in (7.1).

To this point we have being discussing the first moment of the Green function which determines equilibrium characteristics of disordered systems. However, to calculate kinetic characteristics, the conductivity first of all, we need to know the second moment of the Green function. It was recognized long ago [7,28] that the simple decoupling $\mathbf{E}\{G\,G\} = \mathbf{E}\{G\}\,\mathbf{E}\{G\}$, which is in the spirit of any single-site (mean field type) approximation including the CPA, is inconsistent with certain physical and mathematical conditions. The most important is the Hilbert identity which plays the role of the Ward identity here. It was also found that a modification of this decoupling which is free of the above mentioned inconsistencies should only take into account the multiple scattering of two particles by the same site [23].

As we have mentioned above the infinite R, d and n limits do not coincide with the CPA. However these limits have similar properties. In particular, according to Theorem 4.1, all the Green function moments decouple into the products of the first moments in these limits and, according to Section 6, this fact does not contradict the Hilbert identity. The reason is not too fast decay of the respective remainders. As a result we cannot perform these limits inside the sums over the whole lattice entering into the Hilbert identity or in expression (6.5) for the Stieltjes transform of the conductivity. To calculate an expectation containing the mentioned sums we are to consider a larger family of quantities including both the products of the Green functions and certain infinite sums of some of these products and to derive a new infinite system of equations for the expectations of this family. However the common feature of this new system and the systems for the expectation of the products of the Green functions is that in both cases the solutions of the respective limiting systems have a factorized form (see (4.1) and (6.11)) in accordance with the spirit of the single-site approximations.

This should be regarded as the proper mathematical mechanism of these approximations.

It is worth noting that, though the respective corrections (see e.g. (6.12)) are neccessary to guarantee the Hilbert identity, they do not contribute to the conductivity (6.5) due to the antisymmetry of the velocity operator $\hat{v}$ in (6.4). Thus, we can use the simple decoupling $\mathbf{E}\{G\,G\} = \mathbf{E}\{G\}\,\mathbf{E}\{G\}$ in the final calculation of the conductivity. This fact is also well known in the theory of disordered systems (see [7,17,28]).

Vanishing of the simplest two-body corrections to the conductivity seems physically related to vanishing of the backscattering term in a transport equation [23]. As is generally accepted (see e.g. review [15]), the absence of the backscattering contribution is in turn closely related to the absence of localization. This is in agreement with the absence in our formulae of such well known manifestations of localization as exponential tails of the IDS, vanishing of the zero frequency conductivity for the Fermi energies for which the density of states is nonzero, etc.

Since for finite R, d and n the localized states should always be present at least at the edges of the spectrum of the operators (2.1)-(2.3), we have to conclude that the limiting transitions $R,\ d,\ n \to \infty$ "remove" completely the pure point spectrum of these operators. In particular, since according to (6.3) in all the three limits (as well as in the CPA) the support of the zero-frequency conductivity as a function of the Fermi energy coincides with the IDS support , then one may speculate that the edges of this support coincide with the $R,\ d,\ n \to \infty$ limits of the mobility edges of respective operators.

AKNOWLEDGEMENTS
This work was partially supported by Grant N 1/82 of the State Committee for Science and Technology of Ukraine. L.P. would like to thank the Mittag-Leffler Institute for the kind hospitality.

REFERENCES

1. Akhiezer N. The Classical Moment Problem. London: Oliver and Boyd 1964.
2. Berlin, T., Kac, M.: The spherical model of a ferromagnet. Phys. Rev. **86**, 821-825 (1952).
3. Bogachev, L., Molchanov, S., Pastur, L.: On the density of states of random band matrices (in Russian) Mat. Zametki, **50** , 31-42 (1991).
4. Brezin, E., Itzykson, C., Parisi, G., Zuber, J.: Planar diagramms. Commun. Math. Phys. **59** , 35-51 (1978).
5. Casati, G., Molinari, L., Izrailev,F.: Scaling properties of band random matrices. Phys. Rev. Lett. **64**, 1851-1854 (1990).
6. Constantinescu, F., Felder, C., Gawedzki, K., Kupiainen, A.: Analyticity of density of states in a gauge invariant model of disordered systems. J. Stat. Phys. **48** , 365-391 (1987).
7. Elliot, P., Krumhansl, J., Leath, P.: Theory and properties of randomly disordered crystals and related physical systems. Rev. Mod. Phys. **46** , 463-510 (1974).
8. Fernandez, R., Frohlich, J., Sokal, A.: Random Walks, Random Surfaces, Critical Phenomena and Triviality in Quantum Field Theory. Berlin, Heidelberg, New York: Springer 1992.
9. Haake, F.: Quantum Signatures of Chaos. Berlin, Heidelberg, New York: Springer 1991.
10. Kac, M.: Mathematical mechanisms of the phase transitions. In: Chretien, M., Deser, S. (eds.) Statistical Physics, Phase Transitions and Superfluidity, Vol.I, pp. 241-301 . New York: Gordon and Breach 1968.
11. Kac, M., Thompson, C.: Spherical model and the infinite spin dimensionality limit. Physica Norwegica **5** , 163-168 (1971).
12. Khorunzhy, A., Pastur, L.: On the eigenvalue distribution of the deformed Wigner ensemble of random matrices. In: Operator Theory and Related Topics, AMS (in press).

13. Khorunzhy, A., Molchanov, S., Pastur, L.: On the eigenvalue distribution of band random matrices in the limit of their infinite order (in Russian) Teor. Mat. Fiz. **90**, 163-178 (1992).
14. Khorunzhy, A., Khoruzhenko, B., Pastur, L., Shcherbina, M.: The large-n limit in statistical mechanics and the spectral theory of disordered systems. In: Domb, C., Lebowitz, J. (eds.) Phase Transitions and Critical Phenomena Vol. 15, pp. 73-239 . New York: Academic Press 1992.
15. Lee, P., Ramakrishnan, T.: Disordered electronic systems. Rev. Mod. Phys. **57** , 287-337 (1985).
16. Lebowitz, J., Penrose, O.: Rigorous treatment of the van der Waals-Maxwell theory of the liquid-vapour transition. J. Math. Phys. **7** , 98-110 (1966).
17. Lifshitz, I., Gredeskul, S., Pastur, L.: Introduction in the theory of disordered systems. New York: Wiley 1988.
18. Mehta, M.: Random matrices. New York: Academic Press 1967.
19. Pastur, L.: Spectra of random self-adjoint operators. Russ. Math. Surveys, **28** , 1-67 (1973).
20. Pastur, L.: On the spectrum of random matrices (in Russian). Teor. Mat. Fiz. **10** , 102-112 (1973).
21. Pastur, L., Figotin, A.: Spectra of random and almost periodic operators. Berlin, Heidelberg, New York: Springer 1992.
22. Stanley, H.: Spherical model as a limit spin dimensionality. Phys. Rev. **176**, 718-721 (1968).
23. Velicky, B.: Theory of electronic transport in disordered binary alloys: coherent potential approximation. Phys. Rev. **184** , 614-627 (1969).
24. Vlaming, R., Vollhardt, D.: Controlled mean field theory for disordered electronic systems: single particle properties. Rutgers preprint RWTH/ITP - C 6/91.
25. Wegner, F.: Disordered system with n orbitals per site: $n = \infty$ limit. Phys. Rev. **B19**, 783-792 (1979).
26. Wigner, E.: Random matrices in physics. SIAM Review J. **9** , 1-23, (1967).
27. Wegner,F., Opperman, R.: Disordered systems with n orbitals per site: $1/n$ expansion. Z. Phys. **B34** , 327-348 (1979).
28. Yonezawa,F., Morigaki, K.: Coherent potential approximation. Suppl. Progr. Theor. Phys. **53** , 1-76 (1973).

A.M.Khorunzhy, L.A.Pastur, Institute for Low Temperature Physics, Academy of Sciences of Ukraine, 47 Lenin Ave, Kharkiv, 310164, Ukraine

Operator Theory:
Advances and Applications, Vol. 70

Atoms at Finite Density and Temperature and the Spectra of Reduced Density Matrices*

N. Macris and Ph. A. Martin

1 Introduction

Why is matter constituted with atoms and molecules? Everyone who had a first course in quantum mechanics is tempted to answer: because Schroedinger has provided us with his equation (Δ = Laplacian on $\mathbf{R}^3$, $V(x)$ = potential)

$$-\frac{\hbar^2}{2m}\Delta\psi(x) + V(x)\psi(x) = E\psi(x) \tag{1.1}$$

If $V(x)$ is sufficiently attractive there exists square integrable bound states $\psi(x)$ with negative energies $E < 0$. More generally, the Schroedinger equation predicts accurately binding energies, so atoms and molecules exist!

In fact, the situation is not so simple. We observe generally atoms and molecules in thermodynamical phases having non vanishing density and temperature. Consider a quantum particle of mass m in a finite domain Λ with Hamiltonian

$$H_\Lambda = -\frac{\hbar^2}{2m}\Delta_\Lambda + V(x) \tag{1.2}$$

with Dirichlet conditions on the boundaries of Λ. Then, according to the principles of statistical mechanics, the probability

$$p_\Lambda(E) = \frac{e^{-\beta E}}{Z_\Lambda}, \qquad Z_\Lambda = \mathrm{Tr}\, e^{-\beta H_\Lambda} \tag{1.3}$$

to find the particle in a bound state of energy E at inverse temperature β behaves as

$$p_\Lambda(E) = O\left(\frac{\lambda_B^3}{|\Lambda|}\right) \tag{1.4}$$

*work supported by the Fonds National Suisse de la Recherche Scientifique

when $|\Lambda| \to \infty$, where $\lambda_B = (\frac{2\pi\beta\hbar^2}{m})^{1/2}$ is the de Broglie thermal wave length. This is so because the partition function Z_Λ is $O(|\Lambda|)$ as a consequence of the fact that the number of ionized states (forming the continuous spectrum of H as $|\Lambda| \to \infty$) is extensive. The quantity (1.4) is vanishingly small, so the particle has no chance to bind [1]. The way to solve this apparent paradox is to consider a state with an extensive number of particles $N = \rho|\Lambda|$ ($\rho > 0$ is the density) so that the space available per particle remains finite in the thermodynamic limit. The conclusion is that atoms may form only if there is a non vanishing density of them, and only if they have (at least weak) mutual interactions. Thus understanding the formation of an atom requires a study of the many-body situtation.

This problem has been beautifully treated by Ch. Feffermann in ref [2]. In this work, Fefferman considers the quantum mechanical electron-proton gas at thermal equilibrium. He shows that provided that the stability of matter bound holds with an optimal constant, the system is asymptotic to a perfect gas of hydrogen atoms in a suitable low-density low-temperature limit, the "Saha limit" (see section II). Conlon, Lieb and Yau have extended the results to a general system of nuclei and electrons [3]. We present here a simplified model (a single quantum particle interacting with a gas of classical particles by means of short range forces) exhibiting the main features of the problem. In section II, we present the model and describe the formation of an atom in the low-density, low-temperature limit. In section III , we discuss how to formulate the bound and ionized states problem at non vanishing density in terms of the spectral properties of the reduced two-particle density matrix. Section IV is devoted to some aspects of the proofs, and concluding remarks can be found in section V.

2 The Saha limit

We consider a single quantum particle of mass m (the q-particle) in thermal equilibrium with a gas of classical particles of mass M (the c-particles). The q-particle interacts with each of the c-particle by means of an attractive potential $V(x)$. The hamiltonian of the q-particle in presence of a configuration $r_1, ..., r_n$ of c-particles is

$$H^n[r_1, ..., r_n] = -\frac{\hbar^2}{2m}\Delta + \sum_{j=1}^{n} V(x - r_j) \tag{2.1}$$

If there are no c-particles, $H^0 = -\frac{\hbar^2}{2m}\Delta$ is purely kinetic, where Δ is the Laplacian on $\mathbf{R^3}$. Moreover, the c-particles may have a mutual potential interaction energy $U(r_1, ..., r_n)$. We make the following assumptions:

(i) $V(x) \leq 0$, $V(x)$ is locally square integrable and there exists $M > 0$ such that

$$V(x) \leq \frac{M}{|x|^\eta}, \qquad \eta > 3 \qquad \text{as} \qquad |x| \to \infty \tag{2.2}$$

Under the condition (i) the hamiltonian (2.1) has at most a finite number of bound-states of finite multiplicities with negative energies and an absolutely continuous spectrum on $[0, \infty]$.

(ii) we assume that $H^1[r_1 = 0] = -\frac{\hbar^2}{2m}\Delta + V(x)$ has a ground state $\psi_1(x)$ with energy $E_1 < 0$.

(iii) Stability estimate for the total energy: for $n \geq 2$, there exists a constant K, such that $0 < K < |E_1|$ and

$$H^n[r_1, ..., r_n] + U(r_1, ..., r_n) \geq -Kn, \qquad n \geq 2 \tag{2.3}$$

Because of (ii) the inequality (2.3) is only possible with K strictly less then $|E_1|$ for $n \geq 2$. It means that the binding energy per c-particle is the largest when the q-particle binds with only one c-particle.

When all particles (including the q-particle) are in thermal equilibrium at inverse temperature β, the corresponding grand-canonical partition function is

$$\Xi_\Lambda(\beta, \mu) = \sum_{n=0}^{\infty} \frac{z^n}{n!} \int_\Lambda dr_1 ... dr_n \, e^{-\beta U(r_1, ..., r_n)} \mathrm{Tr}\, e^{-\beta H_\Lambda[r_1, ..., r_n]} \tag{2.4}$$

In (2.4), $H_\Lambda[r_1, ..., r_n]$ is the hamiltonian (2.1) with Dirichlet conditions on the boundaries of Λ; moreover

$$z = \left(\frac{M}{2\pi\beta\hbar^2}\right)^{3/2} e^{\beta\mu} \tag{2.5}$$

is the activity and μ is the chemical potential of the classical gas.

A quantity of interest is the average energy distribution $p_\Lambda(dE, \beta, \mu)$ of the q-particle in the gas, conveniently defined by its Laplace transform

$$g_\Lambda(\lambda, \beta, \mu) = \int p_\Lambda(dE, \beta, \mu) \, e^{-\lambda E} = < e^{-\lambda H_\Lambda} >_\Lambda (\beta, \mu) = \frac{\Xi_\Lambda(\lambda + \beta, \mu)}{\Xi_\Lambda(\beta, \mu)} \tag{2.6}$$

where $< - >_\Lambda (\beta, \mu)$ denotes the grand-canonical average.

The following results are obtained in [4] when the classical particles form a gas of hard spheres of diameter d.

Proposition 2.1 *Stability with a good constant*
Given a potential $V(x)$ satisfying (i), the stability bound (2.3) holds for d sufficiently large.

Proposition 2.2 *Low-temperature low-density limit*
(a) Fix $\mu < 0$ and let $\beta \to \infty$ (so that the activity (2.5) tends also to zero). Then

$$\lim_{\beta \to \infty} g(\lambda, \beta, \mu) = \begin{cases} 1, & \text{if } \mu < E_1 \quad (2.7) \\ e^{-\beta E_1}, & \text{if } E_1 < \mu < -K \quad (2.8) \end{cases}$$

(b) Set $\mu(\beta) = E_1 + \sigma\beta^{-1} + o(\beta^{-1})$ for some number σ, $-\infty < \sigma < +\infty$, then

$$\lim_{\beta\to\infty} g(\lambda,\beta,\mu(\beta)) = \alpha + (1-\alpha)e^{-\lambda E_1} \tag{2.9}$$

with

$$\alpha = \left(\left(\frac{M}{m}\right)^{3/2} e^{\sigma} + 1\right)^{-1}, \qquad 0 < \alpha < 1$$

These results have the following interpretation. If $\mu < E_1$ (2.7) implies that $p(E,\beta,\mu) \to \delta(E)$, $\beta \to \infty$ i.e. the energy distribution of the q-particle is concentrated at the edge of the continuous spectrum : the particle does not bind. If μ belongs to the (open) interval $(E_1, -K)$, (2.8) implies that $p(E,\beta,\mu) \to \delta(E - E_1)$, $\beta \to \infty$ i.e. the particle binds with probability one. The case (2.9) ($\mu(\beta)$ tends to E_1 linearly as the temperature vanishes) interpolates between the two limits (2.7) and (2.8): binding occurs with probability $1-\alpha$. When μ is sufficiently negative (case (2.7)), the density is lowered too fast as $\beta \to \infty$ to permit binding : the entropy effect wins. When the decrease of the density is slower, we find successively partial binding (case (2.9), ionization equilibrium phases) and full binding for the specific range $(E_1, -K)$ of the chemical potential. The situation is described in the μ, T phase diagram (fig 1). The same picture is valid in the full quantum mechanical electron proton gas (provided that an appropriate form of the stability of matter holds) [2], [5].

Let us say a few words about the mathematical techniques that are involved in the proof of the proposition 2. The proof relies on an analysis of the low activity expansion of the quantity

$$f(\beta, z) = \lim_{\Lambda\to\infty} \frac{1}{|\Lambda|} \frac{\Xi_\Lambda(\beta,\mu)}{\Xi^o_\Lambda(\beta,\mu)} = \sum_{n=0}^{\infty} z^n f_n(\beta) \tag{2.10}$$

In (2.10), the partition function (2.4) is normalized by that of the purely classical gas $\Xi^o_\Lambda(\beta,\mu)$. Clearly the n-th order Mayer coefficient $f_n(\beta)$ of the series (2.10) will involve combinations of the kernels

$$(x|\exp(-\beta H^k[r_1, ..., r_k])|x), \qquad k = 0, 1, ...n \tag{2.11}$$

integrated on x and the location of the c-particles $r_1, ..., r_k$ (the explicit expressions can be found in [4]). Thus, as $\beta \to \infty$, according to (2.5), the dominant contribution to $z^n f_n(\beta)$ is of the order

$$z^n f_n(\beta) \sim e^{\beta(\mu n - E_n)} \tag{2.12}$$

where

$$E_n = \inf_{|r_i - r_j| \geq d} \operatorname{infspectrum} H^n[r_1, ..., r_n] \tag{2.13}$$

If μ is chosen in the open interval $(E_1, -K)$ (case (2.8) of proposition 2) the stability bound (2.3) implies

$$\mu n - E_n < 0 < \mu - E_1 \tag{2.14}$$

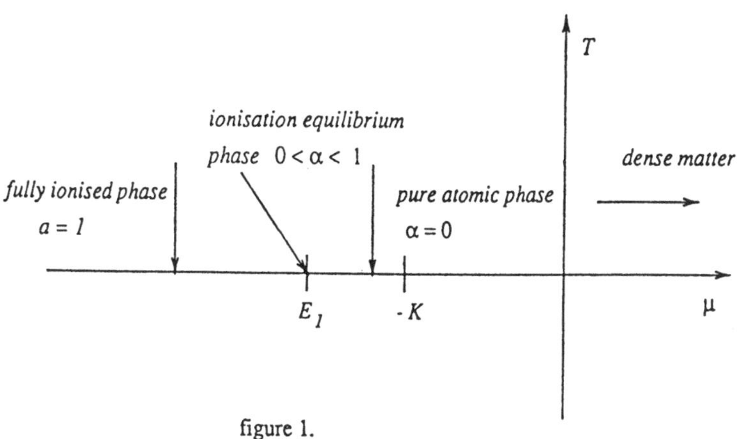

figure 1.

So the term $zf_1(\beta)$ corresponding to the presence of a single c-particle dominates the low activity series. The ionization equilibrium case (part (b) of proposition 2) occurs when the two first terms of the series (2.10) (free q-particle and one c-particle term) are of the same order as $\beta \to \infty$. The technical problem is to show that the rest of the series is negligible. For this, we establish pointwise bounds on the kernels

$$(x|e^{-\beta H[r_1,\dots,r_n]}|y) \leq Ae^{\beta B}(x|e^{-2\beta H_0}|y) \tag{2.15}$$

in terms of the free kernel that are uniform with respect to the location of the c-particles. The theory of Schroedinger semi-groups developed by B. Simon [6] is extensively used in this respect. These estimates in conjunction with standard methods to control the Mayer series enable to prove the proposition. We remark that the hard core repulsion is essential for the validity of (2.3). For a free gas of c-particles, one can show that E_n is a convex function of n, so there are no values of μ such that the Boltzmann weight of any finite aggregate is dominant as $\beta \to \infty$.

It is interesting to investigate the behaviour of the correlations in our model. They are given by the grand-canonical reduced density matrices defined by

$$\rho_{\Lambda,k+1}(x,y|r_1,\dots,r_k) = \frac{1}{\Xi_\Lambda}\sum_{n=0}^{\infty}\frac{z^{k+n}}{n!}\int_\Lambda dr'_1\dots dr'_n \exp(-\beta U(r_1,\dots,r_k,r'_1,\dots,r'_n))$$
$$(x|\exp(-\beta H[r_1,\dots,r_k,r'_1,\dots,r'_n]|y) \tag{2.16}$$

$\rho_{\Lambda,k+1}$ is the density matrix of the q-particle in presence of k c-particles located at $r_1,\dots,r_k$.

The low-temperature low-density limit of $\rho_2(x,y|r)$ corresponding to parts (a) and (b) of the proposition 2 is

$$(a) \qquad \lim_{\beta\to\infty} \rho_2(x,y|r) = \begin{cases} 0, & \text{if } \mu < E_1 \\ \psi_1^*(x-r)\psi_1(y-r) & \text{if } E_1 < \mu < -K \end{cases} \tag{2.17}$$

$$(b) \qquad \lim_{\beta\to\infty} \rho_2(x,y|r) = (1-\alpha)\psi_1^*(x-r)\psi_1(y-r), \qquad \mu(\beta) = E_1 + \sigma\beta^{-1} + o(\beta) \tag{2.18}$$

When $\mu \in (E_1, -K)$, one finds as expected that the correlation between the q-particle and a c-particle at r is given by the ground state wave function $\psi_1(x)$. In the case (b), this correlation is weighted by the probability $1-\alpha$ that the atom forms.

3 Spectra of reduced density matrices

We now come to the study of our model at non vanishing density. If $\rho \neq 0$, the notion of individual atoms looses in principle its meaning and we must find an alternative description of the system. If the pair density matrix $\rho_2(x,y|0)$ is known, it is natural to consider an effective two body Hamiltonian H_{eff} formally defined by

$$\frac{\rho_2(x,y|0)}{\rho} = (x|\exp[-\beta H_{eff}]|y) \tag{3.1}$$

H_{eff} depends on the density ρ and the temperature β^{-1}. As $\rho \to 0$, it should approach the Hamiltonian $H^1[0]$ of the pair formed by the q-particle with a c-particle at the origin in empty space. As ρ increases, H_{eff} progressively embodies the effects of the many particle system on this pair. These effects are seen in the deformation of the spectrum of H_{eff} as ρ varies. If ρ is small, the spectrum of H_{eff} is expected to be close to that of $H^1[0]$, i.e. it consists of a continuous part with a number of isolated eigenvalues. When ρ gets larger, the eigenvalues as well as the treshold of the continuous spectrum are displaced. It may even happen that above a large enough critical density ρ_c all the eigenvalues have merged into te continuum. If this phenomenon occurs, one speaks of pressure induced ionisation, or Mott effect, and ρ_c is the Mott density. In a plasma, the Mott effect results in a strong increase of the electrical conductivity. These properties are reflected in the spectrum of ρ_2 which is simply the exponential of that of H_{eff}. A qualitative picture of the spectrum of ρ_2 as a function of the density when the Mott effect occurs is given in fig 2.

We now state the main results that we obtain for two different models of the system of c-particles: a perfect gas (pg) and the "cell model" of a fluid (cm). In the perfect gas model we simply set $U(r_1, ..., r_n)$ equal to zero and $\rho = z$. In the cell model, one divides the space into cubic lattice cells of volume $\Delta = a^3$ (a is the lattice spacing). The allowed configurations of the c-particles have at most one particle per cell, with uniform distribution in the cell; the probability for the

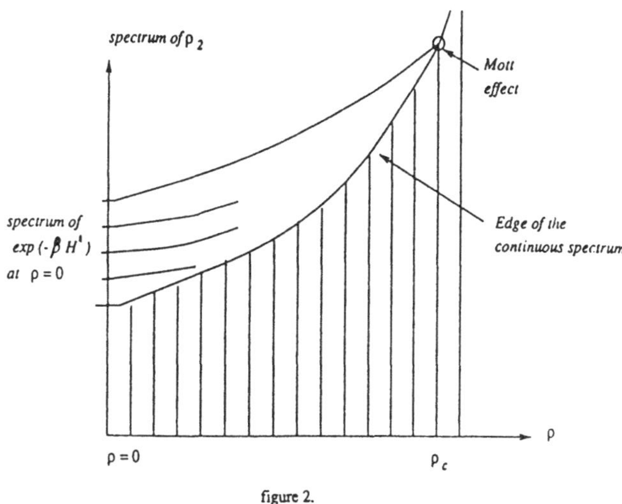

figure 2.

occupation of a cell is $\frac{z}{1+z}$ and the corresponding number density is $\rho = \frac{1}{\Delta}\frac{z}{1+z}$. An additional average over the lattice translations is performed in order to restore the full translational symmetry. The advantage of the cell model in our case is that it mimics a short range repulsion between the particles without allowing for phase transitions such as cristallisation which would occur in a system of hard spheres at high density. Let us note an important difference between the perfect gas and the cell model which will lead to different spectral properties of ρ_2 in the two models. In a perfect gas, the particle fluctuations grow linearly with the density ρ, which corresponds on a microscopic scale to large variations of the potential seen by the q-particle.On the contrary, in the cell model, the particle fluctuations tend to zero as full occupancy is approached, leading to a rather uniform potential at high density.

We introduce the Fourier transform

$$\tilde{\rho}_1(k) = \int dx\, e^{-ik.x} \rho_1(x, 0) \tag{3.2}$$

which physically represents the momentum distribution of the quantum particle in the classical gas. Note that if $\rho = 0$ it reduces to the usual Maxwellian distribution namely $\exp[-\beta |k|^2]$.

The propositions 3.1 and 3.2 hold for both models at any values of the density and the temperature

Proposition 3.1

For any ρ and β, ρ_1 has an absolutely continuous spectrum in the interval $[0, \tilde{\rho}_1(0)]$. Moreover $|\tilde{\rho}_1(k)| < \tilde{\rho}_1(0)$, $k \neq 0$.

Proposition 3.2

ρ_2 has an absolutely continuous part $[0, \Sigma]$, and possibly a finite number of eigenvalues λ_ν, $\lambda_\nu > \Sigma$, with $\Sigma = \rho\tilde{\rho}_1(0)$.

These two propositions validate the general features of fig 2. As far as the low density regime is concerned, we have the more precise statement of proposition 3.3, valid for both models.

Proposition 3.3
Let E_ν be the eigenvalues of $H^1[0]$. Then ρ_2 has isolated eigenvalues λ_ν satisfying $\lim_{\rho\to 0} \frac{\lambda_\nu}{\Sigma} = \exp(-\beta E_\nu)$ and $\Sigma = O((2\pi\beta)^{3/2}\rho)$.

In the high density limit, the perfect gas and the cell model exhibit different spectral properties, stated in propositions 3.4 and 3.5.

Proposition 3.4: perfect gas
For ρ large, ρ_2 has at least one eigenvalue $\lambda > \Sigma$. This holds for arbitrarily weak attractive potentials. Moreover $\Sigma = O(\rho^{1/4})$

Proposition 3.5 : cell model
The discrete spectrum of ρ_2 is empty for ρ large enough and $\Sigma = O(\rho^{3/4})$.

These two propositions show a striking difference between the two models at high density. In the cell model the Mott effect indeed occurs as indicated in fig 2. On the contrary, in the perfect gas there always remains an eigenvalue of ρ_2 above the continuous spectrum.

4 Proofs of the spectral properties

In this section we illustrate the basic ideas of the proofs of the results described in section III. These are based on a functional integral representation of the kernels of ρ_2 and ρ_1 and use tools from the spectral analysis of Schroedinger operators such as the stability of absolutely continuous spectra under trace-class perturbations, regular perturbation theory, the variational principle and the Birman-Schwinger principle.

The functional integral representation of the kernels of the density matrix is based on the Feynman-Kac formula

$$(x|e^{-\beta H_\Lambda[r_1,\ldots,r_n]}|y) = \frac{e^{-\frac{(x-y)^2}{2\beta}}}{(2\pi\beta)^{3/2}} \int D\alpha\, \chi_\Lambda(\alpha_{xy}) \prod_{j=1}^{n} e^{-\beta\int_0^1 ds V(\alpha_{xy}(s)-r_j)} \tag{4.1}$$

where

$$\alpha_{xy}(s) = x + s(y-x) + \sqrt{\beta}\alpha(s) \tag{4.2}$$

and α is the Gaussian Brownian bridge process $\alpha(s) = (\alpha_1(s), \alpha_2(s), \alpha_3(s))$, $0 < s < 1$, $\alpha(0) = \alpha(1) = 0$ with zero mean and covariance $s(1-t)\delta_{ij}$,$s \leq t$. $\chi_\Lambda(\alpha_{xy})$ represents the characteristic function of paths α_{xy} that stay in Λ for all s. Inserting (4.1) in the definition (2.14) for $k = 0$ and $k = 1$ one can derive the following formulas (the reader can find a detailed derivation in apppendix A of ref. [1]) for the density matrices of the perfect gas and cell models

$$\rho_1(x,y) = \frac{1}{N(\rho,\beta)} \frac{e^{-\frac{(x-y)^2}{2\beta}}}{(2\pi\beta)^{3/2}} \int D\alpha\, G_1(\alpha_{xy}) \tag{4.3}$$

$$\rho_2(x,y|0) = \frac{\rho}{N(\rho,\beta)} \frac{e^{-\frac{(x-y)^2}{2\beta}}}{(2\pi\beta)^{3/2}} \int D\alpha\, G_2(\alpha_{xy})\, e^{-\beta \int_0^1 ds V(\alpha_{xy}(s))} \tag{4.4}$$

where $N(\rho,\beta)$ is a normalization factor

$$N(\rho,\beta) = (2\pi\beta)^{-3/2} \int D\alpha\, G_1(\alpha) \tag{4.5}$$

In (4.3) and (4.4) the functionals $G_1(\alpha_{xy})$ and $G_2(\alpha_{xy})$ contain the effect of the environment and appear when one integrates over the coordinates r_j of the classical particles. Their expression is of course different for the two models under consideration. For the perfect gas model (pg) one finds

$$G_1^{pg}(\alpha_{xy}) = G_2^{pg}(\alpha_{xy}) = \exp\left[\rho \int dr \left(e^{-\beta \int_0^1 ds V(\alpha_{xy}(s)-r)} - 1\right)\right] \tag{4.6}$$

whereas for the cell model (cm) of a fluid

$$G_1^{cm}(\alpha_{xy}) = \frac{1}{|\Delta|} \int_\Delta d\tau \exp\left[\sum_{j\in\mathbf{Z}^3} C_{j,\tau}(\alpha_{xy})\right] \tag{4.7}$$

$$G_2^{cm}(\alpha_{xy}) = \frac{1}{|\Delta|} \int_\Delta d\tau \exp\left[\sum_{j\neq 0} C_{j,\tau}(\alpha_{xy})\right] \tag{4.8}$$

with

$$C_{j,\tau}(\alpha_{xy}) = \ln\left[1 + \rho \int_\Delta dr \left(e^{-\beta \int_0^1 ds V(\alpha_{xy}(s)-r-ja-\tau)} - 1\right)\right] \tag{4.9}$$

Remark that as $\rho \to 0$ (4.7) reduces to (4.6). Moreover for $\rho = 0$, G_j, $j = 1,2$ are identically 1 and $\rho_1(x,y)$, $\rho_2(x,y|0)$ reduce respectively to the kernels of $\exp[\beta\Delta]$ and $\exp[-\beta(-\Delta+V)]$. In all cases the kernels define positive self-adjoint integral operators ρ_1, ρ_2 from $L^2(\mathbf{R^3},\mathbf{dx})$ to $L^2(\mathbf{R^3},\mathbf{dx})$.

Part of our analysis of the spectral properties is based in the introduction of the truncated reduced density matrix ρ_T,

$$\rho_2 = \rho\rho_1 + \rho_T \tag{4.10}$$

the kernel of which can easily be deduced from the previous formulas. We consider $\rho\rho_1$ as the "unperturbed operator" (when $\rho = 0$ it becomes proportional to $\exp[\beta\Delta]$) and ρ_T as the "perturbation" (when $\rho = 0$ it becomes proportional to $\exp[-\beta(-\Delta+V)] - \exp[\beta\Delta]$).

Inspection of the kernels of the operators ρ_1, ρ_2, ρ_T leads to useful pointwise bounds, which will be needed later and are stated here without proof (see [7]). Setting

$$v(\beta) = \int dr \left(e^{-\beta V(r)} - 1\right) \tag{4.11}$$

we have

$$\frac{1}{N(\rho,\beta)} \frac{e^{-\frac{(x-y)^2}{2\beta}}}{(2\pi\beta)^{3/2}} \le \rho_1(x,y) \le \frac{e^{\rho v(\beta)}}{N(\rho,\beta)} \frac{e^{-\frac{(x-y)^2}{2\beta}}}{(2\pi\beta)^{3/2}} \tag{4.12}$$

$$\rho_2(x,y|0) \le \frac{e^{[\rho v(\beta)+\beta \sup_x |V(x)|]}}{N(\rho,\beta)} \frac{e^{-\frac{(x-y)^2}{2\beta}}}{(2\pi\beta)^{3/2}} \tag{4.13}$$

These bounds imply that

$$\tilde{M} \equiv \max\left[\sup_x \int dy\, |\rho(x,y)|, \sup_y \int dx\, |\rho(x,y)|\right] < \infty \tag{4.14}$$

and as a consequence [8], ρ_1 and ρ_2 are bounded selfadjoint operators on $L^2(\mathbf{R^3})$. For the kernel of the truncated density matrix we have

$$|\rho_T(x,y)| \le C(\rho,\beta) \frac{e^{-\frac{(x-y)^2}{2\beta}}}{(2\pi\beta)^{3/2}} \int D\alpha \int_0^1 ds\, [1 + \alpha_{xy}(s)|^2]^{-\eta/2} \tag{4.15}$$

where the constant $C(\rho,\beta)$ depends only on the density and the temperature. This estimate is valid for the two models. For the cell model of a fluid we can obtain a better estimate for small a (i.e. high density) which turns out to be important for the high density results,

$$|\rho_T^{cm}(x,y)| \le A e^{[\beta \sup_x |V(x)|]} (z^{-1} + a) \frac{e^{-\frac{(x-y)^2}{2\beta}}}{(2\pi\beta)^{3/2}} \int D\alpha \int_0^1 ds\, [1 + \alpha_{xy}(s)|^2]^{-\eta/2} \tag{4.16}$$

This shows that ρ_T can be made arbitrarily small at high densities (large z, small a) in the cell model of a fluid. Propositions 3.1 and 4.1 (see below) play a key role. We proceed to the proof of proposition 3.1

Proof of Proposition 3.1

By (4.12), $\rho_1(x,0)$ is in $L^2(R^3) \cap L^1(R^3)$. Moreover, inspection of (4.3) shows that $\rho_1(x,y) = \rho_1(x-y)$. Thus ρ_1 acts as a multiplication operator in the Fourier representation, i.e. for $\phi \in L^2(R^3)$, $\widetilde{\rho_1\phi}(k) = \tilde{\rho}_1(k)\phi(k)$. Because of the upper bound in (4.12) $\tilde{\rho}_1(k)$ is an entire function of the components of k, so it cannot be constant on open sets of $\mathbf{R^3}$. Thus ρ_1 has an absolutely continuous spectrum given by the image of the function $\tilde{\rho}_1(k)$. Since $\tilde{\rho}_1(k) \ge 0$ and $\lim_{|k|\to\infty} \tilde{\rho}_1(k) = 0$ the spectrum is $[0, \sup_{|k|} \tilde{\rho}_1(k)]$. Finally since $\rho_1(x,0) \ge 0$ we have

$$\tilde{\rho}_1(0) \le \sup_{|k|} \tilde{\rho}_1(k) \le \int dx\, |\rho_1(x,0)| = \int dx\, \rho_1(x,0) = \tilde{\rho}_1(0) \tag{4.17}$$

Then the continuous spectrum is given by $[0, \tilde{\rho}_1(0)]$. Furthermore since $\rho_1(x,0) = \rho_1(-x,0)$ and $\rho_1(x,0)$ is strictly positive (from the lower bound in (4.12)), for any $k \neq 0$ the integrand of

$$\tilde{\rho}_1(k) = \int dx \, \cos(k.x)\rho_1(x,0) \tag{4.18}$$

is strictly less than $\rho_1(x,0)$ on some open set not containing the points $k.x \in 2\pi\mathbf{Z}$. Hence $\tilde{\rho}_1(k) < \tilde{\rho}_1(0)$. This completes the proof of the proposition.

Proposition 4.1
For any ρ and β, the truncated reduced density matrix $\rho_T = \rho_2 - \rho\rho_1$ belongs to the Trace-Class.

Proof of Proposition 4.1
Let h be the multiplication operator by the function $h(x) = 1$ for $|x| \leq 1$ and $h(x) = |x|^{-(\frac{3}{2}+\epsilon)}$ for $|x| \geq 1$, with $\epsilon > 0$. One can easily check that the operator $K = (-\Delta+1)^{-1}$ is Hilbert-Schmidt. Since $\rho_T = K(K^{-1}\rho_T)$ it is sufficient to prove that $K^{-1}\rho_T h$ is Hilbert-Schmidt, i.e.

$$\int dx \int dy \, |h(x)|^{-2} |(-\Delta_x + 1)\rho_T(x,y)|^2 < \infty \tag{4.19}$$

One can show [7] that $|(-\Delta_x + 1)\rho_T(x,y)|$ satisfies a bound similar to (4.15), implying that (4.19) holds with our choice of h provided ϵ is small enough and $\eta > 6$.

General Properties of the Spectrum
Now we have gathered all the preliminary material needed to sketch the proof of proposition 3.2. The philosophy of the proof is closely related to the methods used for Schroedinger operators as explained after (4.10).

Proof of proposition 3.2
First we state immediate consequences of Propositions 3.1 and 4.1. From standard theorems [8] on the stability of absolutely continuous spectra $\rho_2 = \rho\rho_1 + \rho_T$ has an absolutely continuous part in its spectrum spanning the interval $[0, \rho\tilde{\rho}_1(0)]$. Moreover by the Weyl-von Neumann theorem, this interval coincides with the essential spectrum. Since ρ_2 is a positive operator, outside this interval the spectrum can consist only of isolated eigenvalues all greater than $\Sigma = \rho\tilde{\rho}_1(0)$, with finite multiplicities.

Since ρ_2 is bounded it could have an infinity of eigenvalues only if Σ would be an accumulation point of the discrete spectrum. Using the Birman-Schwiger technique we show that this does not occur, and thus the number of eigenvalues is finite as stated in 3.2. For $\lambda > \Sigma$ we introduce a Birman-Schwinger operator

$$K(\lambda) = |\rho_T|^{1/2}(\lambda - \rho\rho_1)^{-1}|\rho_T|^{1/2} \tag{4.20}$$

which is Hilbert-Schmidt since for $\lambda > \Sigma$, $(\lambda - \rho\rho_1)^{-1}$ is bounded, and $|\rho_T|^{1/2}$ (here $|\rho_T| = \sqrt{\rho_T^* \rho_T}$)is Hilbert-Schmidt from Proposition 2. By the Birman-Schwinger principle [10], [11]

$$\{\text{Number of eigenvalues of } \rho_2 \text{ greater than } \Sigma\} \leq \lim_{\lambda \to \Sigma} \mathrm{Tr} K(\lambda)^* K(\lambda) \tag{4.21}$$

We will show that the limit in (4.21) is in fact finite. We make the following decomposition

$$\mathrm{Tr}\, K(\lambda)^* K(\lambda) = \mathrm{Tr}\, |\rho_T|\, S(\lambda) |\rho_T|\, S(\lambda) + \frac{2}{\lambda} \mathrm{Tr}\, |\rho_T|\, S(\lambda)\, |\rho_T| - \frac{1}{\lambda^2} \mathrm{Tr}\, |\rho_T|^2 \tag{4.22}$$

where $S(\lambda) = (\lambda - \rho\rho_1)^{-1} - \lambda^{-1}$. As $\lambda \to \Sigma$ the last term in (4.22) has a finite limit. Let us control the limit of the second term. From the proof of Proposition 1 we see that $S(\lambda)$ has a kernel defined by the convergent integral

$$S(\lambda)(x, y) = \int dk\, e^{ik.(x-y)} \frac{\rho \tilde{\rho}_1(k)}{\lambda - \rho \tilde{\rho}_1(k)} \tag{4.23}$$

and since $|\rho_T|$ and $S(\lambda)|\rho_T|$ are Hilbert-Schmidt we can represent the trace by multiple integrals involving (4.23) and the kernel of $|\rho_T|$. Then estimating the integrals by the integral of the absolute value we arrive at

$$\mathrm{Tr}\, |\rho_T|\, S(\lambda)\, |\rho_T| \leq \left\{ \int dz \int dy \int dx\, \big| |\rho_T|(z, x)\, |\rho_T|(x, y) \big| \right\} \int dk\, \frac{\rho \tilde{\rho}_1(k)}{\lambda - \rho \tilde{\rho}_1(k)} \tag{4.24}$$

With the help of the estimate (4.15) we can show that the triple integral in large braces converges. Here the non-trivial point is that they involve the kernel $|\rho_T|(x, y)$ whereas (4.15) gives information on the decay of $\rho_T(x, y)$. This problem is circumvented by the following successive applications of Schwartz inequality

$$\begin{aligned} \int dx \int dy \int dz \big| |\rho_T|(x, z)|\, \rho_T|(z, y) \big| &\leq \int dx \left[\int dz \big| |\rho_T|(x, z) \big|^2 \right]^{1/2} \\ &= \int dx h(x) [h(x)]^{-1} \left[\int dz \big| |\rho_T|(x, z) \big|^2 \right]^{1/2} \\ &\leq \left[\int dx h(x)^2 \right]^{1/2} \int dx \int dz h(x)^{-2} \big| |\rho_T|(x, z) \big|^2 \end{aligned} \tag{4.25}$$

In the last term of (4.25) we recognize the Hilbert-Schmidt norm of $h^{-1}|\rho_T|$ which is equal to that of $h^{-1}\rho_T$. One can check that the latter is finite by using the bound (4.15).

It remains to check that

$$\lim_{\lambda \to \Sigma} \int dk\, \frac{\rho \tilde{\rho}_1(k)}{\lambda - \rho \tilde{\rho}_1(k)} = \int dk\, \frac{\tilde{\rho}_1(k)}{\tilde{\rho}(0) - \tilde{\rho}_1(k)} \tag{4.26}$$

is finite. We remark that:

(i) by proposition 1, the integrand has a unique singularity at $k = 0$,

(ii) $\int dx\, x\rho_1(x,0) = 0$ because $\rho_1(x,0) = \rho(-x,0)$ and

(iii) $0 < \int dx\, |x|^2 \rho_1(x,0) < \infty$.

Thus the singularity of the integrand is $O(|k|^2)$, an integrable one in 3 dimensions.

The control of the limit $\lambda \to \Sigma$ for the first term in (4.22) can be achieved by the same type of arguments (see [7]). This completes the proof of 3.2.

Proof of Proposition 3.3 (*Low density limit*)
The regime of fixed temperature and low density is easily treated by regular perturbation theory. Expanding the functionals (4.6) and (4.8) to first order in the density yields

$$\rho_2 = \frac{\rho}{N(\rho,\beta)}(e^{-\beta H[0]} + R) \tag{4.27}$$

where the kernel of R is expressed by a functional integral and the operator norm satisfies

$$||R|| \leq \max\left[\sup_x \int dy\, |R(x,y)|, \sup_y \int dx\, |R(x,y)|\right] \leq e^{[\beta \sup_x |V(x)|]}(e^{\rho v(\beta)} - 1) \tag{4.28}$$

Thus for the two models under consideration for small ρ, $||R|| = O(\rho)$ so for β fixed and ρ small enough ρ_2 has isolated eigenvalues λ_ν satisfying $\lim_{\rho\to 0} \frac{\lambda_\nu}{\Sigma} = e^{-\beta E_\nu}$ (recall that E_ν are the eigenvalues of $H[0]$).

Proof of proposition 3.4 (*High Density Limit for the Perfect Gas Model*)
Our proof of the existence of eigenvalues at high density for the perfect gas model is based on the variational principle. In view of Propositions 3.1 and 3.2, we know that the discrete spectrum is not empty if we can find $\phi \in L^2(R^3)$, $||\phi||_2 = 1$ such that

$$\frac{(\phi, \rho_2\phi)}{\Sigma} > 1 \tag{4.29}$$

The asymptotic behaviour of the left hand side in (4.29) can be computed exactly as $\rho \to \infty$. For a sufficiently smooth normalized function ϕ one finds

$$\lim_{\rho\to\infty} \frac{(\phi, \rho_2\phi)}{\Sigma} = \int dx\, |\phi(x)|^2 e^{-\beta V(x)} \tag{4.30}$$

Thus if we choose ϕ not vanishing on the support of V (4.29) will hold. We remark that it will hold even if V is replaced by λV, $0 < \lambda << 1$, thus the perfect gas model supports bound states even if V does not.

The computations involve the application of Laplace method to functional integrals of the form

$$I(\rho) = \int dx \frac{e^{-\frac{x^2}{2\beta}}}{(2\pi\beta)^{3/2}} \int D\alpha\, g(x,\alpha) \exp\left[\rho \int dr \left(e^{-\beta \int_0^1 ds V(\alpha(s)-r)} - 1\right)\right] \tag{4.31}$$

for appropriate functionals $g(x, \alpha)$. It can be shown that the functional in the bracket of the exponential has a non degenerate absolute maximum at $x = \alpha = 0$. Expanding to second order around this maximum leads to the computation of Gaussian integrals with the modified brownian bridge measure

$$D\alpha \exp\left[-\frac{\gamma^2}{2}\int_0^1 du \int_0^1 dv(\delta(u-v)-1)\alpha(u)\alpha(v)\right] \tag{4.32}$$

where γ is proportional to ρ. It turns out that this measure is intimately connected to the Schroedinger operator of a free particle in a uniform magnetic field and consequently the covariance can be explicitely calculated.

Proof of proposition (3.5) (*High Density Limit for the Cell Model of a Fluid*) The proof of the absence of eigenvalue at high density rests on the Birman-Schwinger principle in the form : ρ_2 has eigenvalue λ if and only if $K(\lambda)$ has eigenvalues -1. We establish that

$$\sup_{\lambda \geq \Sigma} ||K(\lambda)|| < 1 \tag{4.33}$$

where $||.||$ denotes the operator norm. The operator norm could be estimated by the Hilbert-Schmidt norm and then one could make estimates similarly as in the proof of 3.2 but this does not lead to sharp enough inequalities. Instead, we relate $K(\lambda)$ to the kernels of the reduced density matrices by the

Lemma
Let $h(x)$ be the function used in the proof of Proposition 4.1. Then

$$\sup_{\lambda > \Sigma} ||K(\lambda)|| \leq \frac{m_0}{\Sigma} + ||h||_2^2 \frac{\max(m_1, m_2)}{\Sigma} \int dk \, \frac{\rho\tilde{\rho}_1(k)}{\Sigma - \rho\tilde{\rho}_1(k)} \tag{4.34}$$

where

$$m_0 = \sup_x \int dy \, |\rho_T(x,y)| = \sup_y \int dx \, |\rho_T(x,y)| \tag{4.35a}$$

$$m_1 = \sup_y \int dx \, h(x)^{-2} |\rho_T(x,y)| \tag{4.35b}$$

$$m_2 = \sup_x h(x)^{-2} \int dy \, |\rho_T(x,y)| \tag{4.35c}$$

For the proof of the lemma we refer the reader to [7]. Let us now show why it leads to (4.33). From (4.16) and translation invariance, we deduce easily that

$$\int dy \, |\rho_T(x,y)| \leq A\beta e^{\beta \sup_x |V(x)|}(z^{-1} + a)\rho\tilde{\rho}_1(0) \tag{4.36}$$

Thus we see that

$$\frac{m_0}{\Sigma} \leq C(\beta)(z^{-1} + a) \tag{4.37}$$

which can be arbitrarily small for large z and small a.

For $\frac{\max(m_1,m_2)}{\Sigma}$ we have a similar upper bound. The proof however is more complicated. It involves again the inequality (4.16), but also the asymptotic evaluation of functional integrals similar to (4.31) with

$$\frac{\beta^2}{24a}\int dr \Big| \int_0^1 ds \, \nabla V(sx+\sqrt{\beta}\alpha(s)-r)\Big|^2 \tag{4.38}$$

replacing the functional in the exponential of (4.31). This comes about by considering the small a and large z behaviour of (4.7) and (4.8). To leading order the sums over the cells j become Riemann integrals and this leads to (4.38). As for the case of the perfect gas (4.38) has a non degenerate absolute maximum at $x = \alpha = 0$ and the quadratic fluctuations around this point are governed again by the Gaussian measure (4.32).

The same methods permit also to compute the asymptotics of the momentum distribution

$$\frac{\rho\tilde{\rho}_1(k)}{\Sigma} = (1+o(1))\exp\bigg(-\frac{2(3a)^{1/2}}{\sqrt{\beta}W(\beta)}|k|^2\bigg) \tag{4.39}$$

with $(W(\beta))^2 = \frac{1}{3}\int dr \sum_{i=1}^3 |\nabla \frac{\partial}{\partial y_i}V(y)|^2$. Note that to leading order the momentum distribution is a Maxwellian renormalized by the effects of the environment. This yields

$$\int dk \, \frac{\rho\tilde{\rho}_1(k)}{\Sigma - \rho\tilde{\rho}_1(k)} = O(a^{-3/4}) \tag{4.40}$$

Gathering the bounds on $\frac{m_i}{\Sigma}$, $i = 0, 1, 2$ and (4.40) we obtain (4.33). We see that the result holds because for the cell model the density fluctuations are sufficiently reduced, as illustrated by (4.16), (4.37), to overcome the divergence in (4.40).

5 Concluding remarks

Beyond the Saha low density, low temperature regime, there are no generally accepted criteria, to characterize the pressure induced ionization that occurs at high densities. Here we have shown that the study of the spectral properties of the reduced density matrices can provide a precise formulation of the Mott effect at least in some models. This leads to interesting spectral problems which can be treated by the mathematical tools existing for Schroedinger operators, combined with functional integral techniques.

The study of sections III and IV clearly shows that the disapearance of the discrete part of the spectrum (Mott effect) in the cell model is related to a significant reduction of particle fluctuations. On the contrary in the perfect gas model the variance of the number of particle in a given region of space is proportional to the density and for this reason the excess density can take appreciable values leading to the binding of the quantum particle to clusters of classical particles. As a consequence the discrete spectrum is not empty at high density for the perfect gas model.

Finally, we mention that the present problematic can be viewed from an other angle, namely that of disordered systems. One can consider that the quantum particle is in the presence of annealed disorder, i.e. the disorder (the classical gas or fluid) adapts itself to the configurations of the quantum particle. The spectral properties of reduced density matrices of the annealed system have no obvious relationship with those of random Schroedinger operators which model systems with quenched (or frozen) disorder. However, a better understanding of the connections and differences between these two problems would be interesting. Also interesting would be the study of relationships between the spectral analysis presented here and the dynamical properties of systems with annealed disorder. Dynamical properties have been studied in the framework of the time dependent Born-Oppenheimer theory by coupled Schroedinger-Newton algorithms in similar models (see for example [12]).

References

[1] see also R. Peierls, "Surprises in Theoretical Physics", Princeton University Press (1979)

[2] Ch. Fefferman, Rev. Math. Iberoamericana **1**, 1 (1985); Ch. Fefferman, Comm on Pure and Applied Mathematics **67** (1986); For an elementary account of Fefferman analysis, see Ph. A. Maartin, Acta Physica Polonica, **24**, 751 (1993)

[3] J. Conlon, E. Lieb, H. Yau, Comm. Math. Phys. **125**, 153 (1989)

[4] N. Macris, Ph. A. Martin, J. Pulé, Helv. Phys. Acta, **63**, 705 (1990)

[5] N. Macris, Ph. A. Martin, J. Stat. Phys. **60**, 619 (1990)

[6] B. Simon, Bull of the Amer. Math. Soc. Vol. **7**, 447 (1982)

[7] J. L. Lebowitz, N. Macris, Ph. A. Martin, J. Stat. Phys. **67**, 909 (1992)

[8] T. Kato, Perturbation Theory for Linear Operators (Springer-Verlag 1984)

[9] H. L. Cycon, R. G. Froese, W. Kinsch, B. Simon, "Schroedinger Operators with applications to Quantum Mechanics and Global Geometry" (Springer-Verlag 1987)

[10] M. Reed, B. Simon, "Methods of Modern Mathematical Physics", Vol. IV, Analysis of Operators (Academic Press, 1978)

[11] M. Klaus, Helv. Phys. Acta, **55**, 49 (1982)

[12] A. Selloni, P. Carnevali, R. Car, M. Parinello, Phys. Rev. Lett **59**, 823 (1987)

N. Macris, Ph. A. Martin, Institut de Physique Theorique, Ecole Polytechnique Federale de Lausanne, CH-1015, Lausanne

Operator Theory:
Advances and Applications, Vol. 70

Quantum Fluctuations in the Many-Body Problem

A. Verbeure and V.A. Zagrebnov

1 From Micro to Macro [1,2]

1.1 The Micro: Microscopic Dynamical System

Any microscopic dynamical system is a triplet $(\mathcal{A}, \omega, \alpha_t)$ where:

- $\mathcal{A} = \bigcup\limits_{\Lambda} \mathcal{A}_\Lambda$ is the *quasi-local algebra* of observables where Λ are the bounded subsets of $\mathbb{R}^d$ or $\mathbb{Z}^d$, $[\mathcal{A}_{\Lambda'}, \mathcal{A}_{\Lambda''}] = 0$ if $\Lambda' \cap \Lambda'' = \{\emptyset\}$.

- ω is a *state* of $\mathcal{A}$. Denote by τ_x, $x \in \mathbb{R}^d$ or $\mathbb{Z}^d$, the space translation automorphism of the translation over the distance x, i.e., $\tau_x : A \in \mathcal{A}_\Lambda \to \tau_x(A) \in \mathcal{A}_{\Lambda+x}$. We assume that the state ω is *homogeneous*: $\omega \circ \tau_x = \omega$ for all $x \in \mathbb{R}^d$ or $\mathbb{Z}^d$, and *clustering*: $\lim\limits_{|x|\to\infty} (A\tau_x B) = \omega(A)\omega(B)$ for $A, B \in \mathcal{A}$.

- α_t is the *dynamics* described by the set of local Hamiltonians $\{H_\Lambda\}_\Lambda$. As usual we consider α_t as the *norm limit* of the local dynamics: $\alpha_t(A) = \lim\limits_{\Lambda} \exp(it[H_\Lambda, \cdot])(A)$, $A \in \mathcal{A}$, i.e., $\alpha_t : \mathcal{A} \to \overline{\mathcal{A}}$-norm-closure of $\mathcal{A}$. The *local structure* is only conserved for *mean-field* interactions, i.e., $\alpha_t : \mathcal{A}_\Lambda \to \mathcal{A}_\Lambda$.

One assumes, usually, that the space and time translations *commute*: $\tau_x(\alpha_t(\cdot)) = \alpha_t(\tau_x(\cdot))$ and that $\omega \cdot \alpha_t = \omega$ (*time invariance*).

1.2 The Macro I: Weak Law of Large Numbers (WLLN)

Definition: For any $A \in \mathcal{A}$, let us consider the local space-mean $m_\Lambda(A) \equiv |\Lambda|^{-1} \sum\limits_{x\in\Lambda} \tau_x(A)$. Define the mapping $m : \mathcal{A} \to \mathbb{C}$ by the ω-weak limit

$$m(A) = \text{weak-}\lim_{\Lambda} m(A)\,, \tag{1.1}$$

and let $m(\mathcal{A}) = \{m(A) : A \in \mathcal{A}\}$.

Then for the state *homogeneous* and *clustering* (see 1.1) one has:

- $[m(\mathcal{A}), \mathcal{A}] = 0$, i.e., $m(\mathcal{A})$ is a set of *observables at infinity* $Z_\omega^\perp$.
- $m(\mathcal{A})$ is an *abelian algebra* and $m(A) = \omega(A) \cdot \mathbb{1}$, i.e., the algebra $Z_\omega^\perp = \mathbb{C} \cdot \mathbb{1}$ and $\omega|_{m(\mathcal{A})} = \mu$ is a *probability measure.*
- The map $m : \mathcal{A} \to m(\mathcal{A})$ is *not injective* (coarse graining), e.g., $m(\tau_a A) = m(A)$.
- The *macro-dynamics* $\tilde{\alpha}_t m(A) = m(\alpha_t(A))$ induced by the micro-dynamics on $m(\mathcal{A})$ is *trivial:*
$m(\alpha_t(A)) = \omega(\alpha_t(A)) \cdot \mathbb{1} = \omega(A) \cdot \mathbb{1} = m(A)$, i.e., $\tilde{\alpha}_t = id$.

These properties yield that the macro-system I defined by the WLLN (1.1) is the triplet $(m(\mathcal{A}) = \mathbb{C} \cdot \mathbb{1}$, $\mu =$ probability measure, $\tilde{\alpha}_t = id)$.

1.3 The Macro II: Quantum Central Limit (QCL)

Definition: For any $A \in \mathcal{A}_{sa} = \{B \in \mathcal{A} : B = B^*\}$ define the local mapping:

$$F_{k,\Lambda}^{\delta}(A) = \frac{1}{|\Lambda|^{1/2+\delta}} \sum_{x \in \Lambda} (\tau_x(A) - \omega(A)) e^{ikx}, \quad k \in \mathbb{R}^1, \quad \delta \in \mathbb{R}^1; \tag{1.2}$$

i.e., the *local fluctuation operator* for the *mode k*. If $\delta = 0$, then this fluctuation operator is called *normal.*

Central Limit Theorem (CLT) [1,2]

Let

$$\gamma_\omega(d) \equiv \sup_{\Lambda,\Lambda'} \sup_{\substack{A \in \mathcal{A}_\Lambda \\ B \in \mathcal{A}_{\Lambda'}}} \left\{ \frac{\omega(A \cdot B) - \omega(A)\omega(B)}{\|A\| \, \|B\|} ; \; d \le \text{dist}\,(\Lambda, \Lambda') \right\}$$

and

$$\sum_{x \in \mathbb{Z}^d} \gamma_\omega(|x|) < \infty.$$

Then for any $A \in \mathcal{A}_{sa}$

$$\lim_{\Lambda} \omega(e^{itF_\Lambda(A)}) = e^{-\frac{1}{2}t^2 S_\omega(A,A)}, \tag{1.3}$$

where $S_\omega(A, B) = Re \sum_{x \in \mathbb{Z}^d} \omega((A - \omega(A)) \cdot \tau_x(B - \omega(B)))$.

The result (1.3) establishes the meaning of the QCL:

$$\lim_{\Lambda} F_\Lambda(A) = F(A), \tag{1.4}$$

where $F(A)$ is the *normal* fluctuation operator (corresponding to the local operator A) for the *mode* $k = 0$. Now, as in probability theory, we face the problem of the *identification* of the right-hand side of (1.4) as a mathematical object.

The CLT in probability theory says that the limit is again a random variable but of *canonical nature*: a Gaussian random variable. In the QCL the $\{F(A)\}$, $A \in \mathcal{A}_{as}$ can be identified with the *algebra* of the *fluctuation operators* which is also somehow *canonical*.

Let $\mathcal{A}_{sa}$ be considered now as a *vectorspace* with the *symplectic form* $\sigma_\omega(\cdot,\cdot)$ defined by the WLLN:

$$i\sigma_\omega(A,B)\cdot \mathbb{1} = \lim_\Lambda [F_\Lambda(A), F_\Lambda(B)] = \lim_\Lambda \frac{1}{|\Lambda|}\sum_{x\in\Lambda}\tau_x\left(\sum_{y\in\Lambda}[A,\tau_{x-y}B]\right).$$

Consider the *Weyl algebra* $W(\mathcal{A}_{sa},\sigma_\omega)$ generated by the Weyl operators $W(A) = \exp(ib_\omega(A))$, $A \in \mathcal{A}_{sa}$, where $\pi_{\tilde{\omega}}(b_\omega(A))$ is the *Boson field* operator, acting on the *representation space* $\mathcal{H}_{\tilde{\omega}}$ of the *quasi-free state* $\tilde{\omega}$ defined by the form $S_\omega(\cdot,\cdot)$:

$$\tilde{\omega}(W(A)) = e^{-\frac{1}{2}S_\omega(A,A)}. \tag{1.5}$$

The identification (1.3) with (1.5) yields the *identification*: $F(A) = \pi_{\tilde{\omega}}(b_\omega(A))$. Hence, the QCL gives a transition from the *micro system* $(\mathcal{A}_{sa},\omega)$ to the *macro system* of the *fluctuation operators* $(\mathcal{F}_{\tilde{\omega}}(\mathcal{A}_{sa},\sigma_\omega),\tilde{\omega})$, where by $\mathcal{F}_{\tilde{\omega}}(\mathcal{A}_{sa},\sigma_\omega) = \{F(A)\}_{A\in\mathcal{A}_{sa}}$ we denote the *CCR-algebra* on the symplectic space $(\mathcal{A}_{sa},\sigma_\omega)$.

The following properties are consequence of the above construction:

- The map $F : \mathcal{A}_{as} \to \mathcal{F}_{\tilde{\omega}}(\mathcal{A}_{sa},\sigma_\omega)$ is not injective for $k = 0$ (*zero-mode coarse graining*), e.g. $\tilde{\tau}_x(F(A)) \equiv F(\tau_x(A)) = F(A)$; $\tilde{\tau}_x = id$.

- The (non-trivial) *macro-dynamics* $\tilde{\alpha}_t(F(A)) \equiv F(\alpha_t(A))$, induced by the micro-dynamics, exists for the finite-range interactions.

- The macro-dynamics $\tilde{\alpha}_t$ is a quasi-free map:

$$\alpha_t(F(A_1)\dots F(A_n)) = \tilde{\alpha}_t(F(A_1))\dots\tilde{\alpha}_t(F(A_n)).$$

- The $\omega\cdot\alpha_t = \omega$ implies the *$\tilde{\alpha}_t$-invariance* of the *macro state* $\tilde{\omega}$.

- The (α_t,β)-KMS property of the *micro state* ω implies the $(\tilde{\alpha}_t,\beta)$-KMS property of the *macro state* $\tilde{\omega}$.

These properties yield that the macro-system II defined by the QCL (1.3) is the triplet $(\mathcal{F}_{\tilde{\omega}}(\mathcal{A}_{as},\sigma_\omega),\tilde{\omega},\tilde{\alpha}_t)$.

2 Applications: The Luttinger Model

This model describes a one-dimensional fermion system defined by the local Hamiltonian (with the periodic boundary conditions):

$$\begin{aligned} H_L &= \int_{-L/2}^{L/2} dx \left[\psi_1^*(x)\frac{1}{i}\partial_x\psi_1(x) - \psi_2^*(x)\frac{1}{i}\partial_x\psi_2(x) \right] \\ &\quad +2\lambda \int_{-L/2}^{L/2} dx \int_{-L/2}^{L/2} dy\ \psi_1^*(x)\psi_1(x)v(x-y)\psi_2^*(y)\psi_2(y)\,. \end{aligned} \tag{2.1}$$

Here $\{\psi_i^\#(x)\}_{i=1,2}$ are creation/annihilation operators corresponding to the two species of fermions. As in the paper [3] we suppose that the pair interaction between two species of the fermions has a bounded Fourier transform $\tilde{v}(k) = \int_{\mathbb{R}^1} dx\ e^{-ikx}v(x)$.
Remark that the free part H_L^0 of (2.1) is defined and essentially self adjoint on the dense domain $\mathcal{D} = \mathcal{P}(\psi_i^*, i = 1,2)\Omega$, generated by the polynomials $\mathcal{P}(\cdot)$ and the cyclic vacuum Ω of the Fock space $\mathcal{F}$, but H_L^0 is *not semibounded from below* on $\mathcal{D}$.
Denote by the $\omega_{0,L}(\cdot)$ the finite-volume ground-state of H_L^0, which is known to satisfy the inequality [4]:

$$\omega_{0,L}(X^*[H_L^0, X]) \geq 0\,, \quad X \in \mathcal{A}\ \text{(CAR-algebra)}\,. \tag{2.2}$$

Then the solution of (2.2) is uniquely defined by $\omega_{0,L}(a_{1,L}^*(k)a_{1,k}(k)) = \theta(-k)$ and $\omega_{0,L}(a_{2,L}^*(k)a_{2,L}(k)) = \theta(k)$,
where $a_{i,L}(k) = \dfrac{1}{\sqrt{L}}\displaystyle\int_{[-\frac{L}{2},\frac{L}{2}]} dx\ e^{-ikx}\psi_i(x)$. With this state one can construct the GNS representation $\pi_{0,L} : \mathcal{A} \to \mathcal{A}_{\omega_{0,L}}$ in the Hilbert space $\mathcal{H}_{\omega_{0,L}}$ with the cyclic vector $\Omega_{0,L}$ such that $\omega_{0,L}(A \cdot B) = (\Omega_{0,L}, \pi_{0,L}(A)\pi_{0,L}(B)\Omega_{0,L})$, $A, B \in \mathcal{A}$. Hence, $\pi_{0,L}(\cdot)$ is a Fock/anti-Fock representation of CAR, i.e., $\Omega_{0,L}$ is a vacuum for particles
$\pi_{0,L}(a_{1,L}^*(k<0))\Omega_{0,L} = \pi_{0,L}(a_{2,L}^*(k>0))\Omega_{0,L} = 0$ (*filled Fermi sea*).

It is this representation which was exploited in [3] to get a "bosonization" of the Luttinger model: $[\pi_{0,L}(\rho_{i,L}(p)), \pi_{0,L}(\rho_{i,L}^*(p'))] = \left(-\dfrac{Lp}{2\pi}\right)\delta_{p,p'}$, $i = 1,2$. These commutators say that in the Fock/anti-Fock representation $\pi_{0,L}(\cdot)$ (filled Fermi sea) the *density operators* for the *modes* $p \in L^* = \left\{\dfrac{2\pi}{L} \times \mathbb{Z}\right\}$

$$\rho_{i,L}(p) = \int_{[-\frac{L}{2},\frac{L}{2}]} dx\ e^{ipx}\psi_i^*(x)\psi_i(x)$$

behave as *bosons.* (Remark that in the original Fock space $(\mathcal{F},\Omega)$ they *commute.*) It is this property, together with the *Kroning's Identity* for H_L^0, which makes the Luttinger model soluble in the finite volume [3].

For the infinite volume one has to reexamine the model starting on the ground state which is unknown on the global algebra $\mathcal{A}$ even for $L < \infty$ [5].

Theorem 2.1. Let the density fluctuation operators be defined as

$$\begin{aligned} F_L^q(N_i) &= \sqrt{\frac{2\pi}{|q|}}\frac{1}{\sqrt{L}}\int_{-L/2}^{L/2} dx\, e^{iqx}(\tau_x N_i(f) - \omega_0(N_i(f)))\,, \\ N_i(f) &= \psi_i^*(f)\psi_i(f)\,, \quad f \in \mathcal{F}(\mathbb{R}) \end{aligned}$$

where $\omega_0(\cdot) = \lim_L \omega_{0,L}(\cdot)$ is the ground-state for the gauge-invariant subalgebra $\mathcal{A}_e \subset \mathcal{A}$. Then for all $\lambda, \mu \in \mathbb{C}$ and $q > 0$ we have (CLT):

$$\begin{aligned} &\lim_L \omega_0(\exp i\{\lambda F_L^q(N_1)^* + \bar{\lambda} F_L^q(N_1) + \mu F_L^q(N_2) + \bar{\mu} F_L^q(N_2)^*\}) \\ &\qquad = e^{-\frac{1}{2}(|\lambda|^2+|\mu|^2)}\,. \end{aligned} \tag{2.3}$$

By means of the ω_0-CLT (2.3) we realized a transition from the *micro* system $\mathcal{A}_e$ and the *micro* state ω_0 to the *one-mode macro algebra* $\tilde{\mathcal{A}}_q$ of the fluctuations realizing the representation of CCR generated by the bosons $\alpha^{\#}(q) \equiv \lim_L F_L^q(N_1)^{\#}$, $\beta^{\#}(q) \equiv \lim_L F_L^q(N_2)^{\#}$ in the state $\tilde{\omega}_0 : \tilde{\omega}_0(\exp i\{\lambda\alpha(q) + \bar{\lambda}\alpha^*(q) + \mu\beta(q) + \bar{\mu}\beta^*(q)\}) = \exp\left[-\frac{1}{2}(|\lambda|^2 + |\mu|^2)\right]$. Unperterbed *micro dynamics* α_t^0 of the density N_1 (the similar for N_2) is defined by the generator $\delta_0 : \delta_0(N_1(x)) = \lim_L \pi_{0,L}([H_L^0, N_1(x)]) = \pi_0(i\partial_x N_1(x))$. Then the evolution of the fluctuation operator is defined by the generator $\tilde{\delta}_0 : \tilde{\delta}_0(\alpha^*(q)) = \lim_L F_L^q(\delta_0 N_1) = q\alpha^*(q)$ and analogously for $\alpha(q)$ and for $\beta^{\#}(q)$. As a consequence of $\omega_0 \circ \alpha_t^0 = \omega_0$, the *macro-state* $\tilde{\omega}_0$ is *$\tilde{\alpha}_t^0$-invariant*, where $\tilde{\alpha}_t^0(\cdot) = \exp it\tilde{\delta}_0(\cdot)$, $\tilde{\delta}_0(\cdot) = [\tilde{H}_{0,q}, \cdot]$ and $\tilde{H}_{0,q} = q(\alpha^*(q)\alpha(q) + \beta^*(q)\beta(q))$, $q \geq 0$. Hence, on the *macro* level of the CCR algebras $\{\tilde{\mathcal{A}}_q\}_q$ the *free fermion* Hamiltonian H_L^0 is represented by a *free boson* one, i.e., we get the *Kroning's identity* in the *thermodynamic limit* [5].

After the ω_0-QCL lifting to the *macro*: $(\mathcal{A}_e, \omega_0, \alpha_t^0) \to (\tilde{a}_q, \tilde{\omega}_0, \tilde{\alpha}_t^0)$, $q > 0$, we can consider the perturbation of H_0^L by the interaction (2.1): $\delta(N_i) = (\delta_0 + \delta_1)(N_i) = \lim_L \pi_{0,L}([H_L, N_i])$.

Theorem 2.2. The perturbed *micro-dynamics* $\alpha_t(\cdot) = \exp it\delta(\cdot)$ induces on the CCR $\mathcal{A}_{q>0}$ the *macro-dynamics* $\tilde{\alpha}_t(\cdot)$ generated by $\tilde{\delta}$:

$$\begin{aligned} \tilde{\delta}\alpha^*(q) &= \lim_L F_L^q(\pi_{0,L}([H_L, N_1])) = q\left(\alpha^*(q) + \frac{\lambda}{\pi}\tilde{v}(q)\beta(q)\right)\,; \\ \tilde{\delta}\beta(q) &= \lim_L F_L^q(\pi_{0,L}([H_L, N_2])) = -q\left(\beta(q) + \frac{\lambda}{\pi}\tilde{v}(q)\alpha^*(q)\right)\,. \end{aligned} \tag{2.4}$$

Hence, $\tilde{\delta}(\cdot) = [\tilde{H}_q, \cdot]$, where $\tilde{H}_q = \tilde{H}_{0,q} + \frac{1}{\pi}\tilde{v}(q)(\tilde{\alpha}^*(q)\tilde{\beta}^*(q) + \alpha(q)\beta(q))$, defines a *quasi-free* evolution $\tilde{\alpha}_t$ on $\tilde{a}_2$.

Theorem 2.3. The ground state equation (2.2) for the *macro-system* $(\tilde{\mathcal{A}}_q, \tilde{\omega}, \tilde{\alpha}_t)$ has the solution $\tilde{\omega}(\cdot) = \tilde{\omega}_0(U \cdot U^{-1})$, where $U = \exp[\varphi(q)(\alpha(q)\beta(q) - \alpha^*(q)\beta^*(q))]$, $\varphi(q) = \frac{1}{2}\text{arcth}\left(-\frac{\lambda}{\pi}\tilde{v}(q)\right)$. The transformation U diagonalizes $\tilde{H}_q$ which leads to the well-known spectrum $\omega(q) = |q|\left\{1 - \left(\frac{\lambda}{\pi}\tilde{v}(q)\right)\right\}^{1/2}$ [3]. This spectrum has the sense of the collective *zero-sound*-type excitations in the Fermi-system (2.1) [5].

References

[1] D. Goderis, A. Verbeure, P. Vets: Prob. Theory Related Fields **82**, 527 (1989).

[2] D. Goderis, A. Verbeure, P. Vets: Commun. Math. Phys. **128**, 533 (1990).

[3] D.C. Mattis, E.H. Lieb: J. Math. Phys. **6**, 306 (1965).

[4] M. Fannes, A. Verbeure: Commun. Math. Phys. **55**, 125 (1977); **57**, 165 (1977).

[5] A. Verbeure, V.A. Zagrebnov: J. Math. Phys. **34**, 785 (1993).

A. Verbeure, Instituut voor Theoretische Fysica, K.U.Leuven
Celestijnenlaan 200 D, B-3001 Leuven, Belgium

V.A. Zagrebnov, On leave of absence from Laboratory of Theoretical Physics, JINR, Dubna 141980, CIS-Russia

Operator Theory:
Advances and Applications, Vol. 70

General Hamiltonians and Model Hamiltonians of the Theory of Superconductivity and Superfluidity in the Hilbert Space of Translation-Invariant Functions

D. Ya. Petrina

In many exactly solvable models of quantum statistical mechanics, an interaction Hamiltonian is equal to an integral of a product of operators of annihilation and creation and a potential divided by a volume of a system raised to a certain power. For example, the interaction Hamiltonian of the BCS model of superconductivity has the form

$$H_{\mathrm{I,M}} = \frac{g}{2V} \iiiint \Phi(x_1 - x_2, x_1' - x_2')\psi^+(x_1)\psi^+(x_2)\psi(x_1')\psi(x_2')dx_1dx_2dx_1'dx_2', \tag{1}$$

where integration is carried out over the whole three-dimensional Euclidean space $\mathbb{R}^3$, V is the volume of $\mathbb{R}^3$, ψ^+ and ψ^- are operators of creation and annihilation of Fermi particles, and Φ is a potential that satisfies the conditions

$$\begin{gathered} \overline{\Phi(x_1 - x_2, x_1' - x_2')} = \Phi(x_2' - x_1', x_2 - x_1), \\ \Phi(x_1 - x_2, x_1' - x_2') = -\Phi(x_2 - x_1, x_1' - x_2') = \Phi(x_2 - x_1, x_2' - x_1') \end{gathered} \tag{2}$$

and is a test function with respect to $x_1 - x_2$ and $x_1' - x_2'$. The dependence on spin is omitted here.

The interaction Hamiltonian of Bogolyubov's theory of superfluidity contains terms of the following form:

$$\begin{aligned} H_{\mathrm{I,M}} = {} & \frac{1}{2V^2} \int a^+(x_1)a^+(x_2)\Phi(x_1 - x_2)dx_1dx_2\Big(\int a(x_3)dx_3\Big)\Big(\int a(x_4)dx_4\Big) \\ & + \frac{1}{2V^2}\Big(\int a^+(x_1)dx_1\Big)\Big(\int a^+(x_2)dx_2\Big) \int a(x_3)a(x_4)\Phi(x_3 - x_4)dx_3dx_4 + \cdots \\ & + \frac{(-5)}{2V^3}\Phi(0)\Big(\int a^+(x_1)dx_1\Big)\Big(\int a^+(x_2)dx_2\Big)\Big(\int a(x_3)dx_3\Big)\Big(\int a(x_4)dx_4\Big), \end{aligned} \tag{3}$$

where a^+ and a are operators of creation and annihilation of particles, respectively, and Φ is a potential which ensures selfadjointness of both general and model (3) Hamiltonians and is absolutely integrable.

Note that integration in (1) and (3) is carried out over the whole $\mathbb{R}^3, V = V(\mathbb{R}^3) = \infty$. We shall understand expressions (1) and (3) as limits

$$\begin{aligned}
&\lim_{\substack{V(\Lambda)\to\infty \\ (\Lambda\nearrow\mathbb{R}^3)}} \frac{1}{V(\Lambda)} \int_\Lambda\int_\Lambda\int_\Lambda\int_\Lambda \Phi(x_1 - x_2, x_1' - x_2')\psi^+(x_1)\psi^+(x_2) \\
&\qquad\qquad\qquad\qquad \times\psi(x_1')\psi(x_2')dx_1dx_2dx_1'dx_2', \\
&\lim_{\substack{V(\Lambda)\to\infty \\ (\Lambda\nearrow\mathbb{R}^3)}} \frac{1}{V^2(\Lambda)} \int_\Lambda\int_\Lambda a^+(x_1)a^+(x_2)\Phi(x_1 - x_2)dx_1dx_2 \\
&\qquad\qquad\qquad\qquad \times(\int_\Lambda a(x_3)dx_3)(\int_\Lambda a(x_4)dx_4), \\
&\ldots\ldots\ldots\ldots\ldots\ldots\ldots\ldots\ldots\ldots ,
\end{aligned} \tag{4}$$

where Λ is a cube with edge $L, V(\Lambda) = L^3$, centered at the origin of coordinates. This paper is devoted to clarifying in what sense these limits exist.

States of infinite systems for certain densities and temperatures are described by sequences of Green's functions which are equal to the statistical average of a product of Heisenberg operators of creation and annihilation. Equations for these are derived in a standard way; these equations also contain integral operators with inverse powers of volume. Thus, the equations for Green's functions of the BCS model contain the following term

$$\begin{aligned}
\frac{g}{2V}\iiint \Phi(x_1 - y, y_1' - y_2')G_{m+1n+1}(t_1, y_1', t_1, y_2', t_2, x_2, \ldots, t_m, x_m; \\
t_1, y, t_{m+1}, x_{m+1}, \ldots, t_{m+n}, x_{m+n})dydy_1'dy_2'.
\end{aligned} \tag{5}$$

The equations for the superfluidity model contain terms of the following form

$$\begin{aligned}
\frac{1}{V^2}\int \Phi(x_1 - y_1)G_{m+1n+1}(t_1, y_2, t_1, y_3, t_2, x_2, \ldots, t_m, x_m; \\
t_1, y_1, t_{m+1}, x_{m+1}, \ldots, t_{m+n}, x_{m+n})dy_1dy_2dy_3 + \ldots \\
+\frac{-5}{V^3}\int G_{m+1n+1}(t_1, y_2, t_1, y_3, t_2, x_2, \ldots, t_m, x_m; \\
t_1, y_1, t_{m+1}, x_{m+1}, \ldots, t_{m+n}, x_{m+n})dy_1dy_2dy_3.
\end{aligned} \tag{6}$$

Here G_{m+1n+1} is a Green's function for $m+1$ annihilation operators and $n+1$ creation operators.

If a finite system of N particles is considered in the whole space $\mathbb{R}^3$ ($V = \infty$), its state is described by eigenfunctions of the Schrödinger operator, which also contain integral operators with inverse powers of volume. Thus, for the BCS model, the Schrödinger equation for N particles contains the following term

$$\sum_{i<j=1}^{N} \frac{g}{V}\int \Phi(x_j - x_i, x_1' - x_2')\{f_N(x_1, \ldots, x_N)|_{x_i=x_1', x_j=x_2'}\}dx_1'dx_2', \tag{7}$$

where f_N is a wave function. From the viewpoint of statistical mechanics, f_N describes a state of a system with density and temperature zero.

Thus, when investigating states of model systems of quantum statistical mechanics, we arrive at the problem of attaching sense to integral operators of the form (5)–(7) which contain inverse powers of the volume of a system when $V = \infty$ ($\Lambda = \mathbb{R}^3$) according to (4). We shall consider these operators in specific spaces of translation-invariant functions and attach a rigorous mathematical meaning to them in these spaces. Let us proceed now to definition of the spaces.

Consider a translation-invariant function of N variables $f_N(x_1, \ldots, x_N) = f_N(x_1 + a, \ldots, x_N + a)$. It depends on $N - 1$ difference variables $\xi_1, \ldots, \xi_N$. We assume that f_N is square integrable with respect to difference variables. The set of these functions forms a Hilbert space; we denote it by h_N. The scalar product of $f_N \in h_N$ and $g_N \in h_N$ is introduced according to the formula

$$\begin{aligned}(f_N, g_N) &= \int \overline{f_N(\xi_1, \ldots, \xi_{N-1})} g_N(\xi_1, \ldots, \xi_{N-1}) d\xi_1 \ldots d\xi_{N-1} \\ &= \lim_{\substack{V(\Lambda) \to \infty \\ (\Lambda \nearrow \mathbb{R}^3)}} \frac{1}{V(\Lambda)} \int_{\Lambda^N} f_N(x_1, \ldots, x_N) g_N(x_1, \ldots, x_N) dx_1 \ldots dx_N;\end{aligned} \tag{8}$$

it does not depend on the choice of difference variables.

Decompose N points into k subsets (clusters) $\{n_1\} = \{x_{i_1}, \ldots, x_{i_{n_1}}\}, \ldots, \{n_k\} = \{x_{j_1}, \ldots, x_{j_{k_1}}\}$, $n_1 + \ldots + n_k = N$ and denote this decomposition by $\sigma_{k,N}$. With these subsets, we associate translation-invariant functions $f_{n_1}(x_{i_1}, \ldots, x_{i_{n_1}}) \in h_{n_1}, \ldots, f_{n_k}(x_{j_1}, \ldots, x_{j_{n_k}}) \in h_{n_k}$, functions f_N of the following form

$$f_N = \sum_{\sigma_{k,N}} f_{n_1}(x_{i_1}, \ldots, x_{i_{n_1}}) \ldots f_{n_k}(x_{j_1}, \ldots, x_{j_{n_k}}), \tag{9}$$

and their linear combinations. By virtue of linear independence of products $f_{n_1} \ldots f_{n_k}$ which correspond to different decompositions σ, the function f_N belongs to the space

$$f_N \in \sum_{\sigma_{k,N}} \oplus h_{n_1} \otimes \ldots \otimes h_{n_k} = h_N^T. \tag{10}$$

Introduce one more space

$$h^T = \sum_N \oplus h_N^T. \tag{11}$$

In this space, we can select subspaces $h_{N,B}$ and $h_{N,F}$ of symmetric and antisymmetric functions, respectively.

For investigation of equations for Green's functions, we need spaces of functions described above but absolutely integrable.

Theorem 1. *The BCS interaction Hamiltonian acts in the space h_N^T as follows*

$$(H_{\mathrm{I,M}}f)_N = \sum_{\sigma_{k,N}} \sum_{i=1}^{k} (\mathrm{I} \otimes \ldots H_{\mathrm{I}n_i} \otimes \ldots \otimes \mathrm{I}), \tag{12}$$

$H_{\mathrm{I}n_i} = 0$ *for* $n_i > 2$, $H_{\mathrm{I}n_i} = H_{\mathrm{I}2}$ *for* $n_i = 2$,

$$(H_{\mathrm{I}2}f_2)(x_1, x_2) = \int \Phi(x_1 - x_2, x_1' - x_2') f_2(x_1' - x_2') d(x_1' - x_2'). \tag{13}$$

Note that decompositions $\sigma_{k,N}$ in the BCS model are such that $n_i \geq 2$. In a standard Fock space, we have $H_{\mathrm{I}} = 0$.

In the momentum space, general interaction Hamiltonians H_{I} usually contain one δ-function, which expresses the momentum conservation law. In all exactly solvable models, additional δ-functions are inserted in interaction Hamiltonians. Here arises the problem in what sense the model BCS Hamiltonian $H_{\mathrm{I,M}}$ approximates the general Hamiltonian H.

Theorem 2. *The space of pairs*

$$h_{\mathrm{P},N} = h_2 \otimes \ldots \otimes h_2, h_{\mathrm{P}} = \sum_N \oplus h_{\mathrm{P},N}$$

is invariant with respect to $H_{\mathrm{I,M}}$; it is not invariant with respect to H_{I}, but the following equality holds

$$H_{\mathrm{I}} h_{\mathrm{P}} = P_{h_{\mathrm{P}}} H_{\mathrm{I}} h_{\mathrm{P}},$$

where $P_{h_{\mathrm{P}}}$ is a projector onto h_{P}.

For models of superfluidity, decompositions σ_k are such that $n_i \geq 1$. The space

$$h_{\mathrm{C}} = \sum_N \oplus \underbrace{h_1 \otimes \ldots \otimes h_1}_{N}$$

is called the space of condensates.

Theorem 3. *The space of pairs and condensates $h_{\mathrm{C,P}} = h_{\mathrm{C}} \otimes h_{\mathrm{P}}$ is invariant with respect to $H_{\mathrm{I,M}}$; it is not invariant with respect to H_{I}, but the following equality holds*

$$H_{\mathrm{I,M}} h_{\mathrm{C,P}} = P_{h_{\mathrm{C,P}}} H_{\mathrm{I}h_{\mathrm{C,P}}},$$

where $P_{h_{\mathrm{C,P}}}$ is a projector onto $h_{\mathrm{C,P}}$.

Also, the general Hamiltonian in the space h_N^T is investigated.

Theorem 4. *The general Hamiltonian is defined on an everywhere dense set h_N^T. It consists of the sum of a selfadjoint operator and a nilpotent one, but its spectrum is real and consists of the union with respect to $\sigma_{k,N}$ of a sum of spectra of Hamiltonians in h_{n_i}. The spectrum is defined by a component of the eigenvector which corresponds to the lowest decomposition.*

It follows from Theorems 2 and 3 that spectra of general and model Hamiltonians of BCS and superfluidity models coincide for eigenvectors with lowest decompositions in h_{P} and $h_{\mathrm{C,P}}$, respectively. Consequently, the Hamiltonian is a model one in the sense that only this spectrum is taken into account.

Investigating equations for Green's functions in spaces of translation-invariant functions, we obtain the following result.

Theorem 5. *Solutions of equations for Green's functions of BCS and superfluidity models coincide with solutions of equations for Green's functions of systems with corresponding approximating Hamiltonian.*

References

[1] D. Ya. Petrina, Theoret. Math. Phys. **4** (1970), 394.

[2] D. Ya. Petrina and V. P. Yatsishin, Theoret. Math. Phys. **10** (1972), 283.

[3] D. Ya. Petrina, *Exactly Solvable Models of Quantum Statistical Mechanics*, Preprint Dipartimento di Matematica Politecnico di Torino No. 18 / 1992, 115 p.

D. Ya. Petrina, Institute of Mathematics, Kiev, Ukraine

Operator Theory:
Advances and Applications, Vol. 70

Poisson Field Representations in the Statistical Mechanics of Continuous Systems

Roman Gielerak* Alexei L. Rebenko†

Abstract

The grand canonical Gibbs ensembles describing continuous systems of particles are represented as functional integrals with respect to some infinitely divisible generalized random fields. Several applications like the convergence of cluster expansions for a wide class of interactions and the existence of the limiting Gibbs measures are presented.

1 Introduction

The statistical mechanics describing continuous systems of particles is still in a very incomplete state. Mathematical results treating the low temperature/high density behaviour of both classical and quantum continuous systems are very rare. This is to be contrasted with the fact that the most important physical phenomena (see i.e. [1,2]) take place in such systems. One of the main obstacles in the way to controling these systems seems to be the very complicated notions of the corresponding configuration space and Gibbs distributions [3,4]. Although the high temperature/low density region is reasonably well understood [3,4] for such systems a new formalism needs to be worked out in order to study the corresponding low temperature/high density phenomena. It is the main aim of the present contribution to present a new mathematical formalism based on the use of the generalized random fields of the Poisson type of the space of distributions $D'(R^d)$. The basic objects of the equilibrium statistical mechanics of continuous systems are represented as Poisson–like integrals on the space $D'(R^d)$. This creates a uniform picture for both classical and quantum systems, see below. The opportunity of using methods of constructive quantum field theory arises naturally. We shall present this new formalism below and also some preliminary applications. For details and complete proofs we refer to our forthcoming paper [5].

It is worthwihle to mention that Poisson–like integrals appeared in the literature in the context of quantum dynamics problems. For example they have been

*Research supported by the Polish National Committee of Science
†Research supported by Ukrainian State Committee of Science and Technology

used to study Feynman path integrals in [6], to construct examples of nonlinear electrodynamic models in [7] and such integrals have been considered by Klauder in his search for alternatives to the conventional perturbative scheme in quantum field theory [8].

2 Poisson integral representation in the classical statistical mechanics

Let us consider a system of classical particles enclosed in the bounded region $\Lambda \subset R^d$ and interacting throughout the two–body forces described by the potential V. The configuration space for the system is the set $C(\Lambda)$ defined as a subset of 2^Λ consisting of those $\underline{x} \in \bigcup_{n\geq 0} \Lambda^{\otimes n}$ such that $\#\underline{x} < \infty$. The natural topology τ_f in $C(\Lambda)$ is defined by component–wise convergence. The corresponding Borel–σ algebra is denoted as $\mathcal{B}(C(\Lambda))$. The (normalized) Poisson distribution $\Pi^z_{\Lambda,0}$ on $\mathcal{B}(C(\Lambda))$ (see [3,4]) describes the system without an interaction. Let us denote by j_Λ the injection

$$j_\Lambda : \qquad C(\Lambda) \ni \underline{x} = (x_1, \ldots, x_n) \longrightarrow j_\Lambda(\underline{x}) = \sum_{i=1}^{n} \delta(x - x_i) \in D'(\Lambda) \tag{1}$$

It is easy to check that j_Λ is $(\beta(D'(\Lambda)),\ \mathcal{B}(C(\Lambda))$ measurable, therefore we can transport the measure $\Pi^z_{\Lambda,0}$ onto the space $D'(\Lambda)$ by the formula:

$$\int_{D'(\Lambda)} P^z_{\Lambda,0}(dq)F(q) \equiv \int_{C(\Lambda)} \Pi^z_{\Lambda,0}(\underline{x})(F \cdot j_\Lambda)(\underline{x}) \tag{2}$$

where $P^z_{\Lambda,0} \equiv j_\Lambda \Pi^z_{\Lambda,0}$, for any $L_1(D'(\Lambda)),\ P^z_{\Lambda,0}$ function F.

In particular taking $F(q) = e^{i<q,f>}$, for $f \in C_0^\infty(\Lambda)$ we obtain the characteristic functional of $P^z_{\Lambda,0}$:

$$\int_{D'(\Lambda)} dP^z_{\Lambda,0}(q) e^{i(q,f)} \equiv e^{z\int_\Lambda dx(e^{if(x)}-1)} \tag{3}$$

in which we recognize the characteristic functional of some infinitely divisible random field on $D'(\Lambda)$ [9]. It follows from the very construction of $P^z_{\Lambda,0}$ that the set $\Sigma(\Lambda) = \{\sum_{i=1}^n \delta(x - x_i),\ \ n < \infty \text{ and } x_i \neq x_j \text{ for } i \neq j\} \subset D'(\Lambda)$ is measurable and of measure $P^z_{\Lambda,0}$ equal to one.

Assume that the two–body V is translation invariant, stable and such that $V(0) < \infty$. Then we can define the functional $|e^{-\beta/2V_\Lambda(q,q)}|$ on $\Sigma(\Lambda)$ by the formula

$$|e^{-\beta/2V_\Lambda(q,q)}| = e^{-\beta/2\int_\Lambda\int_\Lambda dxq(x)V(x-y)q(y)} e^{\beta/2V(0)\#q} \tag{4}$$

Calculating

$$\int_{D'(\Lambda)} dP^z_{\Lambda,0}(q)|e^{-\beta/2V_\Lambda(q,q)}| = \sum_{n\geq 0}\frac{z^n}{n!}\int_{\Lambda^{\otimes n}} dx_1^n \exp -\beta \sum_{1\leq i<j\leq n} V(x_i-x_j)$$
$$= Z^{cl}_\Lambda(z,\beta) \tag{5}$$

we obtain the integral representation for the partition function describing the grand canonical Gibbs ensemble. By a similar calculation we obtain the following Poisson integral representations for the corresponding (reduced) correlation functions

$$\hat{\rho}^n_\Lambda(z,\beta)(x_1^n) = \exp\beta \sum_{1\leq i<j\leq n} V(x_i-x_j)\rho^n_\Lambda(z,\beta)(x_1^n)$$
$$\equiv \int_{D'(\Lambda)} dP^z_\Lambda(q)\prod_{i=1}^n e^{-\beta V_\Lambda(q)(x_i)} \tag{6}$$

where

$$dP^z_\Lambda(q) = \frac{|\exp -\beta/2V_\Lambda(q,q)|dP^z_{\Lambda,0}(q)}{\int_{D'(\Lambda)}|\exp -\beta/2V_\Lambda(q,q)|dP^z_{\Lambda,0}(q)} \tag{7}$$

$$V_\Lambda(q)(x_i) = \int_\Lambda dy q(y)V(x_i-y). \tag{8}$$

Let us denote by Σ_∞ the Borel subset of $D'(R^d)$ consisting of those $q \in D'(R^d)$, such that $q(x) = \sum_{i=1}^\infty \delta(x-x_i)$ where $x_i \neq x_j$ for $i \neq j$ and such that for any bounded $\Lambda \subset R^d$: $\#\{x_i\} \cap \Lambda < \infty$. Any PBC (probabilistic, Borel, cylindric) measure $\mathcal{P}$ on $(D'(R^d),\ \mathcal{B}(D'(R^d))$ such that

(*i*) $\mathcal{P}(\Sigma_\infty) = 1$

(*ii*) $E_\mathcal{P} \circ \hat{E}_\mathcal{P}(F|\tilde{\Sigma}_{\Lambda^C}) = E_\mathcal{P}(F)$

where $F \in L_1(\mathcal{P})$ is $\tilde{\Sigma}(\Lambda)$ measurable and $\Sigma(\Lambda) \equiv \sigma\{(q,f)|f \in C_0^\infty(\Lambda), q \in \Sigma_\infty\}$ is called a P–Gibbs grand canonical measure describing the system at temperature β and for the chemical activity $z > 0$. $\hat{E}_\mathcal{P}(|\tilde{\Sigma}_{\Lambda^C})$ means the following probabilistic kernel:

$$E_\mathcal{P}(F|\tilde{\Sigma}_{\Lambda^C})(q^*_{\Lambda^C}) = (Z_\Lambda^{q^*_{\Lambda^C}}(z,\ beta))^{-1}$$
$$\int_{D'(\Lambda)} \exp\beta/2V_\Lambda(q,q^*_{\Lambda^C})|\exp -\beta/2V_\Lambda(q,q)|F(q) \tag{9}$$

The set of P–Gibbs measures will be denoted by $\mathcal{PG}(z,\beta)$.

Proposition 2.1

There is a bijection $\hat{j}$ between the set $\mathcal{PG}(z,\beta)$ and the set of the corresponding grand canonical Gibbs ensembles $\mathcal{G}(z,\beta)$ on $(C(R^d),\mathcal{B}(C(R^d)))$. Moreover the bijection $\hat{j}$ restricted to the Martin–Dynkin boundary $\partial\mathcal{PG}(z,\beta)$ of the set $\mathcal{PG}(z,\beta)$ is still a bijection between $\partial\mathcal{PG}(z,\beta)$ and $\partial\mathcal{G}(z,\beta)$.

Informally, the bijection $\hat{j}$ is given by transport of the measurable structures induced by the injection j_∞,

The unphysical assumption $V(0) < \infty$ can be relaxed. For this goal, let us assume that $V \in L^1_{loc}(R^d)$ and let $(\kappa_\epsilon) \in C_0^\infty(R^d)$ be a positive mollifier. The regularized potential V^ϵ is given by $V^\epsilon = (\kappa_\epsilon * V)$. Then it is possible to prove:

Proposition 2.2
Let $V \in L^1_{loc}(R^d)$, V is stable. Then for any $p \geq 1$ any bounded $\Lambda \subset R^d$, the limit

$$\lim_{\epsilon \downarrow 0} |e^{-\beta/2V^\epsilon_\Lambda(q,q)}| \equiv |e^{-\beta/2V_\Lambda(q,q)}| \tag{10}$$

exists in $L_p(dP^z_{\Lambda,0})$ space. Moreover there exists a measure dP^z_Λ defined as a weak limit of $dP^{z,\epsilon}_\Lambda = dP^z_\Lambda(V^\epsilon)$. The measure dP^z_Λ is locally absolutely continuous with respect to $dP^z_{\Lambda,0}$ In particular the integral formulae (5) and (6) are valid in the limit.

The statistical independence of the field $dP^z_{\Lambda,0}$ localized in disjoint regions Λ_1, Λ_2 significantly simplifies the construction of the cluster expansion for the measure dP^z_Λ (see [10,11]). Using Proposition (2.2) the following theorem, generalizing slighty the main result of [10] is proven in [5].

Theorem 2.3
Let $V \in L^1_{loc}(R^d)$ be a stable potential and such that $V \in L_1(R^d)$. Then there exists a region $G \subset \{(z,\beta)|z \geq 0, \quad \beta \geq 0\}$ such that the corresponding cluster expansion for $\hat{\rho}^n$ converges.

Ideas for the proof. We shall start with the regularized potential V^ϵ and supply the correspondig cluster expansion in the finite volume. The assumption $V \in L^1_{loc}(R^d)$ is used to obtain estimates uniform in ϵ (also uniform in Λ) on the radius of convergence of this expansion. Because of uniform convergence we can pass to the $\lim_{\epsilon\downarrow 0}$ term by term thus obtaining the convergent cluster expansion for the limiting case V.

Another application of the Poisson integral representation is given by the following Proposition.

Proposition 2.4
Let V be a superstable, locally $L_1(dx)$, lower regular, translation invariant potential. Then the family of measures $(dP^z_\Lambda)_\Lambda$ is a weakly (pre)–compact set of measures on $(D'(R^d),\ \mathcal{B}(D'(R^d)))$ for any $z \geq 0$ and β.

Idea for the proof. By simple manipulation we obtain

$$\int_{D'(\Lambda)} e^{\xi<q,f>} dP^z_\Lambda(q) = \sum_{i=1}^{\infty} \frac{z^n}{n!} \int_{\Lambda^{\otimes n}} dx_1^n \prod_{i=1}^{n} \left[e^{i\xi f(x_i)} - 1\right] \rho^n_\Lambda(x_1^n) \tag{11}$$

Using the estimates of Ruelle [12] we obtain:

$$\int_{D'(\Lambda)} e^{\xi<q,f>} dP_\Lambda^z(q)| \le \exp|\zeta| z \int_{R^d} |e^{Re\xi f(x)} - 1| dx \tag{12}$$

for some constant ζ.

3 Poisson integral representation in the Quantum Statistical Mechanics

The by now classical work of Ginibre [13,14] uses the Wiener integral representation for the basic objects of quantum statistical systems like partition function, (reduced)–density matrices and multitime Green functions. We shall present below new integral representations of these objects. The mathematical details of our constructions shall be presented in [5].

Let $\Omega_\beta(\Lambda)$ be the space of continuous maps $\omega : [0,\beta] \ni \tau \to \omega(\tau) \in \Lambda$; Λ being bounded region in R^d. Then we define $\Xi_\beta(\Lambda)$ as a subset of $2^{\Omega_\beta(\Lambda)}$ consisting of those $\underline{\omega} \in 2^{\Omega_\beta(\Lambda)}$ such that $\#\underline{\omega} < \infty$. The natural topology τ_f in $\Xi_\beta(\Lambda)$ is introduced by defining the convergence: $\underline{\omega}^n \xrightarrow{\tau_f} \underline{\omega}$ iff for all n, $\#\underline{\omega}^n = \#\underline{\omega}$ and for each i $\omega_i^n \longrightarrow \omega_i$ uniformly on $[0,\beta]$. Let us denote by $dW_{\Lambda,x|x}^\beta$ the conditional Wiener measure on $\Omega_\beta(\Lambda)$,; we then define a measure $dW_\Lambda^\beta(\omega)$ by

$$dW_\Lambda^\beta(\omega) = \int_\Lambda dx \int_{\Omega_\beta(\Lambda)} dW_{\Lambda,x|x}^\beta(\omega) \tag{13}$$

on $\Omega_\beta(\Lambda)$. With the help of the measure dW_Λ^β we can define a Poisson–like distribution $\tilde{P}_{\Lambda,0}^{\beta,z}$ on $(\Xi_\beta(\Lambda); \beta(\Xi_\beta(\Lambda)))$ by the formula

$$\int_{\Xi_\beta(\Lambda)} d\tilde{P}_{\Lambda,0}^{\beta,z}(\underline{\omega}) F(\underline{\omega}) \overset{df}{\equiv} \sum_{n\ge 0} \frac{z^n}{n!} \int dW_\Lambda^\beta(\omega_1^n) F_n(\omega_1^n) \tag{14}$$

for any continuous and bounded F on $\Xi_\beta(\Lambda)$.

Let $\Theta_\beta(\Lambda)$ be the space of continuous maps from $[0,\beta]$ into $D'(\Lambda)$, i.e. $\Theta_\beta(\Lambda) \equiv \{Q | Q : [0,\beta] \ni \tau \longrightarrow Q_\tau \in D'(\Lambda), Q\text{continuous}\}$. As in the classical case we define the injection

$$J_\Lambda : \Xi_\beta(\Lambda) \ni \underline{\omega} = (\omega_1, \dots \omega_n) \longrightarrow \sum_{j=1}^n \delta(x - \omega_i(\tau)) \in \theta_\beta(\Lambda). \tag{15}$$

The image of the Poisson measure $\tilde{P}_{\Lambda,0}^{\beta,z}$ in the space $(\Theta_\beta(\Lambda), \mathcal{B}(\Theta_\beta(\Lambda))$ will be denoted by $\tilde{\Pi}_{\Lambda,0}^{\beta,z}$. Taking

$$F(Q) = e^{i<Q,f>} = e^{i\int_0^\beta <Q_\tau,f> d\tau} \tag{16}$$

we obtain

$$\begin{aligned}\int_{\Theta_\beta(\Lambda)} d\tilde{\Pi}^{\beta,z}_{\Lambda,0}(Q)e^{i<Q,f>} &= \int_{\Xi_\beta(\Lambda)} d\tilde{P}^{\beta,z}_{\Lambda,0}(\underline{\omega})e^{\int_0^\beta d\tau<J_\Lambda(\underline{\omega}),f>} \\ &= \exp(z\int_\Lambda dx \int_{\Omega_\beta(\Lambda)} dW^\beta_{\Lambda,x|x}(\omega)e^{i\int_0^\beta d\tau f(\omega(\tau))}) \quad (17)\end{aligned}$$

Taking V to be a stable and continuous function on R^d and defining

$$F_V(Q) = \exp -\frac{1}{2}\int_0^\beta d\tau < \kappa_\Lambda, V(Q_\tau \otimes Q_\tau)\kappa_\Lambda > \quad (18)$$

where

$$Q_\tau(x) = Q(\tau)(x) \in D'(\Lambda). \quad (19)$$

$$< \kappa_\Lambda, V(Q_\tau \otimes Q_\tau)\kappa_\Lambda >= \int_\Lambda dxdy Q_\tau(x)V(x-y)Q_\tau(y) \quad (20)$$

and

$$|F_V|(Q) = Z_\Lambda \cdot F_V(Q) \quad (21)$$

where

$$Z_\Lambda = \exp \beta/2\#Q \cdot V(0) \quad (22)$$

we obtain:

$$\begin{aligned}&\int_{\Theta_\beta(\Lambda)} d\tilde{\Pi}^{\beta,z}_{\Lambda,0}(Q)|\exp -\frac{1}{2}\int_0^\beta d\tau < \kappa_\Lambda, V(Q_\tau \otimes Q_\tau)\kappa_\Lambda > | = \\ &= \sum_{n\geq 0}\frac{z^n}{n!}\int_{\Omega_\beta(\Lambda)^{\otimes n}} dW^\beta_\Lambda(\omega_1^n)\exp -\frac{1}{2}\int_0^\beta d\tau \sum_{1\leq i<j\leq n} V(\omega_i(\tau)-\omega_j(\tau))\end{aligned} \quad (23)$$

in which we recognize the finite–volume partition function for the system of particles obeying the Maxwell–Boltzmann statistics.

To cover the case of quantum statistics we define for $\epsilon = \pm 1$ the following Poisson–like distributions $dP^{\beta,z}_{\Lambda,\epsilon,0}$ on $\Xi_\beta(\Lambda)$:

$$\begin{aligned}&\int_{\Xi_\beta(\Lambda)} dP^{\beta,z}_{\Lambda,\epsilon,0}(\underline{\omega}))F(\underline{\omega}) = \sum_{n\geq 0}\frac{z^n}{n!}\int_{\Lambda^{\otimes n}} dx_1^n \sum_{\Pi\in S_n}\epsilon^{\#\Pi} \\ &\int_{(\Omega_\beta(\Lambda))^{\otimes n}} \bigotimes_{i=1}^n dW^\beta_{\Lambda,x_i|\Pi(x_i)}(\omega_i)F_n(\omega_1,\ldots\omega_n)\end{aligned} \quad (24)$$

Then we transport $P^{\beta,z}_{\Lambda,\epsilon,0}$ onto the space $\Theta_\beta(\Lambda)$ by the injection J_Λ as above obtaining

$$\int_{\Theta_\beta(\Lambda)} d\Pi^{\beta,z}_{\Lambda,\epsilon,0}(Q)F(Q) = \int_{\Xi_\beta(\Lambda)} dP^{\beta,z}_{\Lambda,\epsilon,0}(\underline{\omega})F(J_\Lambda\underline{\omega}) \tag{25}$$

In particular we have obtained the following representation for the corresponding quantum statistical partition functions:

$$Z^\epsilon_\Lambda(z,\beta) = \int_{\Theta_\beta(\Lambda)} d\Pi^{z,\beta}_{\Lambda,\epsilon,0}(Q)|\exp(-\frac{1}{2}\int_0^\beta d\tau < \kappa_\Lambda, V(Q_\tau \otimes Q_\tau)\kappa_\Lambda >)| \tag{26}$$

The corresponding correlation functionals can be defined by the following formulas

$$\begin{aligned} P^\epsilon_\Lambda(z,\beta)(\omega^n_1) = z^n \exp -\frac{1}{2}\int_0^\beta d\tau \sum_{1\leq i<j\leq n} V(\omega_i(\tau) - \omega_j(\tau)) \\ \cdot \int d\Pi^{z,\beta}_{\Lambda,\epsilon}(Q)\prod_{i=1}^n \exp - \int_0^\beta d\tau V_\Lambda(Q)(\omega_i(\tau)), \end{aligned} \tag{27}$$

where

$$V_\Lambda(Q)(\omega_i(\tau)) = \int_\Lambda dx V(\omega_i(\tau) - x)Q_\tau(x)dx \tag{28}$$

Several applications of the representations (26), including the construction of the cluster expansion, independence of the limiting correlation functionals of the classical boundary conditions etc. will be presented in our forthcoming paper [5].

References

1. Federbush,P. Kennedy,T., Surface effects in Debye screening , Comm. Math. Phys. **43**, 55-97 (1988).

2. Rebenko,A.L., Mathematical Foundations of Equilibrium Classical Statistical Mechanics of charged particles, Russ.Math.Surv. **43**, 55–97 (1980) and refs. therein.

3. Ruelle,D., Statistical mechanics: Rigorous results, New York, Amsterdam: Benjamin 1969.

4. Petrina,D.Ya., Gerasimenko,V.J. Malyshew,P.V. Mathematical Foundation of Classical Statistical Mechanics. Continous System., New York, London, Paris, Gordon and Breach Science, 1989.

5. Gielerak,R., Rebenko,A.L., Poisson Integral representation in the Statistical Mechanics of Continuous Systems, paper in preparation.

6. Ph.Combe, R. Hoegh-Krohn, R. Rodriguez, M. Siruge, M. Siruge-Collin, Comm. Math. Phys. **77**, 269–288.

7. Albeverio,S. Iwata,K. Hoegh-Krohn,R., Random fields as solutions of the inhomogenous quaternionic Cauchy-Riemann equations, Poisson Processes on Groups and Feynman path Integrals, Comm.Math.Phys. **132**, 555-580 (1981).

8. Klauder,J.R., Measures and support in functional integration, pp. 31–56 in "Progress in Quantum Field Theory", Eds: H. Ezawa, S. Kamefuchi, Elsevier Science 1986, and refs, therein

9. Gelfand,J.M., Vilenkin,N.Ya, Generalized functions Vol.4. Applications of harmonic analysis. New York, London: Academic Press 1968.

10. Rebenko, A.L., Poisson Measure representation and Cluster Expansion in Classical Statistical Mechanics, Comm.Math.Phys. **151**, 427-435 (1993).

11. Brydges,D., Fedurbusti,P., The cluster expansion in statistical mechanics, Comm.Math.Phys. **49**, 233–246 (1976)

12. Ruelle,D., Superstable Interactions in Classical Statistical mechanics, Comm. Math. Phys. **18** , 127 (1970)

13. Ginibre,J., Reduced Density Matrices of Quantum Gases, I,II,III: Journ. Math. Phys. **6**(2) 238–251 (1965), **6**(2), 252–262 (1965), **6** 1432–1446 (1965)

14. Ginibre,J., Phase Transitions and Critical Phenomena, vol 1, Domb I.C. and Green M.S. eds pp 11–136, Academic Press, New York 1972.

Author's addresses

Roman Gielerak
Institute of Theoretical Physics
University of Wroclaw
50-205 Wroclaw
Poland

Alexei L. Rebenko
Institute of Mathematics
Acad. Sci. Ukraine
Repin Street
252601 Kiev
Ukraine

Operator Theory:
Advances and Applications, Vol. 70

Exact ground states for quantum spin chains

M. Fannes
Bevoegdverklaard Navorser N.F.W.O. Belgium

It is much harder to construct translation invariant states on a quantum spin system than on a system of classical spins. The reason for this is that non-commutativity greatly complicates positivity conditions. This is already visible in the following toy model: let $\mathcal{H}$ be a finite dimensional Hilbert space and let $\phi \in \mathcal{H} \otimes \mathcal{H}$ be a normalized vector such that for $A \in \mathcal{B}(\mathcal{H})$:

$$\langle \phi, A \otimes \mathbb{1}\, \phi \rangle = \langle \phi, \mathbb{1} \otimes A\, \phi \rangle.$$

This condition expresses translation invariance for a two-site system. Can one construct a vector ϕ_{123} in $\mathcal{H} \otimes \mathcal{H} \otimes \mathcal{H}$ such that $\langle \phi_{123}, \cdot\, \phi_{123} \rangle$ coincides with $\langle \phi_{12}, \cdot\, \phi_{12} \rangle$ on the first two factors and with $\langle \phi_{23}, \cdot\, \phi_{23} \rangle$ on the last two? (ϕ_{12} and ϕ_{23} are copies on sites 12 and 23 of the three spin system.) This is in general impossible. Indeed, let P be the orthogonal projection operator in $\mathcal{B}(\mathcal{H} \otimes \mathcal{H})$ on $\phi^{\perp}$. Such an extension ϕ_{123} would have the property:

$$\langle \phi_{123}, (P_{12} \otimes \mathbb{1}_3 + \mathbb{1}_1 \otimes P_{23})\, \phi_{123} \rangle = 0.$$

Generically however $P_{12} \otimes \mathbb{1}_3$ and $\mathbb{1}_1 \otimes P_{23}$ don't commute and the ground state energy of $P_{12} \otimes \mathbb{1}_3 + \mathbb{1}_1 \otimes P_{23}$ will be strictly larger than the sum of the ground state energies of P_{12} and P_{23}, which is 0. One can show that a common extension of $\langle \phi_{12}, \cdot\, \phi_{12} \rangle$ and $\langle \phi_{23}, \cdot\, \phi_{23} \rangle$ only exists in the almost commutative case where ϕ is an elementary tensor vector: $\phi = \psi \otimes \psi$, $\psi \in \mathcal{H}$.

A more general situation is the following: let $\mathcal{K}$ be a subspace of $\mathcal{H} \otimes \mathcal{H}$. Generically:

$$\mathcal{G}_n = \mathcal{K} \otimes \mathcal{H} \otimes \cdots \mathcal{H} \bigcap \mathcal{H} \otimes \mathcal{K} \otimes \cdots \mathcal{H} \bigcap \cdots \mathcal{H} \otimes \mathcal{H} \otimes \cdots \mathcal{K} \qquad n \text{ factors}$$

will collapse to the zero vector when n is sufficiently large. For particular choices of $\mathcal{K}$ however, the dimension of this intersection will stay positive, independently of n. We can now consider a two-spin interaction P which is the orthogonal projection in $\mathcal{B}(\mathcal{H} \otimes \mathcal{H})$ on $\mathcal{K}^{\perp}$. Any vector Ω in $\mathcal{G}_n$ will be a ground state for the Hamiltonian $H_{[1,n]}$:

$$H_{[1,n]} = P_{12} + P_{23} + \cdots P_{n-1,n}.$$

Indeed, $H_{[1,n]}$ is a positive operator and, by construction, $H_{[1,n]}\Omega = 0$. Such ground states have a very special property, namely $P_{k,k+1}\Omega = 0$ for all $k = 1, \ldots n-1$. Ω therefore locally minimizes the energy. These features are typical of so-called V(alence)B(ond)S(olid)-Hamiltonians.

It is possible to construct translation-invariant states on a quantum spin chain by a transfer-matrix formalism. Let the observables of a single spin be described by the $d \times d$ matrices $\mathcal{M}_d$. The observables of the spins living at the sites $[a, b] \subset \mathbb{Z}$ are then $\mathcal{A}_{[a,b]} = \otimes_a^b \mathcal{M}_d$. The observables $\mathcal{A}_{\mathbb{Z}}$ of the entire chain are obtained by completing the local observables:

$$\mathcal{A}_{\mathbb{Z}} = \overline{\bigcup_{[a,b]\subset\mathbb{Z}} \mathcal{A}_{[a,b]}}^{\|\ \|}.$$

A state ω of $\mathcal{A}_{\mathbb{Z}}$, is completely determined by its expectation values of observables of the form $\omega(j_m(A_m) \otimes j_{m+1}(A_{m+1}) \otimes \cdots j_n(A_n))$, $m < n$ and $A_i \in \mathcal{M}_d$ for $m \le i \le n$. The map j_n is the injection of $\mathcal{M}_d$ at the n-th site. The prescription for computing these expectation values must be compatible with the requirements that:

$$\omega(j_m(A_m) \otimes \cdots j_n(A_n) \otimes j_{n+1}(\mathbb{1})) = \omega(j_m(A_m) \otimes \cdots j_n(A_n)),$$

and:

$$\omega(j_{m-1}(\mathbb{1}) \otimes j_m(A_m) \otimes \cdots j_n(A_n)) = \omega(j_m(A_m) \otimes \cdots j_n(A_n)).$$

The state ω is translation invariant iff for all choices of $m < n \in \mathbb{Z}$ and $A_i \in \mathcal{M}_d$:

$$\omega(j_m(A_m) \otimes \cdots j_n(A_n)) = \omega(j_{m+1}(A_m) \otimes \cdots j_{n+1}(A_n)).$$

If ω is translation invariant, then we can drop the j_n.

A state ω will be called *finitely correlated* if it has the following structure:

$$\omega(j_m(A_m) \otimes j_{m+1}(A_{m+1}) \otimes \cdots j_n(A_n)) = \rho(\mathbb{E}_{A_m} \circ \mathbb{E}_{A_{m+1}} \circ \cdots \mathbb{E}_{A_n}(\mathbb{1}_k)).$$

A number of ingredients of this formula have to be specified:

i) $\mathbb{E}$ is a completely positive, identity preserving map from $\mathcal{M}_d \otimes \mathcal{B}$ to $\mathcal{B}$, where $\mathcal{B}$ is a *-subalgebra of a matrix algebra $\mathcal{M}_k$ that contains the identity $\mathbb{1}_k$. This means that there are (a finite number of) linear maps $V_i : \mathbb{C}^k \to \mathbb{C}^d \otimes \mathbb{C}^k$, such that

$$\mathbb{E}(X) = \sum_i V_i^* X V_i.$$

As $\mathbb{E}$ is unity preserving $\sum_i V_i^* V_i = \mathbb{1}_k$. If $A \in \mathcal{M}_d$, then $\mathbb{E}_A$ denotes the mapping $B \in \mathcal{B} \mapsto \mathbb{E}(A \otimes B) \in \mathcal{B}$. We will also assume that $\mathcal{B}$ is minimal in the sense that $\mathcal{B}$ is the *-subalgebra of $\mathcal{M}_k$ generated by elements of the form $\mathbb{E}_{A_1} \circ \mathbb{E}_{A_2} \circ \cdots \mathbb{E}_{A_n}(\mathbb{1}_k)$, $n = 0, 1, 2, \ldots$ and $A_i \in \mathcal{M}_d$.

ii) ρ is a state on $\mathcal{B}$. If ρ is a density matrix in $\mathcal{M}_k$, we will identify ρ with the functional $X \in \mathcal{M}_k \mapsto \operatorname{Tr} \rho X$. We will assume that ρ is faithful on $\mathcal{B}$, which means that there is no non-zero element $B \in \mathcal{B}$ with $\rho(B^* B) = 0$.

iii) For all $B \in \mathcal{B}$ we have $\rho(\mathbb{E}(\mathbb{1}_d \otimes B)) = \rho(B)$.

We will say that ω is generated by the triple $(\mathcal{B}, \mathbb{E}, \rho)$. Because $\mathbb{E}$ is identity preserving and ρ satisfies the condition iii) the construction of ω is compatible with tensoring identity operators to the left and the right of a local observable. It is also obvious that ω is translation invariant. The positivity can explicitly be checked by computing the restriction of ω to any finite volume. Due to the complete positivity of $\mathbb{E}$ and the positivity of ρ, all such restrictions are given by density matrices and are therefore positive. The positivity requirements on $\mathbb{E}$ and ρ can in principle be relaxed. Doing so however, leads to intractable problems.

The map $\widehat{\mathbb{E}} = \mathbb{E}_{\mathbb{1}}$ plays an essential role in understanding the properties of the state ω. This is because $\widehat{\mathbb{E}}$ controls the clustering of ω. If τ denotes the shift by 1 site to the right:

$$\omega(A\tau^n(B)) = \rho(\mathbb{E}_A \circ (\widehat{\mathbb{E}})^{n-1} \circ \mathbb{E}_B(\mathbb{1}_k)), \qquad A, B \in \mathcal{M}_d.$$

We will restrict our attention here to the "best" case, where the only eigenvector of $\widehat{\mathbb{E}}$ corresponding to an eigenvalue of modulus 1 is the identity $\mathbb{1}_k$ of $\mathcal{B}$. In this case $\widehat{\mathbb{E}}$ has *trivial peripheral spectrum* and ω will be exponentially clustering. The possible cluster rates are completely determined by the spectrum of $\widehat{\mathbb{E}}$. Because of the finite dimensionality of $\mathcal{B}$, only a discrete set of cluster rates can appear. If there exist other peripheral eigenvalues than $\mathbb{1}_k$, then ω can be decomposed into a finite convex combination of exponentially clustering finitely correlated states (which could be periodic instead of translation invariant). Suppose that $\mathbb{E}$ is a non-trivial sum of r maps of the form $X \mapsto V_i^* X V_i$ then the maps $\mathbb{E}_A$ will decompose accordingly. For a finite interval of length n, the state ω restricted to $\mathcal{A}_{[1,n]}$ will decompose into a convex decomposition of r^n states. Such a decomposition can be expected to generate a finite entropy density. This intuitive argument can be turned into a rigorous one and it leads to:

Theorem A:
A finitely correlated state ω is pure if and only if it is generated by a triple $(\mathcal{M}_k, \mathbb{E}, \rho)$, where $V : \mathbb{C}^k \to \mathbb{C}^d \otimes \mathbb{C}^k$ is an isometry, $\mathbb{E}(X) = V^* X V$, $\widehat{\mathbb{E}}$ has trivial peripheral spectrum and ρ, satisfying $\rho \circ \widehat{\mathbb{E}} = \rho$, is faithful.

Let us compute the reduced density matrices $\rho_{[1,n]}$ of a pure, finitely correlated state.

$$\begin{aligned} \operatorname{Tr}_{\otimes^n \mathbb{C}^d} \rho_{[1,n]} A_1 \otimes \cdots A_n &= \omega(A_1 \otimes \cdots A_n) \\ &= \rho\left(\mathbb{E}_{A_1} \circ \cdots \mathbb{E}_{A_n}(\mathbb{1}_k)\right) \\ &= \operatorname{Tr}_{\mathbb{C}^k} \rho V^{(n)*} (A_1 \otimes \cdots A_n \otimes \mathbb{1}_k) V^{(n)} \\ &= \operatorname{Tr}_{\otimes^n \mathbb{C}^d} \left(V^{(n)} \rho V^{(n)*}\right) (A_1 \otimes \cdots A_n \otimes \mathbb{1}_k), \end{aligned}$$

where:

$$V^{(n)} = (\mathbb{1}_d \otimes \mathbb{1}_d \otimes \cdots V) \cdots (\mathbb{1}_d \otimes V)V.$$

As this holds for all choices of the A_i we must have:

$$\rho_{[1,n]} = \operatorname{Tr}_{\mathbb{C}^k} V^{(n)} \rho V^{(n)*}.$$

Let $\{e_1, \ldots e_k\}$ be an orthonormal basis of $\mathbb{C}^k$. It is clear from the computation of above that, for $m < n$, the reduced density matrix matrix $\rho_{[m,n]}$ will live on the subspace of $(\mathbb{C}^d)^{\otimes(n-m+1)}$ spanned by the vectors $\{\phi_1, \ldots \phi_k\}$, where:

$$\phi = \sum_{i=1}^{k} \phi_i \otimes e_i,$$

and ϕ runs trough $(\mathbb{1} \otimes \cdots \otimes \mathbb{1} \otimes V) \cdots (\mathbb{1} \otimes V)V\mathbb{C}^k$. Therefore $\rho_{[m,n]}$ is supported by a subspace $\mathcal{G}_{[m,n]}$ of dimension at most k^2, independently of m and n. Let r be the smallest integer such that $\dim(\mathcal{G}_{[1,r]}) = k^2$. The standard VBS interaction associated to ω is now chosen as the orthogonal projection operator in $\otimes^{r+1}\mathbb{C}^d$ on the orthogonal complement of $\mathcal{G}_{[1,r+1]}$. The (standard) VBS Hamiltonian is then:

$$H_{[m,n]} = \sum_{j=m-1}^{n-r-1} \tau^j(h).$$

Note that ω is a ground state of this Hamiltonian, locally minimizing the energy, because $\omega(h) = 0$.

Theorem B:
A pure finitely correlated state ω is the unique ground state of its associated standard VBS Hamiltonian. Moreover the Hamiltonian exhibits a ground state energy gap.

It is now quite easy to construct a large number of Hamiltonians with exactly computable ground state. Indeed, fix the dimension k of the auxiliary space $\mathbb{C}^k$ and choose an isometry $V : \mathbb{C}^k \to \mathbb{C}^d \otimes \mathbb{C}^k$. Generically $\mathbb{1}_k$ will be the only eigenvector of $\widehat{\mathbb{E}}$, with $\mathbb{E}(X) = V^*XV$, corresponding to an eigenvalue of modulus 1, and the (unique) density matrix ρ on $\mathcal{M}_k$, which satisfies $\rho = \rho \circ \widehat{\mathbb{E}}$, will be faithful. Using Theorems A and B we can then construct the associated VBS Hamiltonian. It is also quite easy to construct models that are invariant under a local symmetry. We start with two unitary representations U and $\tilde{U}$ of a group $\mathcal{G}$ on $\mathbb{C}^d$ and $\mathbb{C}^k$. If there is an isometry V, intertwining $\tilde{U}$ and $U \otimes \tilde{U}$, then we can again construct a finitely correlated state with this V. Provided the conditions of Theorem A are satisfied, we will obtain in this way a $\mathcal{G}$-invariant VBS Hamiltonian with exactly known, unique, $\mathcal{G}$-symmetric ground state. Probably the simplest example of this type is the spin 1 antiferromagnetic chain studied in. In this model $d = 3$ and $k = 2$. The isometry V is the intertwiner between the spin 1/2 representation of SU(2) on $\mathbb{C}^2$ and the product of the spin 1 representation on $\mathbb{C}^3$ with the spin 1/2 representation. The finitely correlated state constructed with this V turns out to be the ground state of the spin 1 chain with interaction

$$h = \frac{1}{2}\vec{S_1} \cdot \vec{S_2} + \frac{1}{6}(\vec{S_1} \cdot \vec{S_2})^2 + \frac{1}{3}.$$

Higher, SU(2) invariant, integer spin chains can be constructed in a similar way. It can be proven that the ground states of some half-integer antiferromagnets, such as the (isotropic) Heisenberg antiferromagnet, are not finitely correlated. It is also possible, with suitable care, to incorporate in this scheme spin chains that are invariant under a quantum group. Another consequence of the construction is that we can approximate an arbitrary translation invariant state by a pure finitely correlated state in the w*-sense. This is in sharp contrast with classical systems, where the pure translation invariant states coincide with the translation invariant configurations, e.g. all spins up or down for an Ising system.

Theorem C:
The pure, translation invariant states on $\mathcal{A}_{\mathbb{Z}}$ are w-dense in the translation invariant states.*

There are still a number of interesting open problems in connection with these states.

- The most obvious one is to extend the construction to the case with an infinite dimensional auxiliary algebra $\mathcal{B}$. Allowing $\mathcal{B}$ to be completely general, will produce any translation invariant state. We should therefore look at $\mathcal{B}$ that are still "sufficiently close" to the finite dimensional case.
- It would also be quite interesting to establish a link with the Bethe Ansatsz. Can one efficiently model the correlation functions of the ground state of the spin 1/2 antiferromagnet with such a transfer matrix like construction?
- It is possible to construct finitely correlated states on a tree. Can one make non-trivial constructions on lattices with closed loops?

Acknowledgement
The results which are presented here have been obtained in a very enjoyable collaboration with B. Nachtergaele and R.F. Werner.

References

Proofs and an extensive list of references can be found in:

M. Fannes, B. Nachtergaele and R.F. Werner
Finitely correlated states on quantum spin chains
Commun. Math. Phys., **144**, 443–490 (1992)

The characterization of pure states (Theorem A) will appear in *J. Funct. Anal.*.
Theorem C was proven in *Lett. Math. Phys.*, **25**, 249–258 (1993)

M. Fannes, Instituut Theoretische Fysica, Celestijnenlaan 200D, B-3001 Heverlee

Operator Theory:
Advances and Applications, Vol. 70

The spectrum of the spin-boson model

Matthias Hübner and Herbert Spohn

Many physical problems can be described as a single (or a few) degrees of freedom interacting with a free field, regarded as a bath or reservoir. It is commonly assumed that a nontrivial coupling to the field serves as a mechanism for the small system to dissipate its energy. This somewhat vague formulation should be made precise by mathematical physics and we take here the point of view that it is reflected in the spectrum of the coupled Hamiltonian.

Mathematical theorems assure that quantum states $|\psi\rangle$ in the absolutely continuous (a.c.) spectral subspace $\mathcal{H}_{ac}$ of a Hamiltonian H decay weakly to zero under the time evolution induced by H,

$$\langle f|e^{-iHt}|\psi\rangle = \langle f|\psi(t)\rangle \to 0 \qquad \text{as} \quad t \to \infty \qquad \forall\, |f\rangle \in \mathcal{H}. \tag{1}$$

Of course the Hilbert space norm of $|\psi(t)\rangle$ is preserved and we interpret the weak convergence as a decay of any excitation measurable in a bounded region of the configuration space R^d. In the models which we shall consider, we expect this decay to hold for any physical state orthogonal to the ground state. The problem is then to prove that for interacting models the Hamiltonian has a unique ground state and that the remainder of the spectrum is purely a.c.. Here we would like to outline how such a property is proved for the spin-boson Hamiltonian.

In the spin-boson model the bath is a free Bose field and the small system is a localized degree of freedom, which corresponds to a particle in a double well potential. If the well is reflection symmetric, then the particle has a symmetric ground state and an antisymmetric first excited state. In the spin-boson Hamiltonian we take only these two lowest states into account and model the interaction with the bath by a linear coupling,

$$H = \frac{\mu}{2}\sigma_z \otimes I + I \otimes H_B + \sigma_x \otimes (a^*(\lambda) + a(\lambda)). \tag{2}$$

The first term is the energy of the localized degree of freedom with (uncoupled) energy separation $\mu > 0$. σ_x, σ_z are Pauli spin matrices and H_B is the selfadjoint Hamiltonian of noninteracting bosons. In the momentum representation, it is given by

$$H_B = \int d^d k \omega(k) a^*(k) a(k) = d\Gamma(\omega). \tag{3}$$

with domain $D(H_B)$ in the standard Fock space $\mathcal{F}$. The annihilation and creation operators $a(k), a^*(k)$ are defined over R^d and satisfy canonical commutation relations $[a(k), a^*(k')] = \delta(k - k')$. The second quantization functor is denoted by $d\Gamma(\cdot)$. For example, the boson number N_B equals $d\Gamma(I)$. The Hilbert space $\mathcal{H}$ for the model is the tensor product $C^2 \otimes \mathcal{F}$ of the spin 1/2 representation space and Fock space. We also use the shorthand $a^*(\lambda) = \int d^d k \lambda(k) a^*(k)$.

In general, the dispersion relation ω and the coupling λ are fixed by the physics of the problem. Here we consider them as free parameters and would like to have a spectral information for a whole class of ω, λ's. We state our technical conditions on ω. The dispersion is a real function on momentum space which induces by pointwise multiplication a 1-particle operator on the Hilbert space $L^2(R^d)$, whose second quantization is just H_B. We require

i) $\omega : R^d \to R$ is spherically symmetric, i.e. only a function of $|k|$, and everywhere positive with the possible exception of the origin $k = 0$. ω is a.c. as a real function, with positive derivative away from the origin.

$$|\nabla_k \omega| = \omega'(|k|) > 0 \quad \text{for} \quad k \neq 0, \qquad \text{and} \quad \lim_{k \to \infty} \omega(k) = \infty. \tag{4}$$

These properties are obviously satisfied by the dispersion $\omega(k) = \sqrt{k^2 + m^2}$ and its limiting cases $\omega(k) = |k|$ or $\omega(k) = k^2/2m$. Condition (4) implies that ω is not constant on sets of positive measure and is in fact an operator with purely a.c. spectrum on the 1-particle Hilbert space $L^2(R^d)$. Its second quantization $d\Gamma(\omega)$ is then purely a.c. as well, apart from the Fock vacuum. Clearly, in order to prove a.c. spectrum for the coupled H, we better assume absolute continuity of H_B.

We turn to the technical conditions on λ.

ii) λ should belong to the 1-particle Hilbert space, i.e. $(\lambda, \lambda) = \int d^d k |\lambda|^2 < \infty$. If $\inf \omega > 0$, this is enough to ensure selfadjointness of H on the natural domain $C^2 \otimes D(H_B)$ by the Kato-Rellich theorem, since $a(\lambda), a^*(\lambda)$ are then H_B-bounded with H_B-bound zero. By a unitary multiplication on momentum space, we can transform an arbitrary complex coupling function to a real nonnegative one. Thus it is no further restriction to assume $\lambda \geq 0$.

iii) In order that the Hamiltonian be bounded from below, we impose $\int d^d k \lambda^2/\omega < \infty$. If $\inf \omega = 0$, this is also the natural condition for selfadjointness of H on $C^2 \otimes D(H_B)$, which can again be established using the Kato-Rellich theorem.

iv) To ensure the existence and uniqueness of the ground state in Fock space, we need the even stronger condition $\int d^d k (\lambda/\omega)^2 < \infty$. The ground state of the spin-boson model was thoroughly investigated in [2-5]. Notice that ii) implies iii),iv) if $\inf \omega > 0$.

Our technical tools to prove the desired spectral properties are *Mourre estimates*, which are discussed in the book [6], following the original papers [7,8]. The idea is to find a so-called conjugate operator $-iA$ which has a positive commutator with the Hamiltonian, up to a compact perturbation,

$$[H, iA] \geq \alpha I - C, \quad \alpha > 0. \tag{5}$$

We say then that a selfadjoint operator obeys a (global) Mourre estimate. Imagine that this conjugate operator shifts the statesΠ continuously and even differentiably to higher energies. Intuitively, this is possible only in the a.c. spectral subspace. Mourre proved his theorems for *selfadjoint* conjugate operators A. The conjugate operators we will use have no selfadjoint extensions, but generate strongly continuous semigroups of isometries. We first give a generalization of Mourre's virial theorem.

Lemma 1: Let H be a selfadjoint operator with domain $D(H)$ and let iA be the closed generator of a semigroup of isometries. We assume:

a) $D(A) \cap D(H)$ is a core for H.

b) $e^{iA\alpha}$ leaves $D(H)$ invariant and for each $\psi \in D(H)$

$$\sup_{0<\alpha<1} ||He^{iA\alpha}\psi|| < \infty. \tag{6}$$

c) The quadratic form $[H, iA] = i(HA - AH)$ defined on $D(A) \cap D(H)$ is bounded below and closable and the selfadjoint operator associated to its closure is H-bounded.

Then for every eigenvector $\psi \in \mathcal{H}_{pp} = P_{pp}\mathcal{H}$ (the closed subspace generated by the eigenvectors)

$$\langle\psi|[H, A]|\psi\rangle = 0. \tag{7}$$

The proof of the virial theorem follows Mourre's proof with minor modifications. Whereever Mourre assumes the resolvent set of his selfadjoint A to contain the complement of the real line, in our case only the lower complex halfplane belongs to the resolvent set of our semigroup generator A. Mourre [6,7] proves that a global estimate like (5) implies that H has only finitely many eigenvalues, and each eigenvalue has finite multiplicity. In particular, this shows that eigenvalues cannot accumulate. Furthermore, he proved absence of singular continuous spectrum. We need here an explicit bound on the number of eigenstates.

Lemma 2: Under the assumptions of Lemma 1, let us suppose that H obeys a global Mourre estimate (5), with $\alpha > 0$ and C a positive operator of trace class. Then the following bound on the number of eigenvalues, counted with their multiplicity, holds

$$\text{dim}P_{pp} \leq \alpha^{-1}\text{tr}C. \tag{8}$$

Proof: We use the virial theorem which states that

$$\langle\psi|[H, iA]|\psi\rangle = 0 \tag{9}$$

for every eigenvector of H. Then

$$0 = \text{tr}P_{pp}[H, iA] \geq \text{tr}P_{pp}(\alpha I - C) \geq \alpha\text{dim}P_{pp} - \text{tr}C, \tag{10}$$

which proves the lemma. We turn to find a suitable conjugate operator for H. On the 1-particle level, a conjugate operator for ω fulfilling (4) is the radial derivation on momentum space, properly multiplied with the group velocity, and symmetrized,

$$D = \frac{1}{2}\left(\frac{1}{|\nabla\omega|^2}\nabla\omega\cdot\nabla_k + \nabla\omega\cdot\nabla_k\frac{1}{|\nabla\omega|^2}\right). \tag{11}$$

It is easy to see that $[D,\omega]=1$. The obvious guess for a suitable operator conjugate to H is then

$$-iA = d\Gamma(D). \tag{12}$$

D and $-iA$ generate semigroups of shifts directed outward to increasing $|k|$ on 1-particle momentum space and Fock space, respectively. These semigroups of isometries have no extensions to unitary groups. Thus our conjugate operators have no selfadjoint extensions. Because of this, Mourre's theorems don't apply directly to our choice (12) of A. Nevertheless it should be possible to generalize Mourre's theorems to generators of isometric semigroups.

It is not hard to verify the conditions of the lemmata for H of (2) provided $\inf\omega > 0$. If $\inf\omega = 0$, we compress H to a subspace $P_N\mathcal{H}$ of maximal boson number N. This means that we replace H with $P_N H P_N$ and P_N is the projection onto the closed subspace with at most N bosons. Then the conditions in Lemma 1 and 2 are satisfied and also the technical conditions d),e) in Mourre's theorem [7,p.392], which we did not mention explicitly.

We now observe that, because of reflection symmetry, the Hamiltonian commutes with the parity operator

$$P = -\sigma_z\otimes(-1)^{N_B} \qquad \text{and} \qquad [P, H_{SB}] = 0. \tag{13}$$

The Hilbert space $\mathcal{H}$ decomposes then into the two eigenspaces $\mathcal{H}_+, \mathcal{H}_-$ of the operator P corresponding to the eigenvalues ± 1. To obtain H restricted to $\mathcal{H}_\pm$ we first apply the unitary transformation

$$U = \exp\left(i\pi\frac{1-\sigma_x}{2}\otimes N_B\right) = U^* = U^{-1} \tag{14}$$

with the result

$$\begin{aligned} &U\sigma_z\otimes(-1)^{N_B}U = \sigma_z,\\ &\qquad UH_{SB}U = \frac{\mu}{2}\sigma_z\otimes(-1)^{N_B} + I\otimes H_B + I\otimes(a^*(\lambda)+a(\lambda)). \end{aligned} \tag{15}$$

Note that the interaction term contains no explicit spin term anymore. The conserved parity has changed to $-\sigma_z$ with the eigenvalues ±1. Thus $\mathcal{H}_\pm$ is isomorphic to $\mathcal{F}$ and the Hamiltonians $H_\pm$ on the eigenspaces reducing H_{SB} are unitarily equivalent to

$$H_\pm = \mp\frac{\mu}{2}(-1)^{N_B} + \int d^dk\,\omega(k)a^*(k)a(k) + a^*(\lambda) + a(\lambda). \tag{16}$$

Our result on the spectral structure on the even subspace is

Theorem 1: Let $\lambda \in D(D)$ and $(D\lambda, D\lambda) < \frac{1}{2}$. The compression $P_N H_+ P_N$ of H_+ to a maximal boson number N has a unique ground state and the rest of the spectrum is purely a.c.. If $\inf \omega > 0$, this holds also for H_+. *Sketch of proof:* Let us compute the commutator

$$[H_+, iA] = N_B + a^*(D\lambda) + a(D\lambda). \tag{17}$$

We displace the harmonic oscillator in the mode $D\lambda$ and obtain a new boson number operator with the old eigenvalues $0, 1, 2, \ldots$, however. This will be bounded from below as a form as

$$\begin{aligned}[H, iA] &= N'_B - (D\lambda, D\lambda)I \\ &\geq I - (D\lambda, D\lambda) - |vac'\rangle\langle vac'| \\ &\geq (1 - (D\lambda, D\lambda))I - |vac'\rangle\langle vac'|.\end{aligned} \tag{18}$$

Here $|vac'\rangle$ denotes a shifted Fock vacuum due to completing the square. Application of Lemma 2 leads then to the estimate for the number of eigenvalues,

$$\dim P_{pp} \leq \frac{1}{1 - (D\lambda, D\lambda)} \quad . \tag{19}$$

The $\mathcal{H}_-$ sector is more difficult, since we have to make sure that the number of eigenvalues is less than one. Here we try a conjugate operator with an undetermined function f,

$$-iA = d\Gamma + a(f) - a^*(f). \tag{20}$$

f will have to be optimized at the end. Of course the commutator is now somewhat more complicated

$$\begin{aligned}[H_-, iA] = N_B &+ a(D\lambda + \omega f) + a^*(D\lambda + \omega f) \\ &+ (f, \lambda) + (\lambda, f) - (-1)^{N_B} a(\mu f) - a^*(\mu f)(-1)^{N_B}.\end{aligned} \tag{21}$$

We remark that $(f, \lambda), (\lambda, f)$ are the good terms which offer a chance that the commutator will be positive at all. For the formulation of the next theorem, we replace λ by $\alpha\lambda$ with λ considered to be fixed, and introduce thereby a real coupling constant.

Theorem 2: For every λ in the domains of D and ω there exists a coupling constant α_c depending on μ and λ such that for $0 < \alpha < \alpha_c$ the spectrum of $P_N H_- P_N$ is purely a.c.. If $\inf \omega > 0$, no cutoff in the boson number is needed. The odd sector is much harder to treat than the even sector. The strategy is to partition the commutator into two summands and to estimate them separately from below. Our results in this direction are not completely satisfory yet. More details will appear in a future publication [9].

References

[1] A.J.Leggett et al.: Rev. Mod. Phys. 59, 1 (1987).

[2] H.Spohn, R.Dümcke: J. Stat. Phys. 41, 389 (1985).

[3] H.Spohn: Comm. Math. Phys. 123, 277 (1989).

[4] M.Fannes, B.Nachtergaele, A.Verbeure: J.Phys. A 21, 1759 (1988).

[5] M.Fannes, B.Nachtergaele, A.Verbeure: Comm. Math. Phys. 114, 537 (1988).

[6] H.L.Cycon, R.G.Froese, W.Kirsch, B.Simon: Schrödinger Operators, Springer, Berlin 1987.

[7] E.Mourre: Comm. Math. Phys. 78, 391 (1981).

[8] R.Froese, I.Herbst: Duke Math. J. 49, 1075 (1982).

[9] M.Hübner, H.Spohn, to be published.

Matthias Hübner and Herbert Spohn, LMU, Theoretische Physik, Theresienstraße 37, D-80333 München, BRD

Operator Theory:
Advances and Applications, Vol. 70

A Survey oh Wigner-Poisson Problems

Horst Lange, Bruce V.Toomire and Paul F. Zweifel

1 Introduction

In 1932, E.Wigner [1] introduced a phase-space method of computing physical observables in quantum (statistical) mechanics; some surveys have been published recently [2], [3]. See also a review article by Carruthers [4]. Here we give a brief synopsis of the method to make this paper self-contained, referring the reader to the references for details, including proofs.

Consider a system of N particles with a hamiltonian given by

$$H = \sum_{i=1}^{N} \frac{\vec{P}_i^2}{2m_i} + V(\vec{X}_1, \ldots, \vec{X}_n) \tag{1.1}$$

where the $\vec{P}_i$ are the momentum operators and the $\vec{X}_i$ the position operators. The system is described quantum mechanically by a state vector $\Psi \in L^2(\mathbb{R}^{3N})$; or more generally by a density matrix ρ [5]. Physical attributes of the system are described by the expectation values of certain self-adjoint operators representing observables. If A is such an operator, then the expectation value can be written as

$$\langle \mathrm{A} \rangle \quad = \mathrm{Tr}\, \rho\, \mathrm{A}\,. \tag{1.2}$$

This equation can be converted to a phase-space representation by defining the Wigner transform, $A_w(x,p)$ of the operator A:

$$A_w(x,p) = \sum_{k,l} A_{kl} \int_{\mathbb{R}^{3N}} e^{\frac{ipz}{\hbar}} u_k(x - \frac{z}{2}) \overline{u}_l(x + \frac{z}{2}) dz \tag{1.3a}$$

where $\{u_k\}$ is a basis for $L^2(\mathbb{R}^{3N})$ and:

$$A_{kl} = (u_k, Au_L)_{L^2(\mathbb{R}^{3N})}\,. \tag{1.3b}$$

One proves [2] for any two operators A and B

$$\mathrm{Tr} AB = (2\pi\hbar)^{-3N} \int_{\mathbb{R}^{3N}\times\mathbb{R}^{3N}} A_w(x,p) B_w(x,p) dx dp\,. \tag{1.3c}$$

Here $x = X_w$ and $p = P_w$ are interpreted as the classical phase-space coordinates. In particular:

$$\langle A \rangle \quad = (2\pi\hbar)^{-3N} \int A_w(x,p)\rho_w(x,p)dxdp\,. \tag{1.4}$$

The function:

$$f_w(x,p) = (2\pi\hbar)^{-3N} \rho_w(x,p) \tag{1.5}$$

is called the "Wigner distribution function" or simply the "Wigner function". In most cases, ρ is given by [5]

$$\rho = \sum \lambda_i \rho_i \tag{1.6}$$

where ρ_i is the projection onto ψ_i and λ_i is the probability of the state i. In such a case one computes

$$\rho_w(x,p) = \sum_n \lambda_n \int_{\mathbb{R}^{3N}} e^{\frac{ip\cdot z}{\hbar}} \psi_n(x - \frac{z}{2})\overline{\psi}_n(x + \frac{z}{2})dz\,; \tag{1.7}$$

we observe the ρ_w is real, but not necessarily positive.

By tranforming the evolution equation for ρ [4],given by the following:

$$i\hbar\frac{\partial \rho}{\partial t} = H\rho - \rho H \tag{1.8}$$

one derives the evolution equation for the Wigner distribution function f_w:

$$\frac{\partial f_w}{\partial t} + v \cdot \nabla_x f_w - \frac{i}{\hbar}\Theta(V) f_w(x,p,t) = 0 \tag{1.9a}$$

where $\Theta(V)$ is the pseudo-differential operator with symbol:

$$\delta V = V(x + \frac{\hbar\tau}{2}) - V(x - \frac{\hbar\tau}{2}) \tag{1.9b}$$

and v is the velocity vector with component $\frac{p_i}{m_i}$. Equation (1.9) is to be solved subject to the initial condition:

$$f_{w,I}(x,p) = \sum_n \lambda_n \int_{\mathbb{R}^{3N}} e^{\frac{ip\cdot z}{\hbar}} \phi_n(x - \frac{z}{2})\overline{\phi}_n(x + \frac{z}{2})dz \tag{1.9c}$$

where the ϕ_n are the initial values of the Schrödinger functions ψ_n which obey:

$$i\hbar\frac{\partial \psi_m}{\partial t} = (\sum_i \frac{\vec{P}_i^2}{2m_i} + V)\psi_m\,. \tag{1.10}$$

We recall that $\vec{P}_i^2 = -\hbar^2 \Delta_{x_i}$.

The above discussion suggests and has been proved by Markowich [6] that the Wigner equation (1.9) is equivalent to a system of Schrödinger equations (1.10), the dimension of the system corresponding to the number of terms in the sum (1.6). (If the dimension is one, we speak of a "pure state", otherwise of a "mixture". We expect, in general the the system is infinite dimensional.)

The quantum Vlasov equation is a mean-field approximation to the Wigner equation. In this approximation we take $N = 1$ (after integrating over the coordinates of the $N-1$ particles) and approximate V by a "self-consistent" field [2]:

$$V(x) = \int_{\mathbb{R}^{3N}} n(y)v(y-x)dy \tag{1.11a}$$

where v is the two-body potential and $n(y)$ is the density:

$$n(y) = \int_{\mathbb{R}^{3N}} f_w(x,p)dp\,. \tag{1.11b}$$

In this paper, v is taken to be a Coulomb potential, hence V obeys:

$$\triangle V = -\epsilon n \tag{1.12a}$$

or alternatively:

$$V = -\frac{\epsilon}{4\pi}\int_{\mathbb{R}^{3N}} \frac{n(y)}{|x-y|}dy\,. \tag{1.12b}$$

In (1.12), ϵ represents the (normalized) electric charge. We consider both the repulsive ($\epsilon = +1$) and attractive ($\epsilon = -1$) cases. Also, for convenience we set $m_i = 1\,(\forall i)$ and, except in Sec. 5 where the classical limit is discussed, we set $\hbar = 1$.

The mean field version of the classical Liouville equation is the classical Vlasov equation; with a Coulomb potential, one speaks of the Vlasov-Poisson (VP) system. The mean-field version of the Wigner equation is the quantum Vlasov equation; with a Coulomb force, one then calls this the Wigner-Poisson (WP) system. From its equivalence to the corresponding Schrödinger system – the Schrödinger-Poisson (SP) system– we see that the (WP) system may also be considered a statistical version of Hartree-Fock, i.e. of a single particle mean-field Schrödinger equation with a Coulomb force [7]. The analysis described here is, to a large extent, applied to the (SP) system, with the results then lifted to the (WP) system. This enables one to use many known results for the non-linear Schrödinger equation [8] as well as the Hartree-Fock system [9].

For convenience, we repeat the (WP) and (SP) systems which we consider in this paper.

Wigner–Poisson:

$$\partial_t f_w(x,p,t) + v \cdot \nabla_x f_w - i\Theta(V) f_w = 0 \tag{1.13a}$$

$$Sym\Theta(V) = V(x + \frac{\tau}{2}) - V(x - \frac{\tau}{2}) \tag{1.13b}$$

$$f_w(x,p,0) = f_{w,I}(x,p) \tag{1.13c}$$

$$\triangle V = -\epsilon n \tag{1.13d}$$

$$n(x,t) = \int_{\mathbb{R}^{3N}} f_w(x,p,t)dp\,. \tag{1.13e}$$

Schrödinger–Poisson:

$$\partial_t \Psi(x,t) = -\frac{1}{2}\triangle\Psi + V(\Psi)\Psi \tag{1.14a}$$

$$\Psi = (\psi_m)_{m\in\mathbb{N}} \tag{1.14b}$$

$$\Psi(x,0) = \Phi = (\phi_m)_{m\in\mathbb{N}} \tag{1.14c}$$

$$\triangle V = -\epsilon n \tag{1.14d}$$

$$n(x,t) = \sum \lambda_m |\psi_m(x,t)|^2 \tag{1.14e}$$

$$0 \le \psi_m \le 1, \sum_m \lambda_m = 1\,. \tag{1.14f}$$

Eq.(1.14e) follows from the relationship between the (WP) and (SP) solutions – see Eq.(1.7). In dealing with the (SP) system we shall use the following direct sum Hilbert Spaces:

$$X := \{\Gamma = (\gamma_m)_{m\in\widetilde{\mathbb{N}}} \mid \gamma_m \in L^2, \|\Gamma\|_X^2 = \sum_m \lambda_m \|\gamma_m\|_{L^2}^2\} \tag{1.15a}$$

$$Y := \{\Gamma = (\gamma_m)_{m\in\widetilde{\mathbb{N}}} \mid \gamma_m \in H^1, \|\Gamma\|_Y^2 = \sum_m \lambda_m \|\gamma_m\|_{H^1}^2\} \tag{1.15b}$$

$$Z := \{\Gamma = (\gamma_m)_{m\in\widetilde{\mathbb{N}}} \mid \gamma_m \in H^2, \|\Gamma\|_Z^2 = \sum_m \lambda_m \|\gamma_m\|_{H^2}^2\} \tag{1.15c}$$

$$\widetilde{\mathbb{N}} := \{n \in \mathbb{N} \mid \lambda_n > 0\}\,. \tag{1.15d}$$

In (1.15) and subsequently, all function spaces are over $\mathbb{R}^3$.

If the (WP) system is considered rather than the (SP), the $f_{w,I}$ Eq.(1.13c) can be chosen arbitrarily but must conform to Eq. (1.9 c). That is, the ϕ_n in that equation must be chosen to be the eigenvectors of a positive, trace-class operator in L^2, the λ_m being the corresponding eigenvalues, guaranteeing that the f_w will be the Wigner transform of a density matrix. This amounts to requiring that the

Fourier transform (with respect to p) of $f_{w,I}(x,p)$ be the kernel of a trace-class, positive operator.

In Sec. 2 we summarize the existence proof of the Cauchy problem in $\mathbb{R}^3$ for both $\epsilon = \pm 1$, and show how to obtain time asymptotics for the case $\epsilon = 1$. In Sec.3 we describe the existence proof for both the Cauchy problem and the stationary case in a finite domain with periodic boundary conditions. In Sec.4, a dissipative version of the (WP)-(SP) system is introduced and in Sec.5 we present some remarks on a problem which we believe is yet unsolved, namely the $\hbar \to 0$ limit of the WP system.

2 Global Existence, Uniqueness and Asymptotics for (WP)-(SP)

The recent work done on the "Wigner-Poisson" problem in ref. [10] is proven by first considering a bounded region $\Omega \subset \mathbb{R}^3_x \times \mathbb{R}^3_v$ and then taking the formal limit as Ω approachs the whole space. Ref. [11], which this section expounds, starts first by considering the problem in the entire space and then through its analysis arrives at both the unique global-time existence and the specific time asymptotics of the solution, $\Psi(x,t)$ or $f_w(x,p,t)$, and its constituent physical quantities.

The following is a sketch of the existence and uniqueness proof; we refer the reader to section 3 of ref. [11] for the details. The first step is to show that a *mild* local-time solution exists for the system (1.14), i.e. that there exists a $\Psi \in C([0,T];Z)$ such that:

$$\Psi(t) = e^{\frac{it}{2}\triangle}\Phi - i\int_0^t e^{\frac{i}{2}(t-s)\triangle} J(\Psi(s))\,ds \tag{2.1}$$

holds; where $J(\Psi) = V(\Psi)\Psi$ is the nonlinear operator defined by the potential V. To proceed we need the following.

Proposition 2.1. *The mapping* $J\ Z \to Z$, *whose* m^{th} *component is* $J(\Psi)_m = V(\Psi)\psi_m$ *where* V *is the Coulomb potential satisfying (1.12), is locally Lipschitz in* Z.

This proposition is proved by showing for $\widetilde{\Gamma}$, $\Gamma \in Z$ that:

$$\|V(\widetilde{\Gamma}) - V(\Gamma)\|_{L^\infty} \le C\|\widetilde{\Gamma} - \Gamma\|_Y \tag{2.2a}$$

$$\|\nabla V(\widetilde{\Gamma}) - \nabla V(\Gamma)\|_{L^2} \le C\|\widetilde{\Gamma} - \Gamma\|_Y \tag{2.2b}$$

$$\|\triangle V(\widetilde{\Gamma}) - \triangle V(\Gamma)\|_{L^2} \le C\|\widetilde{\Gamma} - \Gamma\|_Y \tag{2.2c}$$

where 'C' is some constant which could depend on the X, Y norms of $\widetilde{\Gamma}$ and Γ. These inequalities are easily shown through the use of the well-known Minkowski, Hölder, and Gagliardo-Nirenberg inequalities [12]. Moreover, we note that the free Hamiltonian associated with the system (1.14) generates a continuous unitary group $\{e^{\frac{it}{2}\triangle}\}_{t\in\mathbb{R}}$ on $L^2(\mathbb{R}^3)$. From this fact and Proposition 2.1, we have that by a classical result on the Cauchy problem [13] there exists a $T > 0$ and unique local strong solution Ψ of (1.14) i.e. a function

$$\Psi \in C([0,T], X) \cap C^1([0,T], Z)$$

such that (1.14) is satisfied in the L^2-sense, and $\Psi(x,0) = \Phi(x)$ (for $\Phi \in Z$). Next consider the conservation laws given by:

Proposition 2.2. *For $\Phi \in X$ and Ψ the mild solution in Z we have:*

$$\|\nabla\Psi\|_X^2 + \epsilon\|\nabla V(\Psi)\|_{L^2}^2 = \|\nabla\Phi\|_X^2 + \epsilon\|\nabla V(\Phi)\|_{L^2}^2 \tag{2.3a}$$
$$\|\Phi\|_X^2 = \|\Psi\|_X^2 \qquad (\forall t \in [0,T])\,. \tag{2.3b}$$

Again these are proved in ref. [11]. Clearly from these laws one can show that: $\|\Psi\|_Y \leq C, \quad \|V(\Psi)\|_{L^\infty} \leq C$ and also $\|\nabla V\|_{L^2} \leq C$. With these additional facts and considering (Eq.2.1) one can show using Gronwall's lemma that:

$$\|\Psi(t)\|_Z \leq C_T \qquad \text{for every } t \in [0,T] \tag{2.4}$$

where C_T is a constant dependent upon T only. Hence another classic result indicates that either the Z-norm of the solution goes to infinity in a finite time, or the solution can be continued onto any time interval. Therefore the solution Ψ exists globally in time, that is $\Psi \in C^1([0,\infty]; X) \cap C([0,\infty]; Z)$ and solves system (1.14); i.e. $\Psi(t)$ is a global strong solution. Formally this gives:

Theorem 2.3. *Let $\Phi \in Z$, $\|\Phi\|_X = 1$; then the Schrödinger-Poisson system (1.14) has a unique, global in time strong solution with the following regularity:*

$$\Psi \in C([0,\infty]; Z) \cap C^1([0,\infty]; X) \tag{2.5a}$$
$$n, \Delta V \in C([0,\infty]; Z) \cap C^1([0,\infty]; W^{2,1}) \tag{2.5b}$$
$$V \in C([0,\infty]; L^\infty) \cap L^\infty([0,\infty] \times \mathbb{R}^3) \tag{2.5c}$$
$$\nabla V \in C([0,\infty]; L^2) \cap L^\infty([0,\infty]; L^2)\,. \tag{2.5d}$$

Moreover, by the properties of the Wigner transform described in Sec. 1 the results of Theorem 2.3 imply the following.

Theorem 2.4. *Assume that $\rho_{w,I} \in L^2(\mathbb{R}_x^3 \times \mathbb{R}_v^3)$ is the kernel of a real trace class operator on L^2. Then the Wigner-Poisson system (1.13) has a unique global classical solution (ρ_w, n, V) satisfying:*

$$\rho_w \in C([0,\infty]; L^2(\mathbb{R}_x^3 \times \mathbb{R}_v^3)) \cap L^\infty([0,\infty]; L^s(\mathbb{R}_x^3 \times \mathbb{R}_v^3)); 2 \leq s \leq \infty \tag{2.6a}$$
$$\partial_t\rho_w, v\cdot\nabla\rho_w, \Theta(V)\rho_w \in C([0,\infty]; L^2(\mathbb{R}_x^3 \times \mathbb{R}_v^3))\,. \tag{2.6b}$$

In addition to the existence results,the analysis of ref. [11] finds time asymptotics for the solution of (WP)-(SP). Note that due to the nature of conservation law (2.3 a) asymptotics were only achieved for the replusive case ($\epsilon = +1$) for the systems (1.12) and (1.13).These asymptotics are summarized by the following theorems.

Theorem 2.5. *For the repulsive case, if* $|x|\Phi \in X$ *then* $|x|\Psi \in X, (\forall t \geq 0)$*;also* $|x|e^{\frac{it}{2}\triangle}\Psi \in X, (\forall t \geq 0)$ *and the following decay estimates hold:*

$$\|\nabla V(\Psi)\|_{L^p} = O(t^{\frac{1}{p}-1}) \text{ for } 2 \leq p \leq \infty \tag{2.7a}$$

$$\|\nabla V(\Psi)\|_{L^p} = O(t^{-\epsilon}) \text{ for } \frac{3}{2} < p \leq 2 \ \forall \epsilon \in (0, 1 - \frac{3}{2p}) \tag{2.7b}$$

$$\|V(\Psi)\|_{L^p} = O(t^{\frac{1}{p}-\frac{2}{3}}) \text{ for } 6 \leq p \leq \infty \tag{2.8a}$$

$$\|V(\Psi)\|_{L^p} = O(t^{-\epsilon}) \text{ for } 3 < p \leq 6 \ \forall \epsilon \in (0, \frac{2}{3} - \frac{2}{p}) \tag{2.8b}$$

$$\|n(t)\|_L^p = O(t^{\frac{-3}{2}(\frac{q-1}{q})}), \text{ for } 1 \leq q \leq 3 \tag{2.9a}$$

$$\sum \lambda_m \|\psi_m(t)\|_{L^p}^2 = O(t^{-3(\frac{1}{2}-\frac{1}{p})}), \text{ for } 2 \leq p \leq 6\,. \tag{2.9b}$$

To sketch the proof of Theorem 2.5 we need the following lemma (see also ref. 9).

Lemma 2.6. *Suppose* $|x|\Phi \in X$*. Then the solution* Ψ *satisfies:*

$$|x|\Psi(t) \in X \qquad (\forall t \geq 0)$$

and

$$\sum \lambda_m \|x\psi_m(t)\|_{L^2}^2 \leq 2 \sum \lambda_m \|x\phi_m\|_{L^2}^2 + C\Big(\int_0^t \big(\sum \lambda_m \|\nabla\psi_m\|_{L^2}^2\big)^{\frac{1}{2}} ds\Big)^2. \tag{2.10}$$

From this result the first assertion of Theorem 2.5 follows. The estimates of (2.7) are then derived from the quantity:

$$\frac{\partial}{\partial t}(xG(-t)\psi_m(t)) \quad (\forall m) \tag{2.11}$$

where $G(t) = e^{\frac{it}{2}\triangle}$ is the free Hamiltonian group operator. By summing over (λ_m) and after much manipulation one gets the "Quasi-Conformal Conservation Law"

$$\sum_m \lambda_m \|xG(-t)\psi_m\|_{L^2}^2 + t^2 \int V \cdot n dx = \sum_m \lambda_m \|x\phi_m\|_{L^2}^2 + \int_0^t s \int V(s) n \, dx ds\,. \tag{2.12}$$

Hence, by using Gronwall's Lemma, (2.12) gives

$$\int V(x,t) n(x,t) dx = \int |\nabla V|^2 dx \leq \frac{C}{t} \tag{2.13}$$

and (2.7a) is proved for $p = 2$. The remaining asymptotics can be derived from (2.13) by using standard inequalities (including the Gagliardo-Nirenberg inequality [12]).

Finally, by the Wigner transform, Minkowski, and Hausdorff-Young inequalities Theorem 2.5 gives the following asymptotics of the solution (ρ_w, n, V).

Theorem 2.6. *Let $\Phi \in X$ and $\sum_m \lambda_m \|\nabla\phi_m\|^2_{L^{\frac{6}{5}}} < \infty$; then the following hold:*

$$\|n(t)\|_{L^p} = O(t^{\frac{3}{2}(\frac{1}{p}-1)}) \text{ for } 1 \le p \le 3 \tag{2.14a}$$

$$\|n(t)\|_{L^p} = O(t^{-\frac{1}{2}(1+\frac{3}{p})+\epsilon}) \text{ for } 3 < p \le \infty,\ \epsilon > 0 \text{ (small)} \tag{2.14b}$$

$$\|\rho_w(x,t)\|_{L^p} = O(t^{-\frac{3}{2p}}) \text{ for } 2 \le p \le \infty \tag{2.14c}$$

$$\|\Theta(V)\rho_w\|_{L^p} = O(t^{-\frac{2}{3}}) \text{ for } 2 \le p \le \infty. \tag{2.14d}$$

3 Periodic Solutions

The problem of the existence of space-periodic solutions (of period 1) to the classical Vlasov-Poisson system has been studied by Batt and Rein [14] [15] on $\mathbb{R}^3$ whereas the one-dimensional case for (WP) was treated by Arnold and Markowich [3]. The physical model in refs. 14 and 15 and also for the periodic (WP) and (SP) systems is a plasma of electrons moving in a background of fixed positive charge $q(x)$ whereas the overall plasma is charge neutral. Thus the Poisson equation takes a different form from that of Refs. 10 and 11:

$$\Delta V(x,t) = q(x) - n(x,t)\,. \tag{3.1}$$

Here $n(x,t)$ is the density of negative charge and $q(x)$ is the given time independent density of positive charge; later on we only consider the normalized case $q(x) \equiv 1$ for simplicity; in this case we have the normalization:

$$\int_Q n(x,t)\,dx = 1 \tag{3.2}$$

where $Q = [0,1]^3$. The density, as usual will be expressed in terms of the wave functions ψ_m of the (SP) system as

$$n(x,t) = \sum_{m=1}^{\infty} \lambda_m |\psi_m(x,t)|^2 \tag{3.3}$$

where $\lambda_m \ge 0, \sum \lambda_m = 1$. The Wigner function f_w is,in our case, slightly different from that of Refs. [10] and [11] because we also have to quantize momentum; we set $v_n = 2\pi n, n \in \mathbb{Z}^3$. The analysis of Section 1 and Refs. [16] [17] are transferred *mutatis mutandis* to the periodic case where the main differences are that in the definition of the Wigner transform integration is taken over Q, and integration over v is replaced by a summation over v_n; normalization constants are deleted. We write f_w as the sequence $f_{w,n}$ defined by:

$$f_{w,n}(x,t) = f_w(x, v_n, t) \,. \tag{3.4}$$

The Wigner equation itself is basically unchanged:

$$\partial_t f_{w,n} + v_n \cdot \nabla_x f_{w,n} - i\Theta(V) f_{w,n} = 0 \,. \tag{3.5}$$

Moreover, $\Theta(V)$ is now defined by (where $Q' = [-1, 1]$)

$$\Theta(V) f_{w,k} = \sum_{k'} \int_{Q'} [V(x + \frac{\eta}{2}, t) - V(x - \frac{\eta}{2})] f_{w,k'}(x,t) e^{2\pi i(k-k')\eta} d\eta \tag{3.6a}$$

$$= \sum_m \lambda_m \int_{Q'} e^{2\pi ikz} \psi_m(x - \frac{z}{2}, t) \overline{\psi}_m(x + \frac{z}{2}, t)[V(x + \frac{z}{2}, t) - V(x - \frac{z}{2})] dz \tag{3.6b}$$

$$f_{w,k} = \sum_m \lambda_m \int_Q e^{2\pi ikz} \psi_m(x - \frac{z}{2}, t) \overline{\psi}_m(x + \frac{z}{2}, t) dz \,. \tag{3.7}$$

The equivalent (SP) system has the form:

$$i\partial_t \psi_m = -\frac{1}{2} \Delta\psi_m + V(\Psi)\psi_m \tag{3.8a}$$

$$\Delta V = 1 - n(x,t) \ (x \in Q,\, t \in \mathbb{R}^+) \tag{3.8b}$$

with $n(x,t)$ defined by (3.3) and $\Psi = (\psi_m)$. Equation (3.8) should be solved with periodic boundary data on Q and a 1-periodic initial condition:

$$\psi_m(x, 0) = \phi_m(x) \tag{3.9}$$

The $\Phi = (\phi_m)$ should be an orthonormal system on $L^2(Q)$ satisfying the consistency condition to guarantee a positive trace class operator on $L^2(Q)$ with eigenvalues (λ_m); call this condition (C).

In this section we state global (in time) existence and uniqueness results for strong solutions of the periodic (SP) and (WP) systems. Furthermore, we give results on the existence of an infinite number of periodic stationary states of (SP) and (WP). Again we only sketch the proofs of these results.

By a *strong solution* of the periodic (SP) system (3.8)-(3.9) we mean functions (Ψ, V, n) such that (3.8) is satisfied in the L^2 sense and:

$$\Psi \in C(S, X^2) \cap C^1(S, X) \cap L^\infty(\ Q)$$

$$V \in C(S, H^2) \cap L^\infty(S, L^p), 2 \le p \le \infty, \int_Q V\, dx = 0$$

$$n \in C(S, L^2) \cap L^\infty(S, L^r), 1 \le r \le \infty, \int_Q n\, dx = 1$$

where $S = [0,T]$, $T > 0$, $\widetilde{Q} = Q \times S$ and X, X^2 are the analogously defined spaces as in Section 1 for the periodic case, e.g.

$$X^k = \{\Psi = (\psi_m) \mid \psi_m \ 1-\text{periodic on } Q, \psi_m \in H^k_{\text{loc}}(\mathbb{R}^3)\};$$

where we consider the L^p-spaces to be over Q.

Theorem 3.1. *Let $\Phi \in X^2$; then for any $T > 0$ there is a unique global strong solution (Φ, V, n) of Eq.(3.8) on $S = [0,T]$.*

Remarks on proof. The details may be found in Ref.[18]. We use a Galerkin method to get the solutions as limits of the approximating Galerkin sequence

$$\psi_m^{(N)}(x,t) = \sum_{|k|\le N} d_{m,k}^{(N)}(t) h_k(x) \tag{3.10}$$

(with $h_k(x) = e^{2\pi k\cdot x}$, $k \in \mathbb{Z}^3$); the coefficients $d_{m,k}^{(N)}(t)$ are determined in such a way that

$$(i\partial_t \psi_m^{(N)} + \frac{1}{2}\Delta\psi_m^{(N)} - V^{(N)}\psi_m^{(N)}, h_k) = 0 \qquad (|k| \le N) \tag{3.11a}$$

$$\psi_m^{(N)}(x,0) = \sum_{|k|\le N} (\phi_m, h_k) h_k \tag{3.11b}$$

are satisfied. Here $V^{(N)}$ is an approximation of V defined by

$$V^{(N)}(\Psi^{(N)}) = \frac{1}{4\pi^2} \sum_{0<|k|\le 2N} \frac{n_k^{(N)}}{k^2} h_k$$

$$n^{(N)}(x,t) = \sum_m \lambda_m |\psi_m^{(N)}(x,t)|^2 = \sum_{|k|\le 2N} n_k^{(N)} h_k$$

where one can compute the $n_k^{(N)}$ to be

$$n_k^{(N)} = \sum \lambda_m \sum_{|l|\le N} \sum_{|l-k|\le N} d_{m,l}^{(N)} \overline{d}_{m,l-k}^{(N)}, \quad n_0 = 1.$$

From (3.11) one sees that the $d_{m,l}^{(N)}$ satisfy the system of nonlinear first-order differential equations:

$$\dot{d}_{m,l}^{(N)}(t) = -2\pi^2 i l^2 d_{m,l}^{(N)}(t) - i/4\pi^2 \sum_{\substack{0<|k|\le 2N \\ |l-k|\le N}} \frac{n_k^{(N)}}{k^2} d_{m,l-k}^{(N)}(t) \tag{3.12a}$$

$$d_{m,l}(0) = (\phi_m, h_l) \qquad |l| \le N\,. \tag{3.12b}$$

(3.12) can be solved globally in time by using the conservation laws

$$\|\Psi(\cdot,t)\|_X^2 = const., \ \|\nabla\Psi(\cdot,t)\|_X^2 + \|\nabla V(\cdot,t)\|_{L^2}^2 = const. \tag{3.13}$$

which are also valid for the Galerkin sequence; (3.12) also implies a priori bounds for the Galerkin sequence in X^1. For getting a strong solution of (SP) one needs an X^2 a priori bound which is provided by looking at the time evolution of $\int |\Delta\psi^{(N)}|^2\,dx$ and using Sobolev embedding. This also gives L^∞-bounds on the Galerkin sequence which are needed for the uniqueness proof. The existence proof is finished by using further compact Sobolev embeddings.

Theorem 3.1 transfers to (WP) by defining a strong 1-periodic solution of (WP) to be a sequence $f_w = (f_{w,n})$ which consists of spatially 1-periodic functions satisfying (WP) in the sense of the function space $W = l^2(L^2(Q))$ such that:

$$f_w \in C^1(S,W),\ (v_n \cdot \nabla_x f_{w,n}) \in C(S,W),\ \Theta(V)f_w \in C(S,W).$$

We say that $f_{w,I}$ satisfies condition (C) if

$$\widehat{\rho}_{w,I}(x,\eta) = \sum_m \lambda_m \phi_m(x-\frac{\eta}{2})\overline{\phi}_m(x+\frac{\eta}{2})$$

where ϕ_m satisfies condition (C) from above; and " ^ " means the Fourier series generated by $\rho_{w,I,n}$.

Theorem 3.2. *Let $\rho_{w,I} \in W$ satisfy condition (C); then the Wigner-Poisson system (3.5) has a unique global strong 1-periodic solution such that*

$$n(x,t) = \sum_{n\in\mathbb{Z}^3} f_{w,n}(x,t) \qquad and \qquad f_w(x,0) = f_{w,I}(x).$$

By a *stationary* 1-period solution of (SP) we mean a sequence of functions $\Psi = (\psi_m)$ such that Ψ solves (SP) and

$$\Psi(x,t) = e^{i\omega t}\Phi(x)$$

with 1-periodic real $\Phi = (\phi_m)$.

Theorem 3.3. *There exist a countably infinite number of stationary 1-periodic solutions $\Phi_j(x,t) = e^{i\omega_j t}\phi_j(x)$ such that $\phi_j \in X^2 \cap C^\infty$ and $\omega_j \to \infty$ for $j \to \infty$.*

Sketch of proof. (details may be found in Ref.[19])

We only consider the case of a pure state. With this assumption one obtains the ϕ_j's as critical points of the functional

$$J(\phi) = \int_Q \{|\nabla\phi|^2 + B(\phi^2 - m(\phi^2))\cdot(\phi^2 - m(\phi^2))\}dx$$

where $m(f) = \int_Q f(x)dx$ is the mean value over Q of any function f, $B\, L_0^2(Q) \to H_0^1(Q)$ is the solution operator $\mu = Bf$ of the Poisson problem $\Delta u = f$ on Q with $m(u) = 0$, and the index zero on the space means "mean value zero". It can be shown that B maps $L_0^2(Q)$ injectively and compactly into itself. The main point to prove *Theorem 3.3* now is to show that J satisfies a *Palais-Smale Condition* on the set

$$\{\phi \in H^1_{\text{per}}(Q) \mid m(\phi^2) = 1\}$$

which is done by using compact and continuous Sobolev embeddings.

4 Dissipative Systems

In this section we consider (SP) and (WP) equations which model dissipation. In some approximation, the magneto-fluid dynamics of a quantum plasma can be discribed by adjoining to the usual hamiltonian terms representing dissipation (or friction, or viscosity). In this case the (SP) system is of Ginzburg-Landau type, [20] (SP) namely we introduce as the new Hamiltonian:

$$H = -\frac{1}{2}\Delta + V + iH_1\,, \tag{4.1}$$

$$H_1 = \alpha\Delta + \beta x^2 + \gamma\,; \tag{4.2}$$

here $\alpha, \beta, \gamma \in \mathbb{R}$; the first term models kinetic friction, the second the loss of energy to a reservoir, and the third static friction (see Ref. [21] for similiar models). We first derive the Wigner equation (WP) obeyed by the Wigner function when the corresponding (SP) system is given by:

$$i\partial_t\psi_m = -\frac{1}{2}\Delta\psi_m + V\psi_m + iH_1\psi_m \tag{4.3}$$

$$-\Delta V = \epsilon n \qquad (\epsilon = \pm 1) \tag{4.4}$$

$$\psi_m(x,0) = \phi_m(x) \quad (x \in \mathbb{R}^3) \tag{4.5}$$

$$n(x,t) = \sum_m \lambda_m |\psi_m|^2\,. \tag{4.6}$$

Propsition 4.1. *The (WP) system for a Hamiltonian given by (4.1) is the evolution equation*

$$[\partial_t + v\cdot\nabla_x - i\Theta(V) + \alpha(2v^2 - \frac{1}{2}\Delta_x) + \beta(2x^2 - \frac{1}{2}\Delta_v) + 2\gamma]f_w = 0\,. \tag{4.7}$$

The proof of Prop.4.1 just uses that

$$f_w(x,v,t) = \sum_m \lambda_m \int e^{-ix\cdot\xi}\hat{\psi}_m(v-\frac{\xi}{2},t)\overline{\hat{\psi}}_m(v+\frac{\xi}{2},t)d\xi$$

(where the " ^" means Fourier transform in the x variable). For proving global existence of the solution to (SP) (which also transfers to (WP)) for the Cauchy problem in $\mathbb{R}^3$ we need the same spaces X, Y, Z as defined in the introduction, and furthermore the space $\widetilde{Z} = Z \cap \widetilde{X}$ where

$$\widetilde{X} = \{\Psi = (\psi_m) \mid \Psi \in X, \|x^2\Psi\|_X^2, \|x \otimes \nabla\Psi\|_X^2 < \infty\};$$

$$\text{here} \quad \|x \otimes \nabla\Psi\|_X^2 = \sum \lambda_m \sum_{j,k=1}^{3} \|x_j \partial_{x_k}\psi_m\|_{L^2}^2 .$$

Furthermore, we use the linear operator $T D(T) = \widetilde{Z} \subset X \to X$ defined by $T(\Psi) = \left((-\frac{1}{2}\Delta + iH_1)\psi_m\right)_{m\in\mathbb{N}}$. One can prove that T is closed, and that

$$D(T) = \{\Psi \in X^1 \mid \Delta\Psi \; exists, x\cdot\Psi \in X, T(\Psi) \in X\};$$

this can be done by considering (for $\gamma = 0$)

$$\|T\Psi\|_X^2 = (\frac{1}{2}+\alpha^2)\|\Delta\Psi\|_X^2 + \beta^2\|x^2\Psi\|_X^2 + 2\beta Re\sum_m \lambda_m \int \Delta\psi_m \cdot x^2\bar{\psi}_m dx$$
$$-\beta Im \sum_m \lambda_m \int \Delta\psi_m \cdot \bar{\psi}_m dx$$

and using partial integration and duality arguments. From Hille-Yosida's theorem one can show that $S := -iT$ generates a contraction semigroup, namely one proves that for $\alpha \geq 0, \beta \leq 0, \gamma \leq 0$

$$\|(\mu I + iT)^{-1}\|_X \leq \frac{1}{\mu} \qquad (\mu > 0)$$

by looking at $Re\sum \lambda_m((\mu I + iT)\psi_m, \psi_m)_{L^2}$.

Theorem 4.2. *Let $\Phi = (\phi_m) \in \widetilde{Z}$ and $\alpha \geq 0, \beta \leq 0, \gamma \leq 0$. then the (SP) system (4.3)-(4.6) has a unique global strong solution (Ψ, n, V) on $[0,\infty]$ such that*

$$\Psi \in C([0,\infty);\widetilde{Z}) \cap C^1([0,\infty);X), n, \Delta V \in C^1([0,\infty);L^1) \cap C([0,\infty);W^{2,1}),$$

$$V \in C([0,\infty);L^\infty(\mathbb{R}^3\cdot[0,\infty))), \nabla V \in C([0,\infty);L^2) \cap L^\infty([0,\infty),L^p) \; (2 \leq p \leq \infty).$$

Sketch of proof. (details may be found in Ref. [22])

One first shows that the nonlinear operator $J(\Psi)_m = V(\Psi)\psi_m$ is locally Lipschitz on $\widetilde{Z}$ which is done along the same lines as in Section 2. This implies local existence of strong solutions of (4.3)-(4.6). To get global solutions one proves that for $\alpha \geq 0, \beta \leq 0, \gamma \leq 0$ there is an a priori bound for any local solution $\Psi(t)$ of the form $\|\Psi(t)\|_{\widetilde{Z}} \leq C(T)$ on any time interval $[0,T]$, $T > 0$. The main point when doing this is to look at the time evolution of $x \otimes \nabla\Psi, x^2\Psi$, and $\Delta\Psi$ in the X-norm. Then one uses the commutation relations $[x \otimes \nabla, \Delta] = -2\nabla \otimes \nabla$, $[x \cdot \nabla, x^2] = -2x \otimes x$, $[\Delta, x^2] = 6I - 4x \otimes \nabla$, and the integral version of (SP) ; i.e.

$$\begin{aligned} x \otimes \nabla\Psi(t) =& G(t)(x \otimes \nabla\Phi) + \int_0^t G(t-s)[\sigma\nabla \otimes \nabla\Psi(s) \\ & +2i\beta x \otimes x\Psi(s) - ix \otimes \nabla J(\Psi(s))]ds \end{aligned}$$

where $G(t)$ is the semigroup generated by $-iT$. We also use the evolution law for the "energy"

$$E(t) = E(x,\Psi) = \int_{\mathbb{R}^3} \{|\nabla\Psi|^2 + |\nabla V|^2 + \delta|\Psi|^2\}dx$$

for some positive δ.

Theorem 4.3. *Let $f_{w,I} \in L^2(\mathbb{R}^3 \times \mathbb{R}^3)$ and*

$$f_{w,I}(x,v) = \sum \lambda_m \int e^{ivt}\phi_m(x - \frac{z}{2})\bar{\phi}_m(x + \frac{z}{2})dz$$

with $\phi = (\phi_m) \in \widetilde{Z}$. Then (4.7) has a global strong solution f_w satisfying

$$\begin{aligned} f_w \in& C([0,\infty); L^2(\mathbb{R}^3 \times \mathbb{R}^3)) \cap L^\infty([0,\infty); L^s(\mathbb{R}^3 \times \mathbb{R}^3)), 2 \leq s \leq \infty, \\ \partial_t f_w,\ v \cdot \nabla_x f_w,\ \Theta(V)f_w \in& C([0,\infty); L^2(\mathbb{R}^3 \times \mathbb{R}^3)) \end{aligned}$$

where $f_w(x,v,0) = f_{w,I}(x,v)$.

Remark. For $\gamma < 0$ we can prove that for the "energy functional"

$$F(t) = \int_{\mathbb{R}^3} \{|\nabla\Psi|^2 + \delta|\Psi|^2\}dx$$

there exists a sufficiently large $\delta > 0$ such that

$$F(t) \leq F(0)e^{-ct} \qquad (\forall t \geq 0).$$

5 Remarks on the Classical Limit

There is considerable interest in the $\hbar \to 0$ limit of the (WP) system for a number of reasons. First, the Wigner method was proposed originally [1] as a means of obtaining quatum corrections to quasi - classical systems; such an equation would presumably be solved by expanding f_w in a similiar sense, whose leading term would obey the classical equation! Second, the existence proofs presented in this paper for the (WP) systems are dramatically simpler than the corresponding proof for the Vlasov-Poisson system [23],[24],[25],and there is some hope that undoubtedly this limit might lead to simplified analysis of the (VP) system.

The existence of a classical limit of the Wigner equation has been viewed with skepticism by a number of authors for some time. In particular, Heller [26] has asserted that a power series expansion of f_w in $\hbar$ might not be valid due to the apparent essential singularity at $\hbar = 0$ [cf Eq.(1.7)]. However, the situation is far from clear, since the singularity appears in the form of an exponential kernel (e.g. choosing $\phi_n = e^{ik_n \cdot x/\hbar}$ leads to $f_w(x,p) = \sum \lambda_n \delta(p - k_n)$, independent of x as required by the uncertainty principle). This essential singularity in the integrand of f_w Wigner transforms into a delta function in the $\hbar \to 0$ limit because of the normalization condition

$$\int_{\mathbb{R}^3 \times \mathbb{R}^3} f_w(x,p,t) dp dx = 1; \tag{5.1}$$

that is, f_w is a delta sequence. While we have not proved this in general, it appears plausible, and is certainly true for two cases which can be worked out explicitly. For the free particle represented by an intial wave packet of mean p_0 and variance $\frac{1}{2\alpha}$ we have [7]

$$\phi_0(x) = (\frac{\alpha}{\pi})^{\frac{3}{2}} (\frac{1}{2\alpha\pi\hbar^2})^{\frac{3}{4}} \int_{\mathbb{R}^3} e^{-\alpha(p-p_0)^2} e^{ip\cdot x/\hbar} dp\,; \tag{5.2}$$

using the Schrödinger flow $e^{-\frac{iHt}{\hbar}}$ gives:

$$\psi_0(x,t) = (\frac{\alpha}{\pi})^{\frac{3}{2}} (\frac{1}{2\alpha\pi\hbar^2})^{\frac{3}{4}} \int_{\mathbb{R}^3} e^{-\alpha(p-p_0)^2} e^{-\frac{ip^2 t}{2m\hbar}} e^{ip\cdot x/\hbar} dp\,. \tag{5.3}$$

Using Eq.(1.7) one computes:

$$f_w(x,p,t) = (\frac{1}{\pi\hbar})^3 e^{-\frac{1}{2\alpha\hbar^2}(x-\frac{p}{m}t)^2} e^{-2\alpha(p-p_0)^2} \tag{5.4}$$

and therefore:

$$\lim_{\hbar\to 0} f_w(x,p,t) = (\frac{2\alpha}{\pi})^{\frac{3}{2}} \delta(x - \frac{p}{m}t) e^{-2\alpha(p-p_0)^2} \tag{5.5}$$

This limit satisfies the Vlasov equation, viewed as a relation between distributions.

For the harmonic oscillator in one dimension in its ground state, Heller [26] has calculated f_w. With $m = 1$ he finds

$$f_w(x,p,t) = \frac{1}{\pi\hbar} e^{-\frac{p^2}{\hbar\omega}} e^{-\frac{\omega x^2}{\hbar}} \tag{5.6}$$

where ω is the usual angular frequency. Again, the limit is a distribution. In fact

$$\lim_{\hbar\to 0} f_w = \delta(x)\delta(p) \tag{5.7}$$

These considerations indicate the probable futility of looking for functions as solutions in the $\hbar \to 0$ limit. In fact, in ref. 27 the limit is sought in a space of measures, which seems eminently reasonable.

The following proposition could be useful in forming the classical limit.

Proposition 5.1.

$$\|f_w\|^2_{L^2(\mathbb{R}^3\times\mathbb{R}^3)} = (2\pi\hbar)^{-3} \sum \lambda_i^2$$

Proof. Since f_w is real

$$\|f_w\|^2_{L^2} = \int f_w^2(x,p,t)dpdx = (2\pi\hbar)^{-3} \int \rho_w f_w dqdx = (2\pi\hbar)^{-3} \langle \rho \rangle$$

by (1.4) and (1.5). The result now follows easily from Eqs.(1.2) and (1.6).

Note that this is now unbounded in the $\hbar \to 0$ limit. Sometimes, it is conjectured that $\sum \lambda_i^2$ contains a factor of $\hbar$, but this is physically impossible. In particular, the (WP) theory and its classical limit should remain valid in the case of a pure state, for which $\lambda_k = \delta_{kl}$.

Acknowledgement: This work was supported by NATO Collaborative Research Grant CRG 910979.

References

[1] E. Wigner, Physics Rev. **40** (1932), 749.

[2] R. Illner, Trans. Theor. and Stat. Phys **21** (1992), 753.

[3] P.F. Zweifel, Trans. Theor. and Stat. Phys **22** (1993), 459.

[4] P. Carruthers, Rev. Modern Phys **55** (1983), 245.

[5] J. von Neumann, *Mathematical Foundations of Quantum Mechanics*, Princeton Univ. Press, 1955.

[6] P.A. Markowich, Math. Meth. Appl. Sci. **11** (1989), 459.

[7] L.I. Schiff, *Quantum Mechanics*, Third ed., McGraw-Hill, 1968.

[8] T. Cazenave, *An Introduction to Nonlinear Schrödinger Equations*, Textos de Methodos Matematicos 22 Rio de Janerio, 1989.

[9] J.P. Dias and M. Figueira, C.R. Acad Sci. Paris **290**, (1980), 889; J. Math. Anal. Appl. **84** (1981), 486.

[10] F. Brezzi and P.A. Markowich, Math. Meth. Appl. Sci. **14** (1991), 35.

[11] R. Illner, H. Lange and P.F. Zweifel, *Global Existence, Uniqueness and Asymptotic behavior of the Solution of the Wigner-Poisson and Schödinger-Poisson System*, Math. Meth. Appl. Sci., (to appear).

[12] A. Friedman, *Partial Differential Equations*, Rinehart and Winston, New York, 1969.

[13] A. Pazy, *Semi-groups Linear Operators and Applications to P.D.E.*, Springer Verlag, Berlin, 1983.

[14] J. Batt and G. Rein, *Global Classical Solution of the Periodic Vlasov-Poisson system in Three Dimensions*, Univ.of Munich (Preprint).

[15] J. Batt and G. Rein, *A Rigorous Stability Result for the Vlasov-Possion System in Three Dimensions*, Univ. of Munich (Preprint).

[16] C.S. Bohun, R. Illner and P.F. Zweifel, Le Matimatiche **16** (1991), 429.

[17] H. Lange and P.F. Zweifel, *Periodic Solutions to the Wigner-Poisson Equation*, Non-linear Analysis (Submitted).

[18] O. Kavian and H. Lange, *On Stationary Solution of the Schrödinger-Poisson and Wigner-Poisson System*, Preprint Dept. des Mathematiques Universite de Nancy (1993).

[19] D.R. Tilley and J. Tilley, *Superfluidity and Superconductivity* (1980), Hilgery Brentol.

[20] J. Messer, Act. Phys. Austriaca **50** (1979), 75.

[21] H. Lange and P.F. Zweifel, *Dissipation in Wigner-Poisson Systems*, J. Math. Phys. (to appear).

[22] K. Pfaffelmoser, *Global Classsical Solutions of the Vlasov-Poisson System in Three Dimensions for General Initial Data*, J. Diff. Eqs. (to appear).

[23] E. Horst, *Global Solution of the Vlasov-Poisson System and their Asymptotic Growth*, Paderborn (Preprint).

[24] R.J. Diperna and J.L. Lions, Rend. Sem. Mat. Univ. Politecn. Torino **46** (1988), 259; *Preprint No.8824*, CEREMADE, Univ. Paris-Dauphiné (1988).

[25] E.Heller, J.Chem. Phys. **65** (1976), 1289.

[26] P.A. Markowich and N.J. Mauser, *The Classical Limit of a Self-consistent Quantum-Vlasov Equation in Three Dimensions*, Technical Report Fachbereich Mathematik TU Berlin, March 1992.

[27] P.L. Lions and T. Paul, *Sur les mesures de Wigner*, CEREMADE, Univ. Paris-Dauphiné (Preprint).

Horst Lange, Fachbereich Mathematik, Universität Köln, Weyertal 86–90, 50931 Köln, Germany

Bruce V.Toomire and Paul F. Zweifel, Center for Transport Theory, and Mathematical Physics, Virginia Polytechnic Institute and State University, Blacksburg, Va 24061, USA

Chapter 5

Chaos

Operator Theory:
Advances and Applications, Vol. 70

Classical d'Alembert field in an one-dimensional pulsating region

J. Dittrich, P. Duclos, P. Šeba

Abstract

Classical massless scalar field in a finite one-dimensional space interval is studied. One end point of the interval is fixed while the second one is periodically oscillating. The field satisfies the Dirichlet boundary conditions at the end points. Sufficient conditions for the unbounded increase of the energy of the field are found. A case with the periodic time evolution of the energy is also shown.

Fermi in 1949 proposed the scattering of charged particles on moving inhomogeneities of the magnetic field as a possible mechanism of cosmic rays acceleration [1]. To study the possibility of such acceleration a simplified model [2] was proposed: a particle elastically bouncing between the fixed and the oscillating walls. The velocity of a classical nonrelativistic particle is increasing with the energy. For large energies, the particle is many times bounced within one period of the wall motion and the energy gained within one half-period is lost within the second half-period. So we expect that the energy of the particle remains bounded which was really proved [3]. Similar nonacceleration have been shown for nonrelativistic quantum particle and smooth periodic motion of the wall , however, the acceleration have been seen for some nonsmooth wall motions here [4].

The waves of a classical field spread with the constant velocity c independent of the energy. A ray can be therefore bounced in phase with the wall motion, always meeting the wall moving with the same velocity against the ray. The energy of the ray thus increases by the same factor at each reflection on the moving mirror (moving wall or end point of our interval). If the distance between the walls is x at the time of reflection and T is the period of the wall motion then the condition of repeating bounces after N periods is

$$2x = NTc. \tag{1}$$

Taking into account the maximal and minimal wall distances the condition

$$x_{min} < \frac{N}{2}Tc < x_{max} \tag{2}$$

is obtained for the increase of energy. We put $c = 1$ in the following.

We are going to demonstrate the validity of these heuristic considerations on the simplest field theoretical model with moving boundary. Let us start from the formulation of the model. We consider a real massless scalar field (or small oscillations of a string) φ satisfying 1+1 dimensional d'Alembert equation

$$\frac{\partial^2\varphi(t,x)}{\partial t^2} - \frac{\partial^2\varphi(t,x)}{\partial x^2} = 0. \tag{3}$$

for

$$t > 0, \quad x \in (0, X(t)) \tag{4}$$

with the boundary conditions

$$\varphi(t,0) = 0, \quad \varphi(t, X(t)) = 0 \tag{5}$$

for all $t \geq 0$. We assume

$$X \in C^2(\mathbb{R}), \quad X > 0, \quad X(t+T) = X(t), \quad |X'(t)| < 1$$

for some $T > 0$ and all $t \in \mathbb{R}$. The field φ is assumed to have continuous second derivatives in the domain (4) and to be continuous in its closure. At the time $t = 0$ the Cauchy data

$$\varphi(0,x) = f_0(x), \quad \frac{\partial\varphi(0,x)}{\partial t} = f_1(x)$$

are given. They must satisfy the consistency conditions

$$f_0 \in C^2([0, X(0)]), \quad f_1 \in C^1([0, X(0)]),$$

$$f_0(0) = f_0(X(0)) = 0, \quad f_1(0) = 0,$$

$$f_1(X(0)) + X'(0) f_0'(X(0)) = 0, \quad f_0''(0) = 0,$$

$$f_0''(X(0)) + X'(0) f_1'(X(0)) + \frac{X''(0)}{1 - X'(0)^2}[f_0'(X(0)) - f_1(X(0))] = 0$$

The field φ has the form

$$\varphi(t,x) = f(t+x) - f(t-x). \tag{6}$$

The function f is expressed with the help of f_0 and f_1 in the interval $[-X(0), X(0)]$ and extended to the whole real axis by the relation

$$f = f \circ F \tag{7}$$

where

$$F = k \circ h^{-1}$$

with

$$h(t) = t - X(t), \quad k(t) = t + X(t)$$

for $t \in \mathbb{R}$.

The energy $E(t)$ of the field is

$$E(t) = \frac{1}{2} \int_0^{X(t)} \left[\left(\frac{\partial \varphi}{\partial t} \right)^2 + \left(\frac{\partial \varphi}{\partial x} \right)^2 \right] dx$$

which is rewritten as

$$E(t) = \int_{h(t)}^{k(t)} [f'(y)]^2 \, dy. \tag{8}$$

The following theorem gives sufficient conditions for the energy unlimited increase.

Theorem 1:
Assume that for some $0 < \xi < T$, X is nondecreasing in $(0, \xi)$, decreasing with $X'(t) < 0$ in (ξ, T) and

$$X(0) < \frac{1}{2} T < X(\xi). \tag{9}$$

Assume further more that none of the functions $f_0' + f_1$ and $f_0' - f_1$ vanish identically in any open subinterval of $(0, X(0))$. Then

$$\lim_{t \to \infty} E(t) = \infty. \tag{10}$$

The detailed proof will be published elsewhere. It is based on the relation (7) which together with repeated substitution $z = F(y)$ leads to

$$E(t) = \int_{F^{-n}(h(t))}^{F^{-n}(k(t))} \frac{f'(z)^2}{\prod_{j=0}^{n-1} F'(F^j(z))} dz.$$

Under the conditions of Theorem 1, all factors $F'(F^j(z))$ are smaller than 1 in a part of the integration interval and we are able to proof relation (10) as $t \to \infty$ corresponds to $n \to \infty$ for suitably chosen n.

The assumption on $\mathrm{supp}(f_0' \pm f_1)$ can be weakened but the set where these functions should be nonzero is expressed in a rather complicated way. The assumption excludes the trivial case of a field constant in some region for which $f'(z)$ is zero in the above mentioned part of the integration interval.

The assumption (9) corresponds to $N = 1$ in the heuristic condition (2). In fact, Theorem 1 can be extended to the case $N > 1$ where condition (9) has to be replaced by (2). The choice of time $t = 0$ when X passes its minimum is just conventional and can be removed.

For the special case

$$X(t) = x_0 + \alpha \sin(\omega t), \quad 0 < \alpha < x_0, \quad 0 < \alpha\omega < 1, \tag{11}$$

equation (9) with the corresponding shift of the time zero reads

$$\frac{\pi}{x_0 + \alpha} < \omega < \frac{\pi}{x_0 - \alpha}$$

giving a band of wall frequencies where the energy is unlimited in time.

Numerical computations indicate that the intervals of wall frequencies given by equation (2) are maximal intervals where the energy is unboundedly growing. However, we have not an analytical proof of that. We can only show the following case of periodic and therefore bounded energies.

Theorem 2:
Let X is given by equation (11) with

$$\omega = (N + \frac{1}{2})\frac{\pi}{x_0}, \quad N = 0, 1, ..., \quad N < \frac{x_0}{\pi\alpha} - \frac{1}{2}. \tag{12}$$

Then

$$E(t + 4x_0) = E(t) \tag{13}$$

for any $t \in \mathbb{R}$.

The proof is based on the relation $F(F(t)) = t + 4x_0$ which is easily verified under the assumption (12).

The work is partly supported by the ASCR grant No. 14814.

References

[1] E. Fermi, Phys. Rev. 75(1949), 1169

[2] S. Ulam, in Proc. 4th Berkeley Symp. on Math. and Probabil., Berkeley - Los Angeles, 1961, v.3, p.315

[3] L. D. Pustylnikov, Teor. Mat. Fiz. 57(1983), 128; Dokl. AN SSSR 292(1987), 549

[4] P. Šeba: Phys. Rev. A 41 (1990), 2306

J. Dittrich, P. Šeba, Nuclear Physics Institute, Academy of Sciences of the Czech Republic, 250 68 Řež near Prague, Czech Republic

P. Duclos, Centre de Physique Théorique, CNRS, Marseille-Luminy, France and PHYMAT, Université de Toulon et du Var, Toulon, France

Operator Theory:
Advances and Applications, Vol. 70

Irregular scattering in one-dimensional periodically driven systems.

Petr Šeba

Abstract

We discuss the irregular scattering in a one dimensional and time periodic model. The existence of sharp quantum resonances is demonstrated. We show that these resonances appear due to tunneling between the stability island and the chaotic layer.

The irregular scattering is one of the frontiers of the nowadays research on the chaotic classical and quantum Hamiltonian systems. In the classical case the existence of the irregular scattering has been discussed for instance by Jung and Eckhardt [1] as a typical representation of the transient chaotic behavior (see for instance [2]). In the quantum case the relevant studies have been initialized by Blumel and Smilansky [3]. Using semi classical arguments they were able to show that the presence of the classical irregular scattering implies certain fluctuations of the corresponding quantum S-matrix which are described by the Dyson ensemble of random matrices.

Today the properties of the classical and quantum irregular scattering have been investigated in a number of various models. Their common feature was that the classical chaotic repeller (which is responsible for the appearance of the fractally organized singularities of the classical scattering characteristics) was fully hyperbolic. The hyperbolicity of the repeller implies among others an exponential decrease of the probability density P(t) that a classical trajectory will stay in the interaction region for a time longer then t:

$$P(t) \approx \alpha e^{-\alpha t} \tag{1}$$

with α being connected with the Lyapunov exponent of the repeller [4]. The exponential decay of P(t) has twofold consequence: In the classical case it leads to a *self similar* structure of the scattering singularities (a kind of Cantor set). In the quantum case (1) implies the absence of long-living resonances and consequently the Ericson fluctuations of the quantum mechanical cross section [3].

In the present letter we would like to describe a *one-dimensional and time periodic* (kicked) model for irregular scattering and to investigate its properties. Depending on the parameters of the model, the corresponding repeller will be either hyperbolic leading to (1) or it will display a large elliptic island leading to an algebraic decay of P(t):

$$P(t) \approx t^{-\alpha} \tag{2}$$

Our aim is to show that the tunneling between the chaotic layer and the stability island leads in the quantum case to the existence of long-living resonances. This resonance mechanism is (according to our opinion) suited to explain the existence of the controversial resonances observed in some ion–ion scattering experiments [12]. The very recent classical calculations performed on the ion–ion scattering model demonstrated [13] the presence of large stability islands. Its phase space structure is therefore quite similar to the phase space of the model presented in our letter.

The system to be investigated consists of a one-dimensional particle moving on a line under the influence of a short range time-periodic potential. The classical Hamiltonian is given by

$$H = \frac{1}{2}p^2 + V(x) \sum_{n=-\infty}^{\infty} \delta(t-n) \tag{3}$$

with a short range potential

$$V(x) = -\lambda e^{-x^2} \ ; \quad \lambda > 0. \tag{4}$$

The dynamics of the system is governed by a classical map

$$p_{n+1} = p_n - V'(x_n), \qquad x_{n+1} = x_n + p_{n+1} \tag{5}$$

where x_n, p_n denote the coordinate and the impulse of the particle just after the n-th kick.

The phase space portrait of the above map has in the interaction region all the typical features common to the time periodic dynamical systems. The origin is a fixed point of the mapping. For $\lambda < 2$ this fixed point is stable and is surrounded by a set of integrable trajectories (KAM curves) defining a stable region (the stability island) in which the motion is quasi-periodic. This island is imbedded into a chaotic layer interspersed with smaller islands (See the Figure 1a). For $\lambda > 2$ the fixed point becomes unstable and the stability island disappears. Some smaller secondary island may, however, still be present. For λ big enough the interaction region becomes hyperbolic (see Figure 1b).

The properties of the transport in the phase space depend heavily on its structure. In the hyperbolic case the probability that a trajectory will stay in a given region of the phase space for a time T is exponentially decaying with T. It is, however, well known that the existence of an elliptic domain will spoil this behavior leading to an algebraic decay. This change in the transport behavior has been observed for the first time in [5]. The theoretical explanation [6] is based on the properties of the boundary layer between the stable and the chaotic domain of

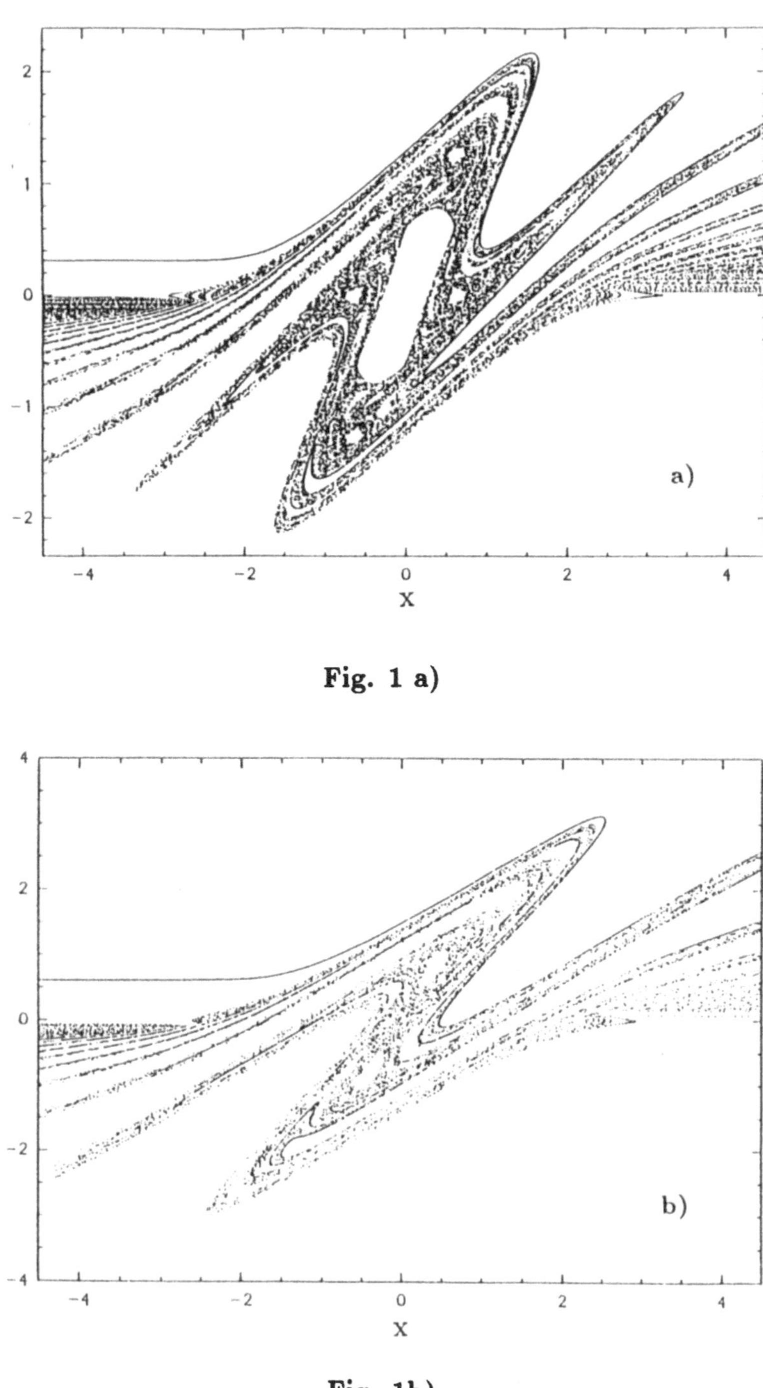

Fig. 1 a)

Fig. 1b)

Fig 1. The phase space portrait of the system. a) $\lambda = 1$, b) $\lambda = 5$

the phase space. The boundary layer contains invariant Cantor sets (the so called Cantori) which are remnants of the KAM curves being destroyed as the coupling constant of the nonlinear term increases. It is the flux through the Cantori which determines the long time behavior of the transport (the trajectory is "sticking" on the boundary where it has to pass through the tiny "holes" in the Cantorus). In the case of a well developed stability island the time decay of the stay-in probability is given by (2) with α roughly equal to 1.5 [7].

As already mentioned the character of the transport in the interaction region is decisive for the structure of the fractal set of the scattering singularities. Very recently Lau, Finn and Ott [8] demonstrated that the customary self similarity of the Cantor set of the classical scattering singularities is a direct consequence of the exponential decay of the probability $P(t)$. In the case of an algebraic decay the situation changes. The authors showed that in this case the fractal dimension is equals to one [8].

In our model the influence of the elliptic island on the behavior of the scattering singularities can be easily investigated. The probability distribution P(t) has been evaluated initializing a large amount of trajectories and inspecting the corresponding time delay. The results are plotted on the Fig. 2. In the case with a large stability island the algebraic decay of P(t) with $\alpha \approx 1.5$ is clearly apparent. The next Figure shows the outgoing impulse p_{out} as a function of the initial position x_{in} of the trajectory. The calculation have been performed for two different coupling constants λ. In both cases the incoming impulse has been fixed and the trajectory has been initialized with x_{in} deep in the asymptotically free domain. In the first case ($\lambda = 5$) the interaction region is hyperbolic leading to exponentially decaying P(t) and to a self-similar structure of the scattering singularities. The second case corresponds to $\lambda = 0.5$ with a large stability island. The difference in the fractal structure is clearly apparent.

In the quantum case is the scattering process governed by the quantum map (Floquet operator)

$$U = e^{-\frac{i}{\hbar}\frac{\hat{p}^2}{2}} e^{-\frac{i}{\hbar}V(x)} \tag{6}$$

where $\hat{p}$ denotes the standard impulse operator

$$\hat{p} = -i\hbar\frac{\partial}{\partial x} \tag{7}$$

The operator U is nothing but the quantized version of the classical map (5).

To define the S - matrix we introduce the Møller operators Ω_-, Ω_+:

$$\Omega_- = \lim_{n\to\infty} U^{-n}U_0^n, \qquad \Omega_+ = \lim_{n\to-\infty} U^{-n}U_0^n \tag{8}$$

with U_0 being the free evolution over one period

$$U_0 = e^{-\frac{i}{\hbar}\frac{\hat{p}^2}{2}}. \tag{9}$$

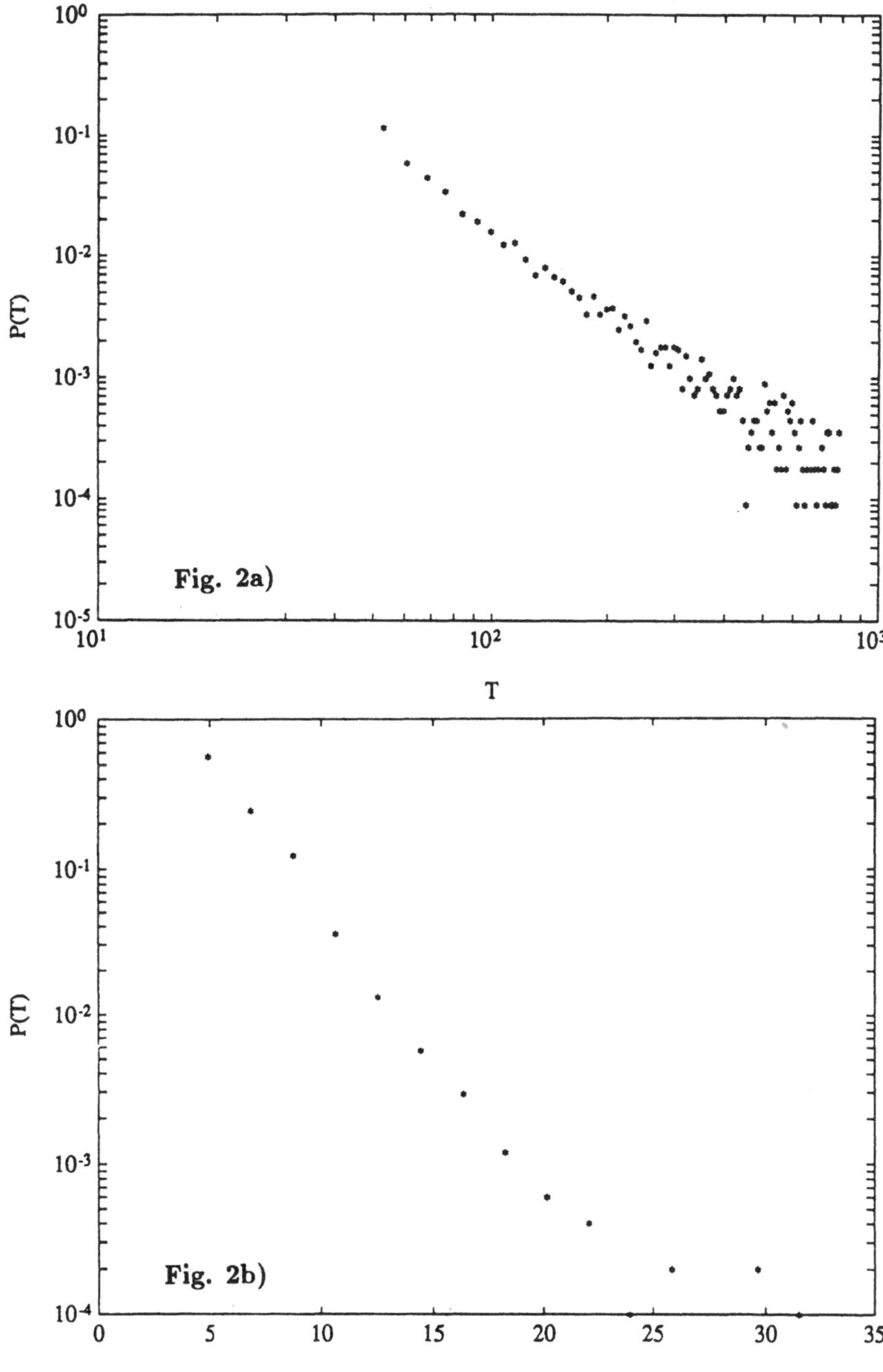

Fig 2. The probability distribution P(T) evaluated for a) $\lambda = 0.5$ b) $\lambda = 5$

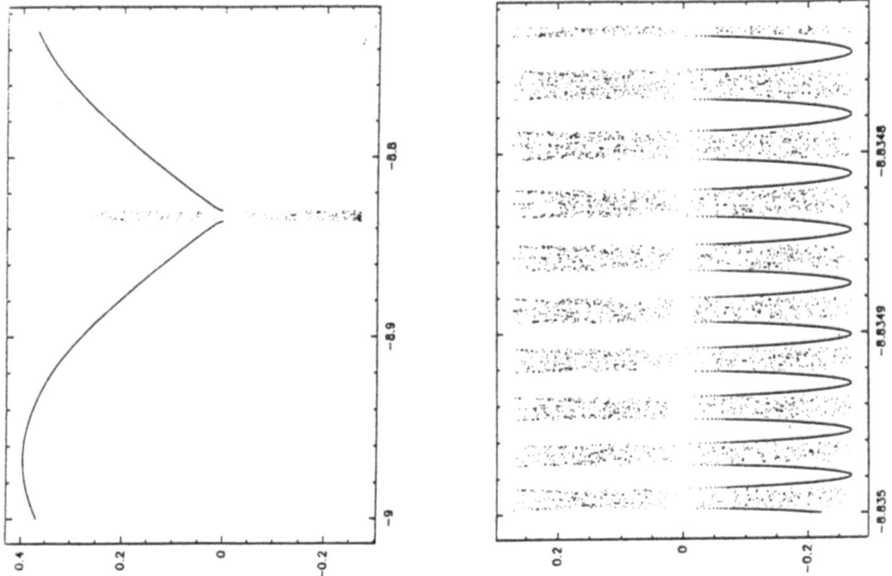

Fig. 3a)

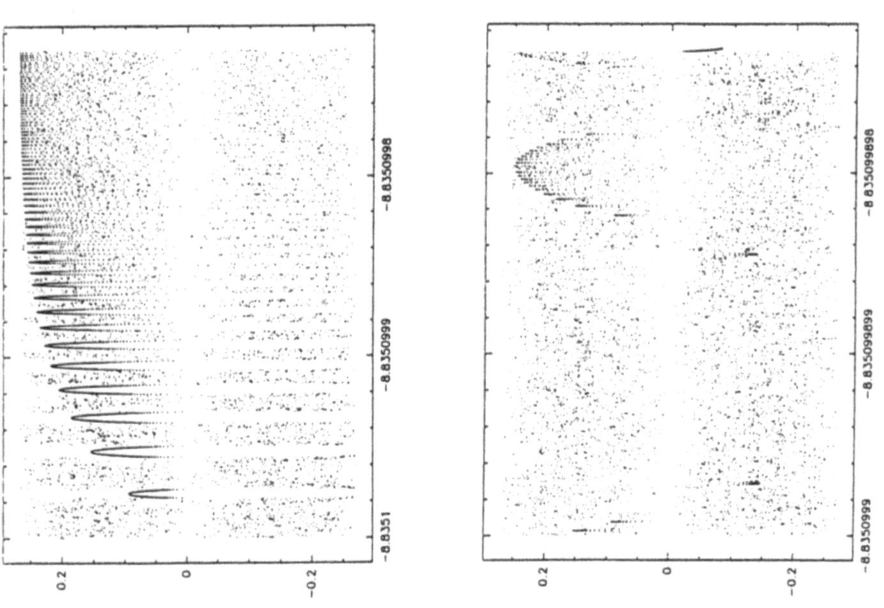

Fig 3. The outgoing impulse of the particle is plotted as a function of x_{in} in a series of subsequently magnified plots. a) $\lambda = 0.5, p_{in} = 0.27$;

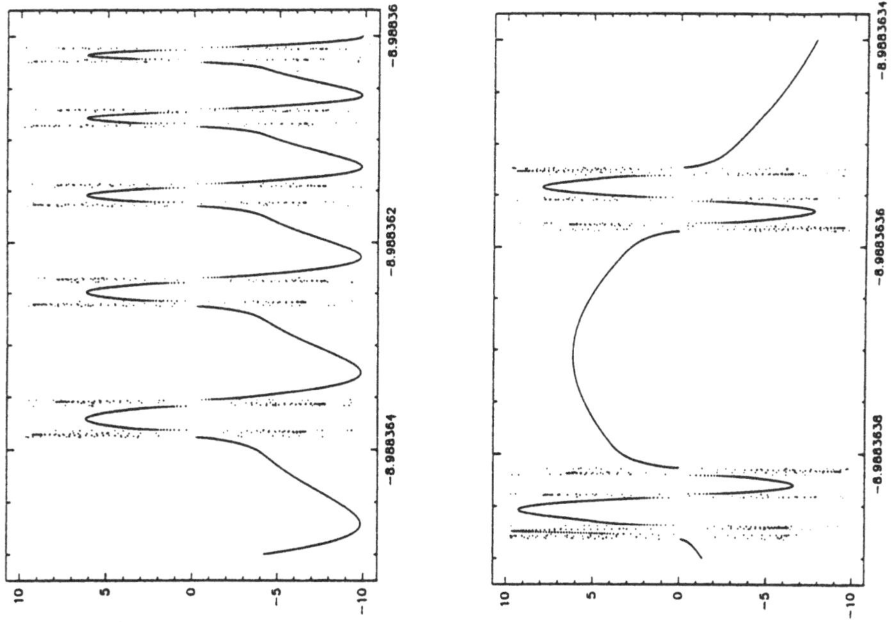

Fig. 3b)

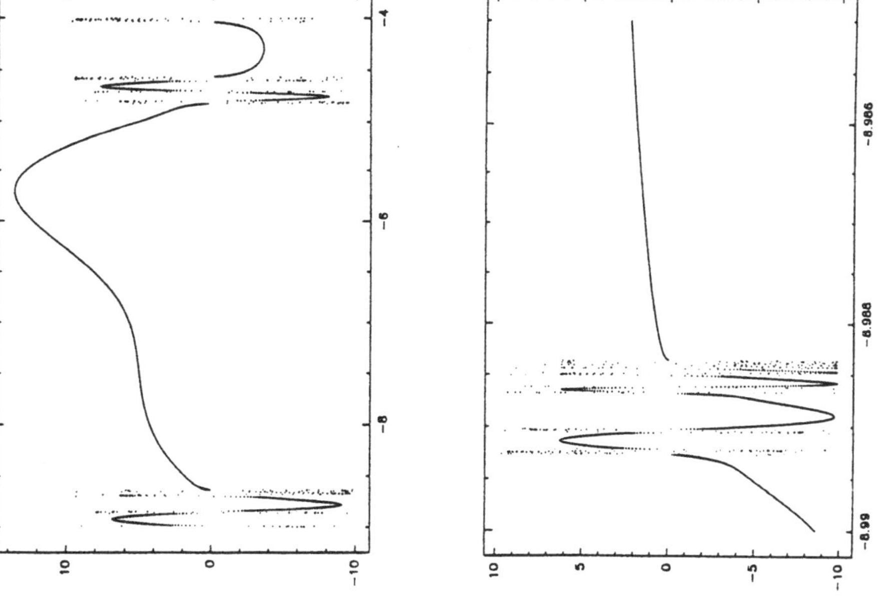

b) $\lambda = 10, p_{in} = 5$.

The quantum S-matrix is then defined in a standard way:

$$S = \Omega_-^{-1}\Omega_+. \tag{10}$$

It is not difficult to prove that the scattering defined by the above equations is complete and the S-matrix is unitary. (This fact can be proved using ideas similar to those sketched in [10].) Moreover the S - matrix commutes with the free Floquet operator U_0: $SU_0 = U_0 S$. This is the precise mathematical formulation of the fact that the free quasi - energy is conserved during the scattering process.

The conservation of the quasi- energy restricts substantially the energy which can be transferred during the quantum scattering. In order to clarify this point let us assume that we have prepared an incoming state $|in>$ with kinetic energy E

$$\frac{1}{2}\hat{p}^2|in>= E|in> \tag{11}$$

After the scattering the state becomes $|out>= S|in>$. The natural question is what is the kinetic energy of $|out>$? From the commutation relations follows

$$U_0|out>= U_0S|in>= SU_0|in>= e^{\frac{i}{\hbar}E}S|in>= e^{\frac{i}{\hbar}E}|out> \tag{12}$$

i.e. the vector $|out>$ has the same quasi- energy as the vector $|in>$. Inasmuch as U_0 is a function of the kinetic energy operator we have

$$\frac{1}{2}\hat{p}^2|out>= E'|out> \tag{13}$$

with E' such that

$$e^{\frac{i}{\hbar}E} = e^{\frac{i}{\hbar}E'}. \tag{14}$$

In other words

$$E' = E + n2\pi\hbar \tag{15}$$

with $n = 0, \pm 1, \pm 2, \ldots.$ Consequently the energy transfer is possible only by *discrete portions of* $2\pi\hbar$.

Let us now say a few word about the structure of the S-matrix defined by (10). In consequence of the quasi- energy conservation we can write the operator S as an direct integral over quasi- energy Θ

$$S = \int S(\Theta)d\Theta \tag{16}$$

with $S(\Theta)$ referring to the scattering on a given quasi energy shell. Taking into account that each quasi-energy state is two times degenerate (the particle can move into two different directions) we can naturally divide the S - matrix elements into four blocks:

$$S(\Theta) = \begin{pmatrix} S^{(-,-)}(\Theta) & S^{(-,+)}(\Theta) \\ S^{(+,-)}(\Theta) & S^{(+,+)}(\Theta) \end{pmatrix} \tag{17}$$

where $S^{(-,-)}(\Theta)$ refers to states coming from $-\infty$ and being finally bounced back to $-\infty$ (reflection) while $S^{(-,+)}(\Theta)$ describes the situation where the particle coming from $-\infty$ moves finally to $+\infty$ (transmission). Due to the time reversal symmetry of the system we find that

$$S^{(-,-)}(\Theta) = \overline{S^{(+,+)}(\Theta)}, \qquad S^{(-,+)}(\Theta) = \overline{S^{(+,-)}(\Theta)} \tag{18}$$

The matrix elements $S_{m,n}(\Theta)$ of $S(\Theta)$ are directly related to the kinetic energies E, E' of the incoming / outgoing particle respectively. From (12) and (15) it follows that one can express E, E' through Θ as

$$E = n2\pi\hbar + \hbar\Theta, \qquad E' = m2\pi\hbar + \hbar\Theta. \tag{19}$$

with n, m being integers corresponding to the indices of the matrix element $S_{n,m}(\Theta)$ and $\Theta \in (0, 2\pi)$. The physical interpretation of the S-matrix elements is straightforward: $|S_{n,m}^{(-,-)}(\Theta)|^2$ is the probability that a particle incoming from left with an energy $E = 2\pi\hbar n + \hbar\Theta$ will be finally reflected back with an energy $E' = 2\pi\hbar m + \hbar\Theta$.

We evaluated the quantum map U using a fast Fourier transform code with 2^{17} elements. This enabled us to obtain reliable results for $\hbar$ as small as 0.01. The evaluated S-matrix has been found to be unitary with high precision.

In the case of large λ (the interaction domain is nearly hyperbolic) the S-matrix elements display Ericson fluctuations [3] similar to those described by Blumel and Smilansky [3-4]. These fluctuations are characterized by an autocorrelation function

$$C(\Delta) = \ll S_{n,m}^{(-,-)}(\Theta)\overline{S_{n,m}^{(-,-)}(\Theta + \Delta)} \gg = \frac{C(0)}{1 - i\frac{\Theta}{\alpha\hbar}} \tag{20}$$

with α referring to the classical probability (1) (see Fig.4.). The situation changes, however, if the interaction region contains a well developed stability island. The quantum tunneling into the regular island leads in this case to the existence of long-living resonances which can be discovered as sharp peaks when plotting the reflection probability as $|S_{n,m}^{(-,-)}(\Theta)|^2$ as a function of the quasi- energy Θ, see Fig. 5.

In order to demonstrate the connection between the quantum resonances and the classical stability island we will suppose for a moment that the outermost KAM curve, which sets the boundary of the island and is impenetrable for the classical particle, is impenetrable for quantum particle as well. In this case some quantum states will be trapped inside the island leading to eigenvalues imbedded into the continuous spectrum of the Floquet operator U. (Formally these eigenvalues are real poles of the S-matrix $S(\Theta)$)

The number $\mathcal{N}$ of them is given roughly by the area $\mathcal{S}$ of the stability island divided by $2\pi\hbar$:

$$\mathcal{N} \approx \frac{\mathcal{S}}{2\pi\hbar} \tag{21}$$

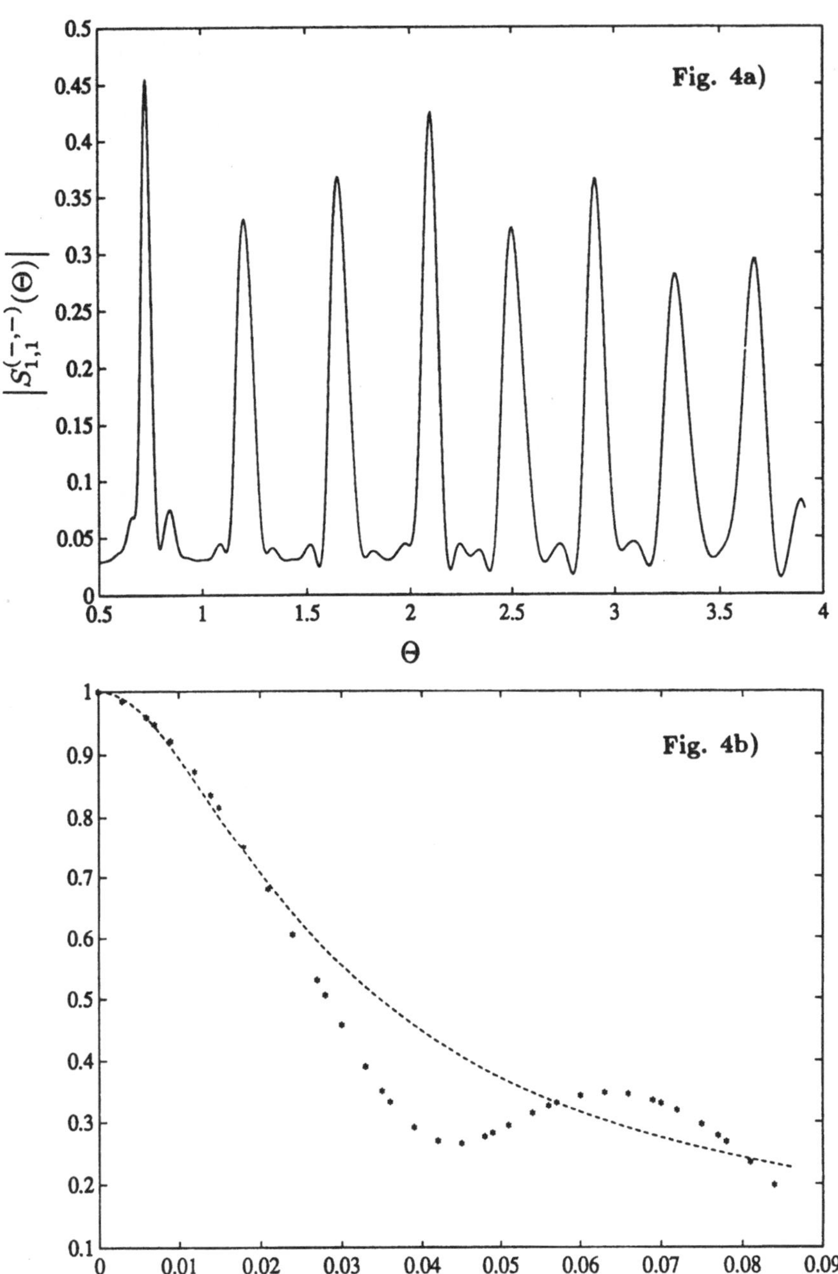

Fig 4. a) The real part of the S-matrix element $S_{1,1}^{(-,-)}(\Theta)$ is plotted as a function of the quasi-energy . $\lambda = 5$ $\hbar = 0.03$

b) The evaluated normalized auto-correlation function $C(\Delta)$ (dots) is compared with the Eriscon shape derived by Blumel and Smilansky. $\lambda = 5$ $\hbar = 0.03$ *and* $\alpha \approx 0.6$.

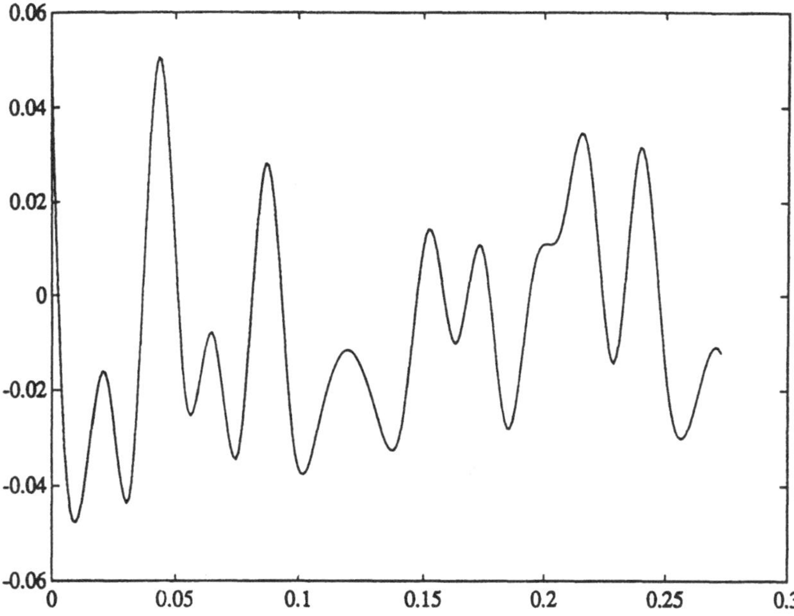

Fig 5. The absolute value of the S-matrix element $S_{1,1}^{(-,-)}(\Theta)$ is plotted as a function of the quasi-energy of the incoming particle. $\lambda = 0.5; \hbar = 0.01$

Moreover in the semi classical regime are the eigenvalues nearly equidistant. (This is a direct consequence of the EBK quantization, see for instance [11].) The distance Δ between the neighboring eigenvalues is easily approximated by

$$\Delta \approx \frac{2\pi}{\mathcal{N}} \tag{22}$$

Inserting (21) into (22) we find

$$\Delta \approx \frac{4\pi^2}{\mathcal{S}}\hbar \tag{23}$$

The quantum tunneling through the outermost KAM curve turns the eigenvalues into resonances and the poles on the S-matrix $S(\Theta)$ will move from the real axis. One can find them, however, by continuing $S(\Theta)$ into the complex plane. The real part of the resonance poles will remain nearly equal to the above described eigenvalues (assuming the tunneling is not very strong). As a result we have to find (for small $\hbar$) a ladder of nearly equidistant resonance peaks when plotting the quantum transition probabilities as a function of the quasi-energy. The distance between the resonance peaks is given by (23). In the case $\lambda = 0.5$ is $\mathcal{S} \approx 0.9$. This leads for $\hbar = 0.01$ to $\Delta = 0.43$ which is in excellent agreement with the mean distance between resonances plotted on Fig.5.

Summarizing we have demonstrated that the existence of a well developed stability island leads in the classical mechanics to a fractal of scattering singularities with fractal dimension equal to one. In the quantum case the tunneling through the outermost KAM curve leads to the appearance of sharp resonances. It is tempting to speculate that the mechanism responsible for their existence (namely the tunneling into the stability domain) can be used to explain some of the recently discovered controversial resonance peaks in the ion–ion scattering experiments [12]. The very recent classical calculations have showed [13] that the phase space portrait of these systems contains stability islands imbedded in the chaotic sea. The quantum tunneling is therefore good candidate for the sought resonance mechanism.

References

[1.] B. Eckhardt, C.Jung: J.Phys.A 19 (1986) L829

[2.] T.Tel: J.Phys.A 22 (1989) L691

[3.] R.Blumel, U.Smilansky: Phys.Rev. Let 60 (1988) 477

[4.] R.Blumel, U.Smilansky: Phys.Rev. Let. 64 (1990) 241

[5.] Ch.F.F.Karney: Physica 8D (1983) 360

[6.] J.D.Meiss, E.Ott: Phys.Rev.Lett. 55 (1985) 2741 W.Bauer at all: Phys. Rev. Lett. 65 (1990) 2213

[7.] B.V.Chirikov, D.L.Shepelyanski: Physica 13 D (1984) 394

[8.] Y.T.Lau, J.M.Finn, E.Ott: Phys.Rev.Lett. 66 (1991) 978

[9.] T.Viczek: Fractal Growth Phenomena, World Scientific, Singapore 1989

[10.] J.Howland: Indiana J. Math. 28 (1979) 471

[11.] L.D.Landau, E.M.Lifshitz: Quantum Mechanics; Chap. 48, Pergamon Press, Oxford 1965

[12.] A.W.Wuasmaa at al: Phys. Rev. C 36 (1987) 1011

[13.] A.Rapisarda, M.Baldo: Phys. Rev. Lett 66 (1991) 2581

Petr Šeba, Nuclear Physics Institute, Czechoslovak Academy of Sciences, 250 68 Řež near Prague, Czechoslovakia

Operator Theory:
Advances and Applications, Vol. 70

Relatively Random Unitary Operators

Karol Życzkowski

Consider unitary operators $\hat{U}_1$ and $\hat{U}_2$ represented by matrices U_1 and U_2 of size N. We shall call operators $\hat{U}_1$ and $\hat{U}_2$ *relatively random*, if

$$\mu = \frac{\mathrm{Re}[\mathrm{Tr}(U_1^\dagger U_2^\dagger U_1 U_2)]}{N} \ll 1. \tag{1}$$

Let $\hat{U}_1' = \hat{U}_2^\dagger \hat{U}_1 \hat{U}_2$ denotes the image of the operator $\hat{U}_1$ transformed by the unitary operation $\hat{U}_2$. It is easy to see that $\mu = \mathrm{Re}[\langle \hat{U}_1' | \hat{U}_1 \rangle]/N$, where $\langle \hat{A} | \hat{B} \rangle = \mathrm{Tr}(A^\dagger B)$ is the scalar product in the space of operators. In other words, an operator $\hat{U}_2$ is relatively random with respect to $\hat{U}_1$, if $\hat{U}_1$ is orthogonal to its image $\hat{U}_1'$. Moreover, the coefficient μ might be used as a measure of commutativity between $\hat{U}_1$ and $\hat{U}_2$, since the norm of commutator reads

$$||[\hat{U}_1, \hat{U}_2]||^2 = 2N(1 - \mu), \tag{2}$$

with the norm $||\hat{A}||^2 = \langle \hat{A} | \hat{A} \rangle$. Relatively random operators do not commute and their eigenbasis are sufficiently different.

The concept of relatively random operators might be used for analysis of quantized chaotic systems. It is well known [1,2] that the statistical properties of quantum chaotic systems are described by ensembles of random matrices [3]. Level spacing distribution which characterizes spectrum of quantum system possessing a generalized time–reversal symmetry is described by the Wigner distribution. Furthermore, according to the theory of random matrices the distribution of components of eigenvectors $y_{ln} = |\langle \psi_l | n \rangle|^2, l = 1, \ldots, N$ of a unitary Floquet operator $\hat{F}$ (or a hermitian Hamiltonian) represented in a suitable basis $|n\rangle, n = 1, \ldots, N$ is given by the Porter–Thomas distribution [4]. We call such a basis *relatively random* with respect to the operator $\hat{F}$.

The distribution of eigenvector components is closely related to the statistics of matrix elements [5,6] of an observable represented in the eigenbasis of the Hamiltonian or the Floquet operator $\hat{F}$. It has been suggested [7] that the statistics of matrix elements of an Hermitian operator $\hat{A}$ is given by the Porter–Thomas distribution, if $\hat{A}$ is relatively random with respect to $\hat{F}$. In this work we conjecture that the statistics of components of eigenvectors of a unitary operator $\hat{F}_1$ describing a chaotic quantum system and represented in the eigenbasis of $\hat{F}_2$, complies to the predictions of random matrices, provided both operators are *relatively random*.

Above mentioned conjecture is supported by a numerical study of the periodically kicked top - a quantum system allowing for chaotic motion [8-10]. Dynamical variables of the system are three components $\hat{J}_l$, $l = 1, 2, 3$ of the angular momentum operator $\hat{J}$. They obey the commutation relation $[\hat{J}_k, \hat{J}_l] = i\epsilon_{kln}\hat{J}_n$. Time evolution of the system is governed by by the Floquet operator

$$\hat{F}(K,p) = \exp(\frac{-iK\hat{J}_x^2}{2j})\exp(-ip\hat{J}_z), \tag{3}$$

where p and K are the parameters of the model. The eigenvalue $j(j+1)$ of the operator $\hat{J}^2$ fixes the dimension of the Hilbert space N as $N = 2j+1$. It is convenient to analyze the system in the eigenbasis of the operator $\hat{J}_z$, $|j,m\rangle$, $m = -j, \dots, j$.

The perturbation operator $\hat{V}$, quadratic in $\hat{J}_x$, does not couple states $|j,m\rangle$ of different parity and the matrix F breaks down into a block diagonal form of size j and $j+1$. Both subspaces are dynamically independent and numerical calculations can be performed separately for each parity. It has been reported [8] that for $p = 1.4$ and the kicking strength $K > 6$ the classical motion is chaotic and the statistical properties of the Floquet operator $\hat{F}$ corresponding to the quantum model can be described by circular orthogonal ensemble (COE) [3].

We are interested in the statistics of eigenvectors of $\hat{F}_1 = \hat{F}(K_1, p_1)$ represented a given orthonormal basis. This basis can be defined as the eigenbasis of a reference operator $\hat{F}_2 = \hat{F}(K_2, p_2)$. In the standard approach to eigenvector statistics one uses the basis of the unperturbed system [11,12], what corresponds to putting $K_2 = 0$ and $p_2 = p_1$. On the other hand, if $K_2 = K_1$ and $p_2 = p_1$, both operators are equal, the statistics of eigenvectors of $\hat{F}_1$ in its eigenbasis is singular and does not contain any information. We put $p_2 = p_1$ and consider arbitrary values of the parameter K_2 determining reference operator $\hat{F}_2$ and study, how large values of the "rotation parameter" $\Delta = k_2 - k_1$ produces COE–like eigenvector statistics described by Porter–Thomas distribution.

Eigenvector statistics may be characterized by the mean entropy of eigenvectors $\langle H \rangle$ [13]

$$\langle H \rangle = -\frac{1}{N}\sum_{l=1}^{N}\sum_{n=1}^{N} y_{ln}\ln(y_{ln}). \tag{4}$$

This quantity varies from zero for totally localized eigenvectors (one component equal to unity and all others to zero) to $\ln(N)$ for a delocalized eigenvector with all components equal to $1/N$. For random matrices representing a member of the orthogonal ensemble the mean entropy can be found analytically [14] and expressed by means of the Digamma Function Ψ [15]

$$H_{OE} = \Psi(\frac{N+2}{2}) - \Psi(\frac{3}{2}). \tag{5}$$

For convenience we use the scaled entropy $\gamma := \langle H \rangle / H_{OE}$ which is equal to unity for matrices pertaining to the orthogonal ensemble.

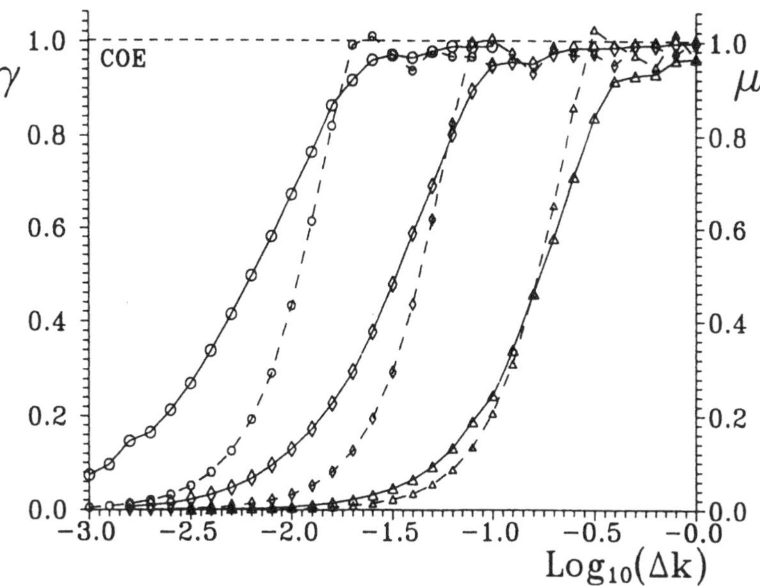

Figure 1. Dependence of scaled entropy of eigenvectors γ on rotation parameter Δ_K for $K = 11.0$, $p = 1.4$ and $j = 400(\circ)$, $j = 100(\diamond)$, $j = 25(\triangle)$. Numerical data are joined by solid lines. Corresponding smaller symbols, connected by dashed lines, represent values of the coefficient μ.

We diagonalized numerically unitary matrices F_1 for values of parameters $p = 1.4$ and $K = 11.0$ corresponding to classically chaotic motion. Obtained eigenvectors where projected onto eigenbasis of $F_2 = F(K + \Delta_K, p)$ and the distribution of eigenvectors was described by the scaled entropy γ. Figure 1 presents the entropy γ as a function of the "rotation parameter" Δ_K (in a logarithmic scale) for $j = 25, 100$ and 400. For small values of Δ_K the reference basis of $\hat{F}_2$ is so close to the eigenbasis of $\hat{F}_1$ that the entropy is negligible. For Δ_K larger than a critical value Δ_c the eigenbasis of F_2 produces eigenvector statistics typical to the orthogonal ensemble and γ achieves unity. Critical value Δ_c is proportional to $1/j$: for larger matrix a smaller value of the rotation parameter Δ_K is sufficient to generate a random basis.

Smaller symbols joined by dashed lines in Figure 1 denote the coefficient μ computed according to equation (1). There are no reasons to expect that for a given spin length j the values of γ and μ would be equal. However, the sudden growth of the coefficient μ coincides, for any j, with the critical value of the rotation parameter Δ_c, for which the scaled entropy tends to unity. Relatively random operator $\hat{F}_2$ generates thus eigenbasis random with respect to the Floquet operator $\hat{F}_1$. Condition (1) might be therefore considered as a simple criterion allowing to select a *random basis*, in which the eigenvector statistics complies with the predictions of ensembles of random matrices.

References

[1] F. Haake, *Quantum Signatures of Chaos*, Springer, Berlin 1991.

[2] L.E. Reichl, *The Transition to Chaos*, Springer, Berlin 1992.

[3] M. L. Mehta, *Random Matrices, 2 ed.*, Academic Press, New York, 1990.

[4] T.A. Brody, J. Flores, J.B. French, P.A. Mello, A. Pandey, S.S.M. Wong, Rev. Mod. Phys. 53, 385 (1981).

[5] M. Feingold, N. Moiseyev, A. Peres, Phys. Rev. A30, 509 (1984).

[6] F. Haake, K. Życzkowski, Phys. Rev. A42, 1013 (1990).

[7] M. Kuś, K. Życzkowski, Phys. Rev. A44, 965 (1991).

[8] F. Haake, M. Kuś, R. Scharf, Z. Phys. B65, 381 (1987).

[9] H. Frahm, H.J. Mikeska, Z. Phys. B65, 249 (1986).

[10] K. Nakamura, Y. Okazaki, A.R. Bishop, Phys. Rev. Lett. 57, 5 (1986).

[11] M. Kuś, J. Mostowski, F. Haake, J.Phys. A21, L1073 (1988).

[12] Y. Alhassid, M. Feingold, Phys. Rev. A39, 374 (1989).

[13] T.H. Seligman, J.J. Verbaarschot, M.R. Zirnbauer, J. Phys. A18, 2751 (1985).

[14] K.R.W. Jones, J. Phys. A23, L1247 (1990).

[15] M. Abramowitz, I.A. Stegun (eds.), *Pocketbook of Mathematical Functions*, Harri Deutsch Verlag, Frankfurt a. Main, 1984.

Karol Życzkowski, Instytut Fizyki, Uniwersytet Jagielloński, ul. Reymonta 4, 30-059 Kraków, Poland

Chapter 6

Operator theory and its application

Operator Theory:
Advances and Applications, Vol. 70

A Trace Formula for Obstacles Problems and Applications

Didier Robert

Abstract

We display here a new trace formula in scattering for obstacles problems. Applications are given for asymptotics of the scattering phase. In particular we prove a Weyl type formula for exterior problems in acoustical scattering for any dimension, extending a result proved by R. Melrose in odd dimension. More generally, we prove that theses results hold for a large class of perturbations of operators elliptic at infinity which may be degenerated in a bounded set.

1 Introduction

One of the most famous result in spectral theory of partial differential equations is the Weyl asymptotic formula for the Laplace-Dirichlet eigenvalues problem on a bounded open set $\Omega \subset \mathbb{R}^n$.
In 1911 [32], H.Weyl proved the following result: let $\{\lambda_j\}_{j\in\mathbb{N}}$ be the nondecreasing serie of eigenvalues for the problem:

$$-\Delta\phi_j = \lambda_j\phi_j; \quad \phi_j|_{\partial\Omega} = 0$$

Let us introduce the counting eigenvalues function:

$$N(\lambda) := Card\{j \in \mathbb{N};\ \lambda_j \le \lambda\}$$

If n=2 and if $\partial\Omega$ is of Lebesgue measure 0 [1] then:

$$\lim_{\lambda\nearrow+\infty} \frac{N(\lambda)}{\lambda} = \frac{Area(\Omega)}{4\pi} \tag{1}$$

After the pionner result by H.Weyl, the asymptotic result (1) was extended in several directions: for any dimension n, for more general boundary value problems, with a remainder estimate and with a second term. If $\Omega \subset \mathbb{R}^n$ is a smooth bounded domain then we have for $\lambda \nearrow +\infty$:

$$N(\lambda) = \gamma_n.vol(\Omega).\lambda^{n/2} + O(\lambda^{(n-1)/2}) \tag{2}$$

[1] see Remark 4.2

The remainder estimate in (2) was a challenging problem during a long time and was proved, in an increasing order of generality, by the following mathematicians: R.Seeley [26], Pham The Lai [19], R.Melrose [14], V.Ivrii [9].
Here we are concerned with obstacle problems where $\Omega = \mathbb{R}^n \backslash K$, K being a compact set of $\mathbb{R}^n$. Let us denote by P the self-adjoint operator $-\triangle$ in $L^2(\mathbb{R}^n)$ with the domain $D(P) = H^2(\Omega) \cap H^1_0(\Omega)$. The spectrum of P is absolutely continuous, equal to the semi real axis: $[0, +\infty[$. In this case, the analogous of the eigenvalues distribution function $N(\lambda)$ is **the scattering phase :** $\theta(\lambda)$ which is defined usually (see Melrose,[15]) by the scattering theory for the wave equation:

$$\frac{\partial^2 u}{\partial t^2} - \Delta u = 0$$

$$\forall x \in \Omega, u(0,x) = f(x);\ u_t(0,x) = g(x);\ \forall x \in \partial\Omega, u(t,x) = 0 \tag{3}$$

Using the Lax-Phillips theory we can define the scattering matrix $S(\lambda)$ acting as a unitary operator in $L^2(\mathbb{S}^{n-1})$; where $\mathbb{S}^{n-1}$ is the unit sphere in $\mathbb{R}^n$. It is well known that $S(\lambda) - \mathbb{1}$ has a smooth kernel (the scattering amplitude) and the scattering phase can be defined by the equality:

$$det\,(S(\lambda)) = \exp(2i\theta(\lambda)) \tag{4}$$

Let us remark here that we can define as well the scattering phase corresponding to the scattering for the Schrödinger equation with an obstacle. Indeed, here we prefer, for technical convenience, to consider that frame.

$$i\frac{\partial \psi}{\partial t} = P\psi \tag{5}$$

$$\forall x \in \Omega, \psi(0,x) = \phi(x);\ \forall x \in \partial\Omega,\ \ \psi(t,x) = 0$$

Let us denote by $s(\lambda)$ the scattering phase associated with the Schrödinger frame. According to the Birman-Krein theory with "two spaces" (see the book [30]) the following trace relation holds, for every $f \in C_0^\infty(]0, +\infty[)$:

$$\int \frac{df}{d\lambda}(\lambda)s(\lambda)d\lambda = tr\,(f(P) - \mathcal{I}f(P_0)\mathcal{I}^*) \tag{6}$$

where $\mathcal{I} : L^2(\mathbb{R}^n) \to L^2(\Omega)$ is the "identification " operator: $\mathcal{I}(\phi) = \phi|_\Omega$.
Moreover we have a simple connection between the scattering phases θ and s, namely: $s(\lambda) = c_n\theta(\lambda^{1/2})$ where c_n is a universal numerical constant. The formula 6 explains in what sense $s(\lambda)$ is the analogous of $N(\lambda)$.

Several mathematicians considered the asymptotic behaviour of $\theta(\lambda)$ as $\lambda \nearrow +\infty$ making some geometrical assumption on the shape of the obstacle K (convex, star-shape, non trapping) (Buslaev[6], Jensen-Kato[10], Majda-Ralston[13], Petkov-Popov[20]. The first general result (without assumption on the shape of the obstacle) was obtained by R.Melrose in 1988 [15]:

Theorem 1.1 *If the dimension n is odd and K is a smooth compact obstacle, then the following asymptotic formula holds:*

$$\theta(\lambda) = \gamma_n.vol(K).\lambda^n + O(\lambda^{n-1}) \tag{7}$$

An essential difference between $N(\lambda)$ and $\theta(\lambda)$ is that the first one is monotone and the second one may be not monotone. So it seems not possible, in general, to apply Tauberian theorem to deduce the behaviour of $\theta(\lambda)$ from the behaviour of some suitable transform of $\frac{d\theta}{d\lambda}(\lambda)$ (Laplace, Fourier, Stieljès,$\cdots$). The other possibility is to get a good approximation of the wave equation for very large times: this is possible if K is non trapping ([20]). In his proof Melrose overcome this difficulty by making a suitable decomposition of $\theta(\lambda)$ using his asymptotic result on the number of scattering poles in ball of radius R ($= O(R^n)$)
The aim of this work is to display a method to solve the following problems:
(*) Is (7) valid for n even ? (it may be surprising that the Weyl formula for the interior problem was proved first for $n = 2$ whereas it is still unproved for two-dimensional exterior problems with general obstacles).
(**) Is (7) valid when Δ is replaced by other elliptic operators, for example for the vibrating plates problem (Δ^2)?

The main point of our method to answer positively to the questions (*), (**), is a suitable trace formula which gives us a decomposition of the scattering phase into pieces which are monotone hence which can be handled with the usual tools of spectral theory for P.D.E.

2 Assumptions and general results in obstacle scattering

Our general frame work is the following. Let us consider a differential operator of order m, $p(x, D) = \sum_{|\alpha|\leq m} a_\alpha(x)D^\alpha$ where $D := -i\nabla$.

We assume that $a_\alpha \in L^1_{loc}(\Omega)$ and the following assumptions:

(H_1) $\exists R_0 > 0$ such that $K \subset \{x \in \mathbb{R}^n;\ |x| < R_0\}$
and $a_\alpha \in C^1(\{x \in \mathbb{R}^n;\ |x| > R_0\})$

(H_2) $\exists R_0 > 0$, $\rho > n$ and an elliptic homogeneous polynomial:

$$p_0(\xi) = \sum_{|\alpha|=m} a_{0,\alpha}\xi^\alpha, \quad a_{0,\alpha} \in \mathbb{C},$$

such that:

$$|\partial_x^\beta(a_\alpha(x) - a_{0,\alpha})| \leq c_{\alpha,\beta} < x >^{-\rho-|\beta|}, \ \forall x,\ |x| > R_0 \tag{8}$$

where $a_{0,\alpha} = 0$ for $|\alpha| < m$ and $< x > := (1 + |x|^2)^{1/2}$

(H_3) $p(x, D)$ is symmetric on $C_0^\infty(\Omega)$

Let us consider a self-adjoint extension P of $p(x,D)$ in $L^2(\Omega)$ with a domain $D(P)$. Assume furthermore:

(H_4) $\forall\zeta \in C_0^\infty(\overline{\Omega})$; $\zeta \equiv 1$ in a neighborhood of K, then we have: $\zeta.D(P) \subset D(P)$.

(H_5) $\exists N_0 \in \mathbb{N}$ such that $\forall\zeta \in C_0^\infty(\overline{\Omega})$ we have: $\zeta.(P+i)^{-N_0}$ is a trace-class operator in $L^2(\Omega)$.

Remark 2.1 Ellipticity for P is assumed only at infinity. In particular P may be degenerate on $\partial\Omega$ (see section 4).

Let us denote by P_0 the self-adjoint extension of $p_0(D)$ in $L^2(\mathbb{R}^n)$ ($D(P_0) = H^m(\mathbb{R}^n)$). As usual, to compare the Hamiltonians P and P_0 we can use the identification operator $\mathcal{I}$ defined above. The general properties given below come from class trace perturbation theory (see [30]). To apply that theory we state the preliminary result:

Proposition 2.2 *Assume (H_1) to (H_5) and $N > \max\{N_0, n/m\}$. Then $\forall z \in \mathbb{C}\backslash\mathbb{R}$, we have the following uniform estimate in trace-norm in $L^2(\mathbb{R}^n)$:*

$$\|\mathcal{I}^*(P-z)^{-N}\mathcal{I} - (P_0-z)^{-N}\|_{tr} \leq C_N \frac{<z>^N}{|\Im z|^{2(N+n)}} \tag{9}$$

Sketch of Proof: We use standard estimates for elliptic operators and arguments from perturbation theory ∎

Corollary 2.3 *(i) $f \to tr\,(f(P) - \mathcal{I}.f(P_0).\mathcal{I}^*)$ is a Schwartz temperate distribution on $\mathbb{R}$.*
(ii) There exists $s : \mathbb{R} \to \mathbb{R}$ such that $\int_{\mathbb{R}} \frac{|s(\lambda)|}{(1+|\lambda|)^N} d\lambda$ and:

$$tr\,(f(P) - \mathcal{I}.f(P_0).\mathcal{I}^*) = \int_{\mathbb{R}} s(\lambda)\frac{df}{d\lambda}(\lambda)d\lambda \tag{10}$$

Furthermore, for the triplet $(P, P_0, \mathcal{I})$, the wave operators:

$$W_\pm := \lim_{t\to\pm\infty} (\exp(itP)\mathcal{I}\exp(-itP_0)) \tag{11}$$

exist and are complete. The scattering matrix, $S(\lambda)$, is a trace-class perturbation of identity on $L^2(\Sigma)$, where $\Sigma = \{\xi \in \mathbb{R}^n, p_0(\xi) = 1\}$ with the Euclidean measure. We have: $det(S(\lambda)) = \exp(-2i\pi s(\lambda))$

3 A trace formula

Let us introduce the notations: $Q = P - P_0$; $A = \frac{1}{2}(x.D + D.x)$. p_0 being homogeneous, we have by the Euler identity:

$$\frac{1}{i}[A, p_0(D)] = m.p_0(D) \tag{12}$$

We fix ζ as in (H_1), (H_2) , with R_0 large enough. Now we can state our main technical result:

Theorem 3.1 *For every $f \in C_0^\infty(\mathbb{R}\backslash(0))$ we have:*

$$\int_{\mathbb{R}} \frac{df}{d\lambda}(\lambda).s(\lambda)d\lambda = tr\,(\mathbb{1}_K.f(P_0)) + tr\,(\zeta^2.f(P))$$
$$+tr\left(\frac{A.[\zeta^2, P]}{im}.P^{-1}.f(P)\right) + tr\left((Q - \frac{[A,Q]}{im})(1-\zeta^2)P^{-1}.f(P)\right) \tag{13}$$

Outline of the proof: By definition, we have:

$$\int \frac{d}{d\lambda}(\lambda.f(\lambda)).s(\lambda)d\lambda = tr\,(P.f(P) - \mathcal{I}P_0.f(P_0)\mathcal{I}^*) \tag{14}$$

Let us begin by a formal munipulation. We have first:

$$P = \frac{1}{im}[A,P] + Q - \frac{1}{im}[A,Q] \tag{15}$$

So we get:

$$tr\,(P.(1-\zeta^2).f(P) - \mathcal{I}P_0.(1-\zeta^2).f(P_0)\mathcal{I}^*) = \tag{16}$$
$$tr\left((Q - \frac{1}{im}[A,Q])(1-\zeta^2).f(P)\right)$$
$$+tr\left(\frac{1}{im}[A,P](1-\zeta^2)f(P) - \mathcal{I}\frac{1}{im}[A,P_0](1-\zeta^2)f(P_0)\mathcal{I}^*\right)$$

The first term of the right hand side is in the trace class by our assumptions H_2. The second term can be checked formally as follows, using cyclicity of traces:

$$\begin{aligned} tr\,([A,P](1-\zeta^2)f(P)) &= tr\,(AP - PA)(1-\zeta^2)f(P)) \\ &= tr\,((A.P(1-\zeta^2) - A(1-\zeta^2)P).f(P)) \\ &= tr\,(A.[\zeta^2, P].f(P)) \end{aligned} \tag{17}$$

The same computation is valid for P_0 with $\zeta \equiv 0$ which gives:

$$tr\,(\mathcal{I}P_0f(P_0).\mathcal{I}^*) = tr\,(\mathbb{1}_K P_0 f(P_0))$$

To make rigorous these computations we proceed as follows:

Let us consider the differential operators with compact support coefficients: $A_R = \chi(\frac{x}{R}).A.\chi(\frac{x}{R})$ where $\chi \in C_0^\infty(\mathbb{R}^n)$ satisfies: $\chi(x) = 1$ for $|x| \leq 1$ and $\chi(x) = 0$ for $|x| \geq 2$.
The calculus is clearly rigorous for A_R in place of A. To end the proof, it is sufficient to apply a uniform trace norm estimate (which will be not proved here) and to take $R \to +\infty$:

Lemma 3.2 *There exists some $C > 0$ such that we have:*

$$\|[A_R, P](1-\zeta^2).f(P) - \mathcal{I}[A_R, P_0](1-\zeta^2)f(P_0)\mathcal{I}^*\|_{tr} \leq C, \ \ \forall R \geq 1 \tag{18}$$

■

4 Applications

We shall give here three applications of our trace formula.
Application I: obstacle problems for second order elliptic operators
Let us consider the differential operator:

$$p(x, D) = \sum_{1 \leq j,k \leq n} D_j(a_{jk} D_k) + \sum_{1 \leq j \leq n} b_j D_j + b_0$$

where we denote : $D_j := i^{-1} \frac{\partial}{\partial x_j}$
Let us assume the following conditions:
(1) $p(x, D)$ is symetric and elliptic:

$$\exists E > 0; \sum a_{jk}(x) \xi_j \xi_k \geq E.|\xi|^2$$

(2) $\exists \rho > n$; $\forall \alpha$; $\forall (j, k)$

$$|D^\alpha(a_{jk}(x) - \delta_{jk})| + |D^\alpha b_j(x)| \leq < x >^{-\rho - |\alpha|} \tag{19}$$

(3) P is a self-adjoint extension of $p(x, D)$ in $L^2(\Omega)$; $\Omega = \mathbb{R}^n \backslash K$, K being a smooth compact obstacle, defined by Dirichlet or Neumann or Robin boundary condition on $\partial\Omega$.

Theorem 4.1 *Under the above conditions, for every dimension $n \geq 1$, the spectral shift function $s(\lambda)$ for the scattering pair (P, P_0), where P_0 is the self-adjoint extension of $-\Delta$ in $L^2(\mathbb{R}^n)$, satisfies the following Weyl formula:*

$$s(\lambda) = \gamma_n.\overline{Vol}(K).\lambda^{n/2} + O(\lambda^{(n-1)/2}); \ \lambda \nearrow +\infty \tag{20}$$

where

$$\overline{Vol}(K) := Vol(K) + \int_{\mathbb{R}^n \backslash K} ((det[a(x)])^{n/2} - 1) dx$$

and $\gamma_n = \frac{(4\pi)^{-n/2}}{\Gamma(n/2+1)}$

Remark 4.2 For n odd and $p(x, D) = -\Delta$ we recover the Melrose's result[15]. Always for n odd and $p(x, D) = -\Delta$ outside a compact set, our result can also be obtained from the estimate of the scattering poles by Vodev, Sjöstrand-Zworski [27] using Melrose's method. But this method can not be used in even dimension or for non compact support perturbations (see also [23, 24, 25])

Remark 4.3 As it was pointed to us during the conference by M.S. Birman and by M. van den Berg, the first term asymptotic of the Weyl formula for the Laplace-Dirichet problem is valid for any bounded domain (without any smoothness condition). We can see easily that this remark can be extended for the exterior problem, using our trace formula and heat kernel techniques to control the boundary contribution.([11, 3, 17])

Application II: Exterior problem for the biharmonic operator

Let us consider $P_0 = \Delta^2$ as a self-adjoint operator in $L^2(\mathbb{R}^n)$ and $K \subset \mathbb{R}^n$ a smooth obstacle as above. Let us consider P a sef-adjoint realization of Δ^2 in $L^2(\Omega)$ ($\Omega = \mathbb{R}^n \backslash K$); and $s(\lambda)$ the scattering phase for the pair $(P_0; P)$. We assume that the domain $D(P)$ of P is determined by elliptic boundary conditions.(in particular Shapiro-Lopatinski conditions hold). Using Vasil'ev's technics [28, 29] we get:

Theorem 4.4 *For* $\lambda \nearrow +\infty$ *we have:*

$$s(\lambda) = \gamma_n.Vol(K).\lambda^{n/4} + O(\lambda^{(n-1)/4})$$

Remark 4.5 With less refined but more general and more flexible technics, ([1]) we can get the above asymptotic formula with a less acurate remainder term $O(\lambda^{(n-\delta)/4})$ for some $\delta > 0$.

Application III: Elliptic degenerate operators

Let us consider the second order operators:

$$p(x, D) = -\sum \partial_j(\varphi a_{jk}\partial_k)$$

the a_{jk} satisfying the assumptions (A) of section 2; $\varphi : \mathbb{R}^n \to [0, +\infty[$ is a smooth function such that:

$$\partial K = \{x : \varphi(x) = 0\};\ \varphi(x) > 0 \Leftrightarrow x \in \Omega$$
$$x \in \partial K \Rightarrow \nabla J\varphi \neq 0$$
$$\exists \rho > n \text{ suchthat} |\partial^\alpha (\varphi(x) - 1)| \leq C_\alpha < x >^{-\rho-|\alpha|} \tag{21}$$

For example we could have: $\varphi(x) = \left(\frac{|x|^{2k}-1}{|x|^{2k}+1}\right)$ with $k \in \mathbb{N}$; $k > \frac{n}{2}$.

Let us consider the realization P of $p(x, D)$ generated by the obvious variationnal form in the Hilbert space:

$$H_1^{1/2}(\Omega) := \{u \in \mathcal{D}'(\Omega);\ \varphi^{1/2}\partial^\alpha u \in L^2(\Omega);\ |\alpha| \leq 1\}$$

It is known that for every $\zeta \in C_0^\infty(\mathbb{R}^n)$ and for $k > n$, $\zeta(P+i)^{-k}$ is a trace class operator in $L^2(\Omega)$.
The Weyl formula for degenerated elliptic operators on bounded domains was proved in different contexts by several authors: Baouendi-Goulaouic [2] Nordin [18], Pham The Lai [19] Bolley-Camus-Pham The Lai [4], Menikoff-Sjöstrand [16]. Using their methods and our trace formula we can get the asymptotics for the scattering phase of the pair (P_0, P)

Theorem 4.6 *We have, for some $\theta > 0$:*

$$If n > 2;\ s(\lambda) = d_n\lambda^{n-1} + O(\lambda^{n-1-\theta}) \tag{22}$$

$$If n = 2;\ s(\lambda) = d_2\lambda\log(\lambda) + O(\lambda^{1-\theta}) \tag{23}$$

The constant d_n is supported by the boundary ∂K and is the same as for the Weyl asymptotics in bounded domains ([18])

References

[1] S. Agmon: Asymptotic formulas with remainder estimates of eingenvalues of elliptic operators. Arch. Rational Mech. Anal. 28, 165-183, (1968)

[2] M.S. Baouendi et C. Goulaouic: Régularité et théorie spectrale pour une classe d'opérateurs elliptiques dégénérés. Arch. Rat. Mech. Anal., vol.34, 361-379, (1969).

[3] M.S. Birman and M.Z. Solomyak: Leading term in the asymptotic spectral formula for "nonsmooth" elliptic problems. Functonnal Anal. Appl. 4, 265-275 (1970)

[4] P. Bolley, J. Camus, Pham The Lai: Noyau, Résolvente et Valeurs propres d'une classe d'opérateurs elliptiques et dégénérés. Lecture Notes in Math. N^0 660 "Equations aux dérivées partielles" Proceedings, Saint Jean de Monts (1977)

[5] C. Bardos, J.C. Guillot, J. Ralston: La relation de Poisson pour l'équation des ondes dans un ouvert non borné. Application à la théorie de la diffusion. Comm. in P.D.E, 7(8), 905-958 (1982)

[6] V.S. Buslaev: On the asymptotic behaviour of the spectral caracteristics of exterior problems for the Schrödinger operator. Math.USSR Izvestija. Vol.9 $N^0$1 (1975)

[7] G. Grubb: Remarks on trace estimates for exterior boundary problems. Comm. in P.D.E, 9(3), 231-270 (1984)

[8] L.Hörmander: The spectral function of an elliptic operator, Acta. Math. 121 (1968) 193-218

[9] V. Ivrii: Book in preparation and preprints of Ecole Polytechnique (1990-92)

[10] A. Jensen, T. Kato: Asymptotic behaviour of the scattering phase for exterior domains. Comm. in P.D.E, 3(12), 1165-1195 (1978).

[11] M. Kac: Can we hear the shape of a drum? American Math. Monthly, 73S, 1-23 (1966)

[12] P. Lax and R. Phillips: Scattering theory, Academic Press (1967)

[13] A. Majda, J. Ralston: An analogue of Weyl's formula for unbounded domains. Duke Math. J. ; **45** 183 (1978); **45** 513 (1978); **46** 725 (1979)

[14] R. Melrose: Weyl's conjecture for manifolds with concave boundary. Proc. Symp. on Pure Math. A.M.S, 257-273, (1980)

[15] R. Melrose: Weyl asymptotics for the phase in obstacle scattering. Commun. in P.D.E, 13 (11), (1988) 1431-1439.

[16] A. Menikoff, J. Sjöstrand: On the eigenvalues of a class of hypoelliptic operators. Math. Ann. 235, 55-84, (1978)

[17] G. Métivier: Valeurs propres de problèmes aux limites irréguliers. Bull. Soc. Math. France, mémoire 51-52, 125-219 (1977)

[18] C. Nordin: The asymptotic distribution of the eigenvalues of a degenerate elliptic operator. Ark för Math., vol.10, 3-21, (1972)

[19] Pham The Lai: Comportement asymptotique du noyau de la résolvente et des valeurs propres d'une classe d'opérateurs elliptiques dégénérés non nécessairement auto-adjoint. J. Math. pures et appl., 55 , 1-42, (1976)

[20] V. Petkov and G. Popov: Asymptotic behaviour of the scattering phase for non trapping obstacles. Ann. Inst. Fourier, Grenoble 32, 3, 111-149 (1982)

[21] G. Popov: Asymptotic behaviour of the scattering phase for the Schrödinger operator. Publication of the Academy of Sciences Sofia-Bulgaria (1982)

[22] M. Reed, B. Simon: Scattering theory. Academic Press (1979)

[23] D. Robert: Asymptotique à grande énergie de la phase de diffusion pour un potentiel, Asymptotic Analysis 3, 301-320 (1991)

[24] D. Robert: Asymptotique de la phase de diffusion à haute énergie pour des perturbations du second ordre du Laplacien. Ann.scient.Ec. Norm. Sup 4e série,t.25,1992, p.107 à 134.

[25] D.Robert: Relative Time Delay and Trace Formula for Long Range perturbations of Laplace Operator. Operator Theory; Advances and Applications, Vol.57 (1992), Birkhäuser Verlag Basel.

[26] R. Seeley: A sharp asymptotic remainder estimate for the eigenvalues of the Laplacian in a domain of $\mathbb{R}^3$. Advances in Math. 29, 244-269 (1978)

[27] J. Sjöstrand, M. Zworskii: Complex scaling and the distribution of scattering poles. J. of Amer. Math. Soc. 4(4), 729-769, (1991)

[28] D.G. Vassil'ev: Two-term asymptotics of the spectrum of a boundary value problem for inside reflection of a general type. Functional Anal. Apll. 18, 1-13 (1984)

[29] D.G. Vassil'ev: Asymptotics of the spectrum of a boundary value problem. Trans. Moscow Math. Soc. 167-237 (1987)

[30] D. Yafaev: Mathematical Scattering Theory. General Theory. AMS Rhode Island, Vol. 105 (1992)

[31] G. Vodev: Sharp polynomial bounds on the number of scattering poles for perturbations of Laplacian. Comm. Math. Phys. 146(1), 205-216, (1992).

[32] H. Weyl: Das asymptotische Verteilungsgesetz der eigenwerte linearer partieller Differentialgleichungen, Math. Ann. 71. 441-469 (1911)

Didier Robert, Department of Mathematics; CNRS-URA 758, University of Nantes, 2, rue de la Houssinière, 44072-Nantes-Cédex 03-France

Operator Theory:
Advances and Applications, Vol. 70

Propagation in Irregular Optic Fibres

F. Bentosela

Introduction

In a recent work (see [B]) we described how to calculate the attenuation in an optical fiber which presents random microdefects.We considered defects which can be described by functions depending only on the longitudinal coordinate, z. For instance, it can be a fiber whose core radius is given by one random process: $\rho_\omega(z)$ or it can be a fiber which has constant core radius but presents microbends, in this case one considers that its axis coordinates are given by two random processes: $x^0_\omega(z)$ and $y^0_\omega(z)$.

In this conference we propose a resume of that work. We showed that the system of coupled differential equations which describe the behavior of the mode amplitudes can be, up to a small error, converted in a system of N decoupled differntial equations plus a coupled system (we call this process, the diagonalisation process).

Each of these second order differential equations gives rise to two symetric Lyapunov exponents. The largest negative one, which correspnds to the solution with slower exponential decrease, can be considered as the attenuation coefficient of general signals propagating in the fiber. We will indicate rapidly the way how to deduce the Lyapunov from the fiber caracteristics and in particular from the distribution of the defects.

One of the problems which arises comes from the fact that our "diagonalisation process" needs differentiability for the coefficients entering the differential system. This means that we have to suppose that the functions which describe the defects are not simply described by ramdom processes of diffusion type but by differentiable processes. Then we cannot directly use the previous results [APW or PF] on Lyapunov estimates which use the former hypothesis.

Diagonalisation

Field components, ψ, satisfy inside the fiber, Helmholtz type equations:

$$H\psi = -\Delta\psi(x,y,z) - (\frac{2\pi}{\lambda_0})^2 n(x,y,z)^2\psi(x,y,z) = 0 \tag{1}$$

λ_0, represents the wavelength in the air, it is a small quantity of the order of 10^{-6}, so, it will play the role of a small parameter, n(x,y,z) is the refraction index at point (x,y,z). We have some boundary condition at the fiber surface.

Call S_z, the fiber section at z and H_z, the restriction of H to S_z acting on the Hilbert space $L^2(S_z)$. Let us denote by $\phi_j^z(x,y)$ and E_j^z resp. its eigenvectors and corresponding eigenvalues:

$$H_z\phi_j^z(x,y) = E_j^z\phi_j^z(x,y) \tag{2}$$

E_j^z are proportional to λ_0^{-2}, then they will be written as $\lambda_0^{-2}\epsilon_j(z)$. The $\phi_j^z(x,y)$ form a basis for $L^2(S_z)$, so, we can express on this basis, a function ψ defined in the volume of the fiber:

$$\psi(x,y,z) = \sum_{j=1}^{\infty} a_j(z)\phi_j^z(x,y)$$

Plugging this expression in equation (1) and using equation (2), we get an infinite set of differential equations:

$$-\frac{d^2a_m}{dz^2} - 2\sum_{j=1}^{\infty} K_{mj}^{(1)}(z)\frac{da_j}{dz} - \sum_{j=1}^{\infty} K_{mj}^{(2)}(z)a_j + \lambda_0^{-2}\epsilon_m(z)a_m = 0 \tag{3}$$

$m = 1, 2\ldots$ where $a_m(z)$ is the amplitude of the mth mode at distance z from the origin and the K are coupling random variables given by:

$$K_{mj}^{(1)}(z) = (\phi_j^z, \frac{\partial\phi_j^z}{\partial z}), \qquad K_{mj}^{(2)}(z) = (\phi_j^z, \frac{\partial^2\phi_j^z}{\partial^2 z})$$

where $(*,*)$ means scalar product in $L^2(S_z)$.

Multiplying the left hand side by λ_0^2 and writing this system as a first order differential system we obtain:

$$\lambda_0\frac{d}{dz}\begin{pmatrix} a_1 \\ \lambda_0\frac{da_1}{dz} \\ a_2 \\ \lambda_0\frac{da_2}{dz} \\ \vdots \end{pmatrix} =$$

$$\begin{pmatrix} 0 & 1 & 0 & 0 & \ldots \\ \epsilon_1(z)+\lambda_0^2K_{11}^{(2)}(z) & -2\lambda_0K_{11}^{(1)} & -\lambda_0^2K_{12}^{(2)} & -2\lambda_0K_{12}^{(1)} & \ldots \\ 0 & 0 & 0 & 1 & \ldots \\ -\lambda_0^2K_{21}^{(2)} & -2\lambda_0K_{21}^{(1)} & \epsilon_2(z)+\lambda_0^2K_{22}^{(2)} & -2\lambda_0K_{22}^{(1)} & \ldots \\ \ldots & \ldots & \ldots & \ldots & \ldots \end{pmatrix}\begin{pmatrix} a_1 \\ \lambda_0\frac{da_1}{dz} \\ a_2 \\ \lambda_0\frac{da_2}{dz} \\ \vdots \end{pmatrix}$$

Let us call this differential system, "system A" and write it in the form:

$$\lambda_0 \frac{dA}{dz} = \left(D_0 + \lambda_0(D_1 + V_1) + \lambda_0^2(D_2 + V_2)\right) A \tag{4}$$

where D_i are "diagonal" matrices i.e constituted of 2x2 blocks on the diagonal. and V_i are off "diagonal". For instance,

$$D_0 = \begin{pmatrix} 0 & 1 & 0 & 0 & \dots \\ \epsilon_1(z) & 0 & 0 & 0 & \dots \\ 0 & 0 & 0 & 1 & \dots \\ 0 & 0 & \epsilon_2(z) & 0 & \dots \\ \dots & \dots & \dots & \dots & \dots \end{pmatrix}$$

Let us write:

$$\lambda_0 \frac{dC}{dz} = \left(D_0 + \lambda_0 D_1 + \lambda_0^2 D_2\right) C$$

and call it, "system C". Each of the decoupled second order differential equations of "system C" gives rise to a pair of symetric Lyapunov exponents whih can be ordered. In general, positive Lyapunov increase as E increases so the smallest positive Lyapunov exponent would correspond to the first mode. If not, we shall select the mode which gives the smallest one and write it at the first place.

We gave an algorithm which constructs a transform $A \to B$ in such a way the two first components of the new function B are partially decoupled (up to order n), i.e. B satisfies:

$$\lambda_0 \frac{dB}{dz} = \left(\sum_{s=0}^{n} \lambda_0^s \widetilde{D}_s\right) B + \lambda_0^{n+1} R(\lambda_0) B \tag{5}$$

where the $\widetilde{D}_s$ are of the form

$$\widetilde{D}_s = \begin{pmatrix} * & * & 0 & 0 & 0 & 0 & 0 & \cdots \\ & * & 0 & 0 & 0 & 0 & 0 & \cdots \\ 0 & 0 & * & * & * & * & * & \cdots \\ 0 & 0 & * & * & * & * & * & \cdots \\ 0 & 0 & * & * & * & * & * & \cdots \\ 0 & 0 & * & * & * & * & * & \cdots \\ 0 & 0 & * & * & * & * & * & \cdots \\ 0 & 0 & * & * & * & * & * & \cdots \\ .. & .. & .. & .. & .. & .. & .. & .. \\ .. & .. & .. & .. & .. & .. & .. & .. \end{pmatrix}$$

We shall call this differential system, "system B".

The transform $A \to B$ will be such that its inverse, T, is expressed as $T = \sum_{r=0}^{n} \lambda_0^r T_r$ with $T_0 = \mathbf{1}$ and T_r, $r \neq 0$ are of the form

$$T_r = \begin{pmatrix} 0 & 0 & * & * & * & * & * & \cdots \\ 0 & 0 & * & * & * & * & * & \cdots \\ & & * & 0 & 0 & 0 & 0 & 0 & \cdots \\ & & * & 0 & 0 & 0 & 0 & 0 & \cdots \\ & & * & 0 & 0 & 0 & 0 & 0 & \cdots \\ & & * & 0 & 0 & 0 & 0 & 0 & \cdots \\ & & * & 0 & 0 & 0 & 0 & 0 & \cdots \\ .. & .. & .. & .. & .. & .. & .. & .. \\ .. & .. & .. & .. & .. & .. & .. & .. \end{pmatrix}$$

Plugging $A = \left(\sum_{r=0}^{n} \lambda_0^r Tr \right) B$ in (4) and using (5) we get the following commutator equations for the unknowns: T_r and $\widetilde{D}_s$

$$[T_1, D_0] = D_1 + V_1 - \widetilde{D}_1 \tag{5.1}$$

$$[T_2, D_0] = -\overset{\circ}{T}_1 - T_1\widetilde{D}_1(-T_0\widetilde{D}_2) + (D_1 + V_1)T_1 + (D_2 + V_2)T_0 \tag{5.2}$$

$$\cdots\cdots$$

$$\begin{aligned}[T_n, D_0] = -\overset{\circ}{T}_{n-1} - T_{n-1}\widetilde{D}_1 \ldots T_0\widetilde{D}_n + (D_1 + V_1)T_{n-1} \\ + (D_2 + V_2)T_{n-2} + \ldots + (D_n + V_n)T_0\end{aligned}$$

The rest is given by:

$$\begin{aligned}R(\lambda_0) = (1 + \lambda_0 T_1 + \lambda_0^2 T_2 + \ldots + \lambda_0^n T_n)^{-1} \cdot [-\overset{\circ}{T}_n - (T_n\widetilde{D}_1 + \ldots + T_1\widetilde{D}_n) \\ + (D_1 + V_1)T_n + (D_2 + V_2)T_{n-1} + \ldots + (D_n + V_n)T_1 \\ - \lambda_0(T_n\widetilde{D}_2 + \ldots + T_2\widetilde{D}_n) + \lambda_0(D_2 + V_2)T_n + \ldots + (D_n + V_n)T_2 \\ \ldots \\ - \lambda_0^{n-1}T_n\widetilde{D}_n + \lambda_0^{n-1}(D_n + V_n)T_n]\end{aligned}$$

Commutator equations can be solved step by step.We are mainly interested in the upper left part of $\widetilde{D}_s$. For instance if,

$$D_0^{ii} = \begin{pmatrix} 0 & 1 \\ \epsilon_i & 0 \end{pmatrix}, \quad D_1 = 0, \quad D_2^{ii} = \begin{pmatrix} 0 & 1 \\ K_{ii}^{(2)} & 0 \end{pmatrix}$$

$$V_1^{ij} = \begin{pmatrix} 0 & 1 \\ 0 & -2K_{ij}^{(1)} \end{pmatrix}, \quad V_2 = 0$$

Then expressions for the upper left part of $\widetilde{D}_s$, denoted $\widetilde{D}_s|_1$ are:

$$\widetilde{D}_1|_1 = 0\,, \qquad \widetilde{D}_2|_{(1)} = \begin{pmatrix} 0 & 0 \\ 1 & 0 \end{pmatrix} \cdot \left(\sum_i 4 \frac{K_{1i}^{(1)} K_{i1}^{(1)}}{\epsilon_1 - \epsilon_i} + K_{11}^2 \right)$$

$\widetilde{D}_3|_1$ has a long expression containing the first derivatives of $K_{1i}^{(1)}$, $K_{i1}^{(1)}$, ϵ_1 and ϵ_i.

Lyapunov calculus

To study the smallest positive Lyapunov exponent of "system A" we use general theorems on stability of Lyapunov spectra, (see [DK]):
- Transform T and its inverse are bounded, then "systems A and B" are said to be kinematicaly similar and as a consequence they have the same Lyapunov spectra.
- "System B" has a Lyapunov spectrum which is close to the Lyapunov spectrum of the partially diagonalised system (obtained from system B neglecting the rest and called B_D^1), the distance between the two is proportional to the norm of the rest, so, it is of order n in λ_0.

To study the Lyapunov spectrum of A we have first to study the Lyapunov spectrum of B_D^1. It is composed of two parts: there are two symetric Lyapunov γ_1 and $-\gamma_1$ coming from the decoupled second order differential equation and the second part which is coming from the remaining coupled equations. Is the smallest positive Lyapunov coming from the second order differential equation? To answer, we have to control the Lyapunov spectrum of the remaining coupled equations. To do that, we neglect in it the off diagonal terms, estimate the Lyapunov of the obtained diagonal system and the corrections introduced by the off diagonal. If the Lyapunov spectrum of the remaining coupled equations is contained in an interval upper than γ_1, then, the answer is YES. If not, we start the diagonalisation procedure as many times as necessary, to know with certitude from what differential equations are coming the first two smallest positive Lyapunov exponants. Suppose we do it N times and call B_D^N the system obtained, it is composed of N decoupled differential equations and of a coupled system. Two decoupled equations will give us the first two positive Lyapunov exponents γ_1^N and γ_2^N for system B_D^N.

Now it is necessary to compare $\gamma_2^N - \gamma_1^N$ and the rests. The smallest positive Lyapunov of "system A" will be well approximated by γ_1^N only if the rests are much smaller than $\gamma_2^N - \gamma_1^N$.

Remark. Rests are of order n in λ_0, and to be small it is necessary that $\epsilon_1(z)$ is not to close to $\epsilon_2(z)$. This condition is not sufficient since in the rests enter also the derivatives.

In conclusion, if $\epsilon_1(z)$ differs from $\epsilon_2(z)$ and if we add artificially a small multiplicative parameter κ in front of the V_s, for small κ it exists a second order differential equation which will give us an approximation of the smallest positive Lyapunov up to order n in λ_0. Is it true for $\kappa = 1$? This will depend on the concrete fiber we are considering.

The problem is now reduced to the calculus of the Lyapunov exponents of an equation of type:

$$-\frac{d^2a_1}{dz^2} + [E_1(z) + W(z)]\, a_1(z) = 0$$

where $W(z)$ is a small random term which can be expressed as a finite series,whose number of terms depend on the order of derivability of the random variables describing the imperfections. If they are C^r:

$$W(z) = W_0(z) + \lambda_0 W_1(z) + \lambda_0^2 W_2(z) + \ldots + \lambda_0^{r-1} W_{r-1}(z)$$

For instance if the random has two derivatives

$$W(z) = E_1(z) - K_{11}^{(2)}(z) + 4E_1(z)\sum_i \frac{K_{1i}^{(1)}K_{i1}^{(1)}}{E_1(z) - E_i(z)} + \lambda_0 W_1(z)$$

Expressions for the Lyapunov were given (see [PF] or [APW]) supposing W(z) of the form $F(Y_\omega(z))$, where F is a continuous function from a compact manifold, $\mathbf{M}$, to $\mathbf{R}$ and $Y_\omega(z)$ a diffusion process on $\mathbf{M}$ (the role of time being played by the space coordinate).

In our case, except for the last one, terms in the series have derivatives, so they cannot be expressed in terms of functions of diffusion process. In [B] is developped an other kind of Lyapunov calculus which can be applied to this case, for instance to the case the process is the convolution of the previous process by a C^r function. It is seen that all W_i with $i < r-2$ dont appear in the expression for the Lyapunov,and as a consequence it appears $\lambda_0^{2(r-2)}$in front of it. In particular if the random variables describing the imperfections are C^2 we recover the usual situation, the Lyapunov has no λ_0 in front and in this case it is given by expression:

$$\gamma_1 \simeq \frac{1}{4\pi}\int_0^\infty b(z) cos 2\sqrt{\lambda} z dx$$

where $b(z) = \mathbf{E}\,[W_0(0)W_0(z)]$.

References

[APW] Arnold L., Papanicolaou G.,Wihstutz V.: SIAM J.Apply.Math., **46** (1986).

[B] Bentosela F.: CPT Preprint**2925** (1993).

[BP] Bentosela F.,Piccoli P.,J.: Phys.France, **49** (1988).

[DK] Daleckii Ju., Krein M.: *Stability of Solutions of Differential Equations in Banach Space*, AMS vol 43, (1974).

[PF] Pastur L.,Figotin J.: *Spectra of Random and Almost-Periodic Operators*, Springer- Verlag (1992).

F. Bentosela, CPT CNRS Case 907, 163 Av. de Luminy, F-13288 Marseille, Cedex 9

Operator Theory:
Advances and Applications, Vol. 70

Singular perturbations, regularization and extension theory

H. Neidhardt
V.A. Zagrebnov

For nonpositive singular potentials in quantum mechanics it can happen that the Schrödinger operator is not essentially self-adjoint on a natural domain of definition or not semibounded from below. In this case we have a lot of self-adjoint extensions each of them is a candidate for the right physical Hamiltonian of the system. Hence the problem arises to single out the right physical self-adjoint extension. Usually this problem is solved as follows. At first one has to approximate the singular potential by a sequence of bounded potentials (cut-off approximation). After that one has to show that the arising sequence of Schrödinger operators converges in the strong resolvent sense to one of the self-adjoint extensions if the cut-off approximation tends to the singular potential. The so determined self-adjoint extensions is regarded as the right physical Hamiltoninan. Very often the right physical Hamiltonian coincides with the Friedrichs extension.

With respect to the Schrödinger operator in $L^2(\mathbf{R}^2)$ this problem was discussed by [3], [4], [5], [9] and [10]. An operator-theoretical investigation of this problem was started by Nenciu in [8] and continued by the authors in [7]. In the following we continue those abstract investigations. We assume that a semibounded symmetric operator admits a monotonously decreasing sequence of semibounded symmetric operators such that the corresponding sequence of Friedrichs extensions converges in the strong resolvent sense to the Friedrichs extension of the symmetric operator with which we have started. The problem will be to find necessary and sufficient conditions that any other sequence of semibounded self-adjoint extensions of the decreasing sequence of symmetric operators converges to this Friedrichs extension too. Unfortunately, we are unable to solve ths problem in full generality. This means we have found a necessary condition which must be satisfied in order to have the desired convergence. However, we can prove the converse only for special sequences of self-adjoint extensions but not for all.

In more detail the problem can be described as follows. Let A and V be two nonnegative self-adjoint operators on the separable Hilbert space $\mathcal{H}$. Further, let $\mathcal{D} \subseteq dom(A) \cap dom(V)$ a dense subset of $\mathcal{H}$ such that

$$(Vf, f) \leq a(Af, f) + b\|f\|^2, \qquad f \in \mathcal{D}, \qquad 0 < a, b. \tag{1}$$

We introduce the abstract operator $\dot{H}_\alpha$

$$\dot{H}_\alpha f = Af - \alpha V f, \quad f \in \mathrm{dom}(\dot{H}_\alpha) = \mathcal{D}, \quad \alpha > 0. \tag{2}$$

If the coupling constant α, $\alpha > 0$, obeys $\alpha < 1/a$, then the operator $\dot{H}_\alpha$ is symmetric, closable and semibounded with lower bound $-\alpha b$. However, the operator $\dot{H}_\alpha$ is in general not esssentially self-adjoint.

Example 1 Let $\mathcal{H} = L^2(\mathbf{R}^1)$ and let A be the usual Laplace operator on $L^2(\mathbf{R}^1)$, i.e. $A = -d^2/dx^2$. By V we denote the multiplication operator arising from the real potential $V(x)$,

$$V(x) = \frac{1}{4\kappa}\frac{1}{|x|^\beta}, \qquad 1 \le \beta \le 2, \qquad \kappa > 0. \tag{3}$$

Let $\mathcal{D} = C_0^\infty(\mathbf{R}^1 \setminus \{0\})$. If $1 \le \beta < 2$, then for every $\kappa > 0$ there are real numbers $a < 1$ and $b \ge 0$ such that

$$\int_{-\infty}^{\infty} \frac{1}{4\kappa}\frac{1}{|x|^\beta}|f(x)|^2 dx \le a \int_{-\infty}^{\infty} |f'(x)|^2 dx + b \int_{-\infty}^{\infty} |f(x)|^2 dx. \tag{4}$$

for $\kappa > 0$. If $\beta = 2$, then this is only true for $\kappa > 1$.

Example 2 Let $\mathcal{H} = L^2(\mathbf{R}^2)$ and let A be the usual Laplace operator on $L^2(\mathbf{R}^2)$, i.e. $A = -\Delta$. Further, let Γ be a smooth curve in $\mathbf{R}^2$ which is parameterized by

$$\Gamma = \{(x, y) \in \mathbf{R}^2 : x = \rho(\varphi)\cos\varphi, y = \rho(\varphi)\sin\varphi, 0 \le \varphi < 2\pi\} \tag{5}$$

where $\rho(\varphi) > 0$ is a smooth function. Again V is the multiplication operator arising from

$$V(x) = \frac{1}{5\kappa}\frac{1}{||x| - \rho(\varphi)|^\beta}, \qquad 1 \le \beta \le 2, \quad |x| = \sqrt{x^2 + y^2}. \tag{6}$$

We set $\mathcal{D} = C_0^\infty(\mathbf{R}^2 \setminus \Gamma)$. If $1 \le \beta < 2$, then for every $\kappa > 0$ there are real numbers $a < 1$ and $b \ge 0$ such that

$$\int_{\mathbf{R}^2} \frac{1}{5\kappa}\frac{1}{||x| - \rho(\varphi)|^\beta}|f(x)|^2 dx \le a \int_{\mathbf{R}^2} |\nabla f(x)|^2 dx + b \int_{\mathbf{R}^2} |f(x)|^2 dx. \tag{7}$$

For $\beta = 2$ this is true only for $\kappa > 1$.

Let us assume that the $\dot{H}_\alpha$ is not essentially self-adjoint. Since $\dot{H}_\alpha$ is semibounded the Friedrichs extension $\hat{H}_\alpha$ exists. Moreover, denoting by $\hat{A}$ the Friedrichs extension of $\dot{A} = A|\mathcal{D}$ it is not hard to see that $\hat{H}_\alpha$ coincides with the form sum of $\hat{A}$ and $-\alpha V$, i.e.

$$\hat{H}_\alpha = \hat{A} \dotplus (-\alpha V). \tag{8}$$

In the above examples the Friedrichs extension corresponds to the Dirichlet boundary condition at $x = 0$ for the first example and on Γ for the second one.

Next let us introduce a regularizing sequence for the singular perturbation.

Definition 3 A sequence $\{V_n\}_{n\geq 1}$ of bounded non-negative self-adjoint operators is called a regularizing sequence of V if

(i) $V_1 \leq V_2 \leq \ldots \leq V_n \leq \ldots \leq V$

(ii) $lim_{n\to\infty}(V_n f, f) = (Vf, f)$, $f \in \mathcal{D} \subseteq dom(V)$.

Example 4 In the Examples 1 and 2 the sequence V_n is given as multiplication operators with the cut-off potentials

$$V_n(x) = \inf_{x\in\mathbf{R}^l}\{n, V(x)\}, \qquad l = 1, 2. \tag{9}$$

With the regularizing sequence $\{V_n\}_{n=1}^{\infty}$ we associate the following sequence of self-adjoint operators $H_{\alpha,n}$,

$$H_{\alpha,n} = A - \alpha V_n, \qquad n = 1, 2, \ldots. \tag{10}$$

The problem is now to find conditions which guarantee that the approximating sequence $\{H_{\alpha,n}\}_{n=1}^{\infty}$ tends to the Friedrichs extension $\hat{H}_\alpha$, i.e.,

$$s - \lim_{n\to\infty}(H_{\alpha,n} - z)^{-1} = (\hat{H}_\alpha - z)^{-1}, \qquad \Im\mathrm{m}(z) \neq 0 \tag{11}$$

However, from the mathematical point of view this setup seems to be unnatural. To explain this we remark that for any $n = 1, 2, \ldots$ the operator $H_{\alpha,n}$ is a self-adjoint extension of the semibounded symmetric operator $\dot{H}_{\alpha,n} = H_{\alpha,n}|\mathcal{D} = \dot{A} - \alpha V_n$, i.e. $\dot{H}_{\alpha,n} \subseteq H_{\alpha,n}$. Taking another semibounded self-adjoint extension $\tilde{A}$ of $\dot{A}$ we get another sequence $\tilde{H}_{\alpha,n}$,

$$\tilde{H}_{\alpha,n} = \tilde{A} - \alpha V_n, \qquad n = 1, 2, \ldots, \tag{12}$$

which naturally implies the question: why we should to investigate the convergence for $H_{\alpha,n}$ and why not for $\tilde{H}_{\alpha,n}$? So in the following we shall search for conditions which guarantee that

$$s - \lim_{n\to\infty}(\tilde{H}_{\alpha,n} - z)^{-1} = (\hat{H}_\alpha - z)^{-1}, \qquad \Im\mathrm{m}(z) \neq 0. \tag{13}$$

for any semibounded self-adjoint extension $\tilde{A}$ of $\dot{A}$. In particular, this would be clarified the uniqueness problem of the limit (13) for the two "extreme cases": the sequence of Friedrichs extension $\hat{H}_{\alpha,n}$,

$$\hat{H}_{\alpha,n} = \hat{A} - \alpha V_n, \qquad n = 1, 2, \ldots, \tag{14}$$

where $\hat{A}$ is the Friedrichs extension of $\dot{A}$, and of the sequence of Krein extensions $\check{H}_{\alpha,n}$

$$\check{H}_{\alpha,n} = \check{A} - \alpha V_n, \qquad n = 1, 2, \ldots, \tag{15}$$

where $\breve{A}$ is the Krein extension (soft extension) [1], [2], [6] of $\dot{A}$ with respect to a given lower bound $\eta < 0$, i.e. $\breve{A} \geq \eta I$.

In general we cannot expect that the sequence $\tilde{H}_{\alpha,n}$ tends to $\hat{H}_\alpha$ assuming only that $\{V_n\}_{n\geq 1}$ is a regularizing sequence. Actually we need a little bit more. Only if $\tilde{A}$ is the Friedrichs extension $\hat{A}$ of $\dot{A}$, i.e. $\tilde{A} = \hat{A}$, then we obtain

$$s - \lim_{n\to\infty} (\hat{H}_{\alpha,n} - z)^{-1} = (\hat{H}_\alpha - z)^{-1}, \qquad \Im\mathrm{m}(z) \neq 0, \tag{16}$$

without any additional assumptions [7]. How to find this additional assumptions? An essential hint comes from the following proposition.

Proposition 5 *Let $\{V_n\}_{n\geq 1}$ be a regularizing sequence of V. If for every self-adjoint extension $\tilde{A}$ of $\dot{A} = A|\mathcal{D}$ obeying $\tilde{A} \geq \eta$, $\eta < 0$, the convergence (13) takes place, then*

$$\sup_{n\geq 1}(V_n h, h) = +\infty \tag{17}$$

for every nontrivial h of $\mathcal{N}_\eta = ker(\dot{A}^* - \eta)$.

By this proposition it seems to be natural to introduce the following notation.

Definition 6 Let $\{V_n\}_{n\geq 1}$ be a regularizing sequence of V. The sequence is called admissible with respect to $\dot{A} = A|\mathcal{D}$ if there is a $\eta < 0$ such that for every nontrivial $h \in \mathcal{N}_\eta = ker(\dot{A}^* - \eta)$ the condition (17) is satisfied.

Remark 7 It can be shown that if (17) is satisfied for one $\eta < 0$, then it holds for every $\eta' < 0$. So the property (17) is independent on $\eta < 0$.

Example 8 It can be shown that the regularizing sequences of Example 4 for the Examples 1 and 2 are admisssible with respect to $\dot{A} = -\frac{d}{dx^2}|C_0^\infty(\mathbf{R}^1 \setminus \{0\})$ and $\dot{A} = -\Delta|C_0^\infty(\mathbf{R}^2 \setminus \Gamma)$.

Hence, the optimal way to solve our problem would be to show that the converse to Proposition 5 is true, i.e., if $\{V_n\}_{n\geq 1}$ is an admissible regularizing sequence of V with respect to $\dot{A} = A|\mathcal{D}$, then for every semibounded self-adjoint extension $\tilde{A}$ of $\dot{A}$ we have that the convergence (13) is valid. Till now we cannot prove this conjecture in full generality. However, if we restrict the set of semibounded self-adjoint extensions $\tilde{A}$ of $\dot{A}$, then we can do it. To describe these restrictions we use a description of all semibounded self-adjoint extensions which goes back to [1]. Let $\tilde{A}$ be any semibounded self-adjoint extension of $\dot{A} = A|\mathcal{D}$ with lower bound greater than $\eta < 0$, i.e. $\tilde{A} \geq \eta$. By $\tilde{\nu} \geq \eta$ we denote the closed quadratic form which corresponds to $\tilde{A}$, i.e.

$$\begin{aligned} \tilde{\nu}(f,f) &= ((\tilde{A}-\eta)^{1/2}f, (\tilde{A}-\eta)^{1/2}f) + \eta(f,f), \\ f \in \mathrm{dom}(\tilde{\nu}) &= \mathrm{dom}((\tilde{A}-\eta)^{1/2}). \end{aligned} \tag{18}$$

In particular, by $\hat{\nu} \geq 0$ we denote the closed quadratic form which corresponds to the Friedrichs extension $\hat{A}$ of $\dot{A}$. In accordance with [1] we have an one-to-one correspondence between the set of all semibounded self-adjoint extensions $\tilde{A}$ of $\dot{A}$ obeying $\tilde{A} \geq \eta$ and all non-negative closed quadratic forms $\tilde{q}$ on the deficiency subspace $\mathcal{N}_\eta = ker(\dot{A}^* - \eta)$, where the form $\tilde{q}$ is not necessarily densely defined on $\mathcal{N}_\eta$. The correspondence is given by the formulas

$$\mathrm{dom}(\tilde{\nu}) = \mathrm{dom}(\hat{\nu}) \dot{+} \mathrm{dom}(\tilde{q}), \tag{19}$$

where $\dot{+}$ means $\mathrm{dom}(\hat{\nu}) \cap \mathrm{dom}(\tilde{q}) = \{0\}$, and

$$\tilde{\nu}(g+h, g+h) = \hat{\nu}(g,g) + \tilde{q}(h,h) + 2\eta\Re(g,h) + \eta(h,h), \tag{20}$$

$g \in \mathrm{dom}(\hat{\nu}), h \in \mathrm{dom}(\tilde{q}) \subseteq \mathcal{N}_\eta$. Therefore, starting with extension $\tilde{A}$ which obeys $\tilde{A} \geq \eta$ we can find a unique non-negative closed quadratic form $\tilde{q}$ on $\mathcal{N}_\eta$ such that (19) and (20) holds. Conversely, if we have a non-negative closed quadratic from $\tilde{q}$ on $\mathcal{N}_\eta$, then we can define by (19) and (20) a semibounded extension $\tilde{A}$ of $\dot{A}$ obeying $\tilde{A} \geq \eta$. The domain of $\tilde{q}$ may be a closed subspace of $\mathcal{N}_\eta$ or not. The Friedrichs extension $\hat{A}$ corresponds to the trivial form $\hat{q}$, i.e., $\mathrm{dom}(\hat{q}) = \{0\}$. Very often this is expressed by $\hat{q} = +\infty$. The Krein extension (soft extension) [1], [2], [6] $\check{A}$ with respect to the lower bound $\eta < 0$, i.e. $\check{A} \geq \eta I$, is given by the form $\check{q}$ which is zero on the whole deficiency subspace $\mathcal{N}_\eta$, i.e., $\check{q} = 0$. All other forms $\tilde{\nu}$ are between $\check{\nu}$ and $\hat{\nu}$ which yields $\check{A} \leq \tilde{A} \leq \hat{A}$.

Of course the description is only unique if we fix some $\eta < 0$. Changing η we get different quadratic forms $\tilde{q}_\eta$ for the same semibounded self-adjoint extension $\tilde{A}$ of $\dot{A}$. However, there are some invariants which do not depent on η. For instance, if $\mathrm{dom}(\tilde{q}_\eta)$ is a closed subspace in $\mathcal{N}_\eta$, then $\mathrm{dom}(\tilde{q}_{\eta'})$ is a closed subspace for $\eta'(< 0)$, too.

Using this description our main theorem can be formulated now as follows.

Theorem 9 *Let $\{V_n\}_{n\geq 1}$ be an admissible regularizing sequence of V with respect to $\dot{A}$ and let $\tilde{A}$ be a self-adjoint extension of $\dot{A}$ obeying $\tilde{A} \geq \eta$ for some $\eta < 0$. If $\tilde{A}$ corresponds to a closed quadratic form $\tilde{q}$ on $\mathcal{N}_\eta = ker(\dot{A}^* - \eta)$ and the domain $dom(\tilde{q})$ is a closed subspace of $\mathcal{N}_\eta$, then for sufficiently small coupling constants $\alpha > 0$ we have*

$$s - \lim_{n\to\infty} (\tilde{H}_{\alpha,n} - z)^{-1} = (\hat{H}_\alpha - z)^{-1}, \qquad \Im m(z) \neq 0, \tag{21}$$

where $\hat{H}_\alpha$ is the Friedrichs extension of $\dot{H}_\alpha = (A - \alpha V)|\mathcal{D}$.

In particular, if $\check{A}$ denotes the Krein extension of $\dot{A}$ with respect to the lower bound $\eta < 0$, then for sufficiently small $\alpha > 0$ we have

$$s - \lim_{n\to\infty} (\check{H}_{\alpha,n} - z)^{-1} = (\hat{H}_\alpha - z)^{-1}, \qquad \Im m(z) \neq 0. \tag{22}$$

If the deficieny indices are finite, then the theorem admits a strengthening.

Theorem 10 *If the deficieny indices of $\dot{A}$ are finite, then for any self-adjoint extension $\tilde{A}$ of $\dot{A}$ and any coupling constant $\alpha < 1/a$ we have (21).*

The Theorem 10 improves the results of Section 3 of [7]. Moreover, the theorem can be slightly generalized.

Corollary 11 *If $\tilde{A}$ is a semibounded self-adjoint extension of $\dot{A}$ such that*

$$dim(dom(\tilde{\nu})/dom(\hat{\nu})) < +\infty, \tag{23}$$

then for $\alpha < 1/a$ (21) is valid.

The theorems and corollary admit an application to our examples.

Example 12 Since in Example 1 the deficiency indices of $\dot{A} = -\frac{d}{dx^2}|C_0^\infty(\mathbf{R}^1\setminus\{0\})$ are finite by Corollary 10 we always have the desired convergence (21).

In Example 2 we have the desired convergence (21) only for a special set of self-adjoint extensions of $\dot{A} = -\Delta|C_0^\infty(\mathbf{R}^2 \setminus \Gamma)$. The set includes the Krein extension (the corresponding boundary condition can be found in [1]) and extensions which are characterized by Corollary 11. However, it remains an open question: whether the sequence of usual Schrödinger operatos $H_{\alpha,n} = -\Delta - \alpha V_n$, where $-\Delta$ denotes the usual Laplace operator in $L^2(\mathbf{R}^2)$ convergences to the Friedrichs extension of the symmetric operator $(-\Delta - \alpha V)|C_0^\infty(\mathbf{R}^2)$? The problem is that the domain of the closed quadratic form, which by (18) - (20) corresponds to the usual Laplace operator $-\Delta$ in $L^2(\mathbf{R}^2)$ regarded as a self-adjoint extension of $-\Delta|C_0^\infty(\mathbf{R}^2)$, is not a closed subspace in $\mathcal{N}_\eta$.

Remark 13 If the deficiency indices are finite, then the strong resolvent convergence (21) can be replaced by the operator-norm convergence [7]. However, if the deficiency indices are infinite this is not true in general. For instance, let in Example 2 the curve Γ be the unite circle. Then one can show that for any interval $\delta \subseteq (-\infty, 0)$ and any integer N there is a greater integer $n \geq N$ such that $H_{\alpha,n}$ has an eigenvalue in δ. Consequently, this excludes the operator-norm convergence for the operators $\{H_{\alpha,n}\}_{n\geq 1}$.

References

[1] Alonso, A.; Simon, B.: The Birman-Krein-Vishik theory of self-adjoint extensions of semibounded operators. J.Operator Theory **4**, 251-270 (1980).

[2] Brasche, J.F.; Neidhardt, H.: Some Remarks to Krein's extension theory. To appear in *Mathematische Nachrichten.*

[3] Gesztesy, F.: On the one-dimensional Coulomb Hamiltonian. J.Phys. A.: Math. Gen. **13**, 867-875 (1980).

[4] Harell, E.M.: Singular perturbation potentials. Ann. Phys. **105**, 379-406 (1977).

[5] Klaus, M.: Removing cut-offs from one-dimensional Schrödinger operators. J. Phys. A: Math. Gen. **13**, L295-L298 (1980).

[6] Krein, M.G.: The theory of self-adjoint extensions of semibounded Hermitian operators and its applications. I. Math. Sbornik **20**, No. 3, 431-495 (1947) (in Russian).

[7] Neidhardt, H.; Zagrebnov, V.: Regularization and convergence for singular perturbations. Commun. Math. Phys. **149**, 573-586 (1992).

[8] Nenciu, G.: Removing cut-offs from singular perturbations: an abstract result. Lett. Math. Phys. **7**, 301-306 (1980).

[9] Schechter, M.: Cut-off potentials and forms extensions. Lett. Math. Phys. **1**, 265-273 (1976).

[10] Simon, B.: Quadratic forms and Klauder's phenomenon: a remark on very singular perturbations. J. Funct. Anal. **14**, 295-298 (1973).

Addresses:

H.Neidhardt
Technische Universität Berlin
Fachbereich Mathematik MA 7-2
Straše des 17. Juni 136
D-10623 Berlin
Germany

V.A.Zagrebnov
Universite d'Aix-Marseille II
Centre de Physique Theorique
CNRS-Luminy-Case 907
13288 Marseille Cedex 9
France

Operator Theory:
Advances and Applications, Vol. 70

Adiabatic Reduction Theory. Semiclassical S-matrix for N-state one-dimensional systems.

G. Nenciu

Abstract

The reduction scheme, which is the standard tool of the analytic theory of perturbations [Ka], has been recently extended to the time time dependent case; for the case when the small parameter lies in the front of the time derivative (the adiabatic case) [Ne4] and also for the case $H(t) = H_0 + \varepsilon V(t)$ [MN1].

In what follows, we shall first review, following [Ne4], [Ne5] the main facts about adiabatic reduction theory, and then apply it to obtain [MN2] the semi-classical (Born-Oppenheimer) behaviour of the S matrix for the n-state one-dimensional Schrödinger operator. On the way a rigorous derivation of the so-called "trajectory model" is given.

1 Adiabatic Reduction Theory

1.1 Invariant Subspace Problem: Heuristics

Consider the evolution equation

$$i\varepsilon \frac{d}{ds} U_\varepsilon(s, s_0) = H(s) U_\varepsilon(s, s_0); \quad U_\varepsilon(s_0, s_0) = 1. \tag{1.1}$$

in a Hilbert (or more general Banach) space $\mathcal{H}$, in the limit $\varepsilon \to 0, \varepsilon > 0$.

The first step in building up the reduction theory for (1.1) is to find (without integrating it) subspaces which are (aproximately) invariant under the evolution U. Let us remind that in the time independent case, the invariant subspaces are provided by the spectral subspaces of H, so that the problem of finding the invariant subspaces is reduced to the spectral analysis of H.

Let us describe more precisely the problem: one has to find decompositions

$$\mathcal{H} = \mathcal{L}(s) + \mathcal{M}(s) \tag{1.2}$$

such that

$$\begin{aligned} U_\varepsilon(s, s_0)\mathcal{L}(s_0) &\approx \mathcal{L}(s) \\ U_\varepsilon(s, s_0)\mathcal{M}(s_0) &\approx \mathcal{M}(s). \end{aligned} \tag{1.3}$$

If the decomposition (1.2) is given in terms of the projection operator

$$P_\varepsilon(s) = P_\varepsilon(s)^2 \tag{1.4}$$

i.e. $\mathcal{L}(s) = P_\varepsilon(s)\mathcal{H}$, then (1.3) can be written as

$$P_\varepsilon(s) \approx U_\varepsilon(s, s_0) P_\varepsilon(s_0) U_\varepsilon(s, s_0)^{-1}.$$

or in differential form

$$i\varepsilon \frac{d}{ds} P_\varepsilon(s) \approx [H(s), P_\varepsilon(s)]. \tag{1.5}$$

Supposing (at the formal series level)

$$P_\varepsilon(s) = \sum_{j=0}^{\infty} E_j(s) \varepsilon^j \tag{1.6}$$

and inserting (1.6) into (1.4) and (1.5) one obtains

$$E_j(s) = \sum_{m=0}^{j} E_m(s) E_{j-m}(s), \tag{1.7}$$

$$i E_{j-1}^{(1)}(s) = [H(s), E_j(s)]. \tag{1.8}$$

The technical problems to be solved are:

i. Find the solutions of (1.7), (1.8).

ii.If the series in (1.6) is not convergent, construct out of $E_j(s)$ a projection $P_\varepsilon(s)$ satisfying (1.5).

1.2 Invariant subspace problem: the setting and the results

For the sake of simplicity we assume $\mathcal{H}$ to be a Hilbert space and $H(s)$ to be bounded (in the unbounded case one has to consider $H(s)$ defined on $\mathcal{D} \subset \mathcal{H}$ and to formulate the smoothness condition $\mathbf{S}_\alpha$ below in the graph norm topology, or alternatively in terms of the resolvent of $H(s)$; see e.g. [Ne4],[JP1]).

Let $s \in (a, b) \subset \mathbf{R}$ and suppose that $H(s)$ is a family of bounded operators in a separable Hilbert space satisfying:

G

$$\sigma(H(s)) = \sigma_0(s) \cup \sigma_1(s),$$
$$dist(\sigma_0(s), \sigma_1(s)) = d(s) > 0.$$

$\mathbf{S}_\alpha$; $\alpha \geq 0$

$H(s)$ is infinitely norm differentiable and

$$\|\frac{d^k}{ds^k} H(s)\| \leq h(s) c(s)^k (k!)^{1+\alpha}, \tag{1.9}$$

with $h(s), c(s) < \infty$

In what follows **G** and $\mathbf{S}_\alpha$ will be assumed to hold true.

Remarks:

i. If $(a,b) = \mathbf{R}$ and

$$\sup_{\mathbf{R}} c(s) < \infty, \quad \int_{-\infty}^{\infty} h(s)ds < \infty$$

then $H(s)$ has limits $H_\pm$ as $s \to \pm\infty$.

ii. The case $\alpha = 0$ corresponds to the case when $H(s)$ is holomorphic in a neighbourhood of the real axis, and $\alpha > 0$ correspond to Gevrey clases which "interpolate" between the $\mathcal{C}^\infty$ case and the holomorphic case.

Let $\Gamma(s)$ be the contour enclosing $\sigma_0(s)$ such that for all $z \in \Gamma(s)$

$$dist(z, \sigma_0(s)) = d(s)/2.$$

Denote

$$r(s) = sup_{z \in \Gamma(s)} \parallel R(s; z) \parallel$$

where

$$R(s; z) = (H(s) - z)^{-1}$$

Remark: If $H(s)$ is self-adjoint, then by the spectral theorem $r(s) = 2/d(s)$. Although, in general $r(s) > 2/d(s)$, in what follows we shall take $r(s)$ as the parameter describing the gap.

In what follows we shall use the standard notation for the derivatives e.g.

$$E_j^{(k)}(s) = \frac{d^k}{ds^k} E_j(s).$$

The answer to the problem i. stated at the end of Section 1.1. is given by the following basic result, proved in [Ne4].

Theorem 1 *Let*

$$P_0(s) = (2\pi i)^{-1} \oint_{\Gamma(s)} R(s; z)dz$$

and $E_j(s)$ *be given by the following recurrent relations*

$$E_0(s) = P_0(s)$$

$$E_j(s) = (2\pi)^{-1} \oint_{\Gamma(s)} R(s; z)\{Q_0(s)E_{j-1}^{(1)}(s)P_0(s) - P_0(s)E_{j-1}^{(1)}(s)Q_0(s)\}$$

$$R(s; z)dz + S_j(s) - 2P_0(s)S_j(s)P_0(s)$$

where

$$Q_0(s) = 1 - P_0(s)$$

$$S_j(s) = \sum_{m=1}^{j-1} E_m(s) E_{j-m}(s).$$

Then $E_j(s),\ j = 0, 1, \ldots,$ *are the unique solution of*

$$E_j(s) = \sum_{m=0}^{j} E_m(s) E_{j-m}(s),$$

$$iE_{j-1}^{(1)}(s) = [H(s), E_j(s)]$$

satisfying

$$E_0(s) = P_0(s).$$

In order to go beyond the formal series level one needs a good control on $\|E_j(s)\|$. The main technical estimate is contained in the following lemma [Ne4],[Ne5].

Lemma 1 *For* $k = 0, 1, 2, \ldots;\ j = 1, 2, \ldots$

$$\|E_j^{(k)}(s)\| \leq \frac{h(s)}{M(s)} A(s)^j a(s)^{j+k} \frac{((j+k)!)^{1+\alpha}}{(1+j)^2(1+k)^2}$$

with

$$M(s) = \frac{1}{4q(s)} + f^2 h(s) + 2f^4 q(s)^2 H(s)^2 h(s),$$

$$A(s) = 16 r(s) f^4 q(s)^4 H(s)^4 M(s)$$

where

$$a(s) = 4c(s) max\{h(s);\ h(s)r(s);\ 2\},$$

$$H(s) = 1 + h(s) \geq 1$$

$$q(s) = r(s) \frac{length \Gamma(s)}{2\pi} \geq 1$$

and f *is the (absolute) constant appearing in the inequality*

$$\sum_{l=0}^{k} \frac{1}{(l+1)^2(k+1-l)^2} \leq f \frac{1}{(k+1)^2}.$$

We are now ready to construct $P_\varepsilon(s)$ satisfying (1.4) and

$$\|i\varepsilon \frac{d}{ds} P_\varepsilon(s) - [H(s), P_\varepsilon(s)]\| \leq \delta(s; \varepsilon)$$

where $\delta(s;\varepsilon)$ depends upon s via $h(s)$, $c(s)$, $r(s)$, and $q(s)$. More exactly, let

$$g(s) = A(s) a(s)$$

$$N_\varepsilon + 1 = [(\frac{1}{g(s)\varepsilon})^{1/1+\alpha}]$$

$$T_\varepsilon(s) = \sum_{j=0}^{N_\varepsilon} E_j(s)\varepsilon^j.$$

Then $P_\varepsilon(s)$ is given by:

$$P_\varepsilon(s) = \frac{1}{2\pi i}\oint_{|z-1|=1/2}(T_\varepsilon(s) - z)^{-1}dz.$$

Using Lemma 1, one obtains the following result [Ne4],[Ne5] :

Theorem 2 *There exist constants: $\varepsilon_0(s) > 0$ depending upon $h(s)$, $c(s)$, $q(s)$, $r(s)$; and $k_1(s) < \infty$, $k_2(s) > 0$ depending upon $h(s)$, $q(s)$, $r(s)$ such that for $0 < \varepsilon < \varepsilon_0(s)$, $P_\varepsilon(s)$ is well defined,*

$$\|i\varepsilon\frac{d}{ds}P_\varepsilon(s) - [H(s), P_\varepsilon(s)]\| \leq k_1(s)h(s)\exp[-\frac{k_2(s)}{(\varepsilon c(s)r(s))^{1/1+\alpha}}] \equiv \delta(s;\varepsilon)$$

and in the sense of asymptotic series

$$P_\varepsilon(s) = \sum_{j=0}^{\infty} E_j(s)\varepsilon^j.$$

Let us discuss now in more detail the self-adjoint case. Since $U_\varepsilon(s, s_0)$ is unitary, one obtains at once

$$\| (1 - P_\varepsilon(s))U_\varepsilon(s, s_0)P_\varepsilon(s_0) \| \leq \varepsilon^{-1}\int_{s_0}^{s}\delta(u;\varepsilon)du \tag{1.10}$$

which contains as particular cases all variants of the adiabatic theorem of Quantum Mechanics. In particular if $\alpha = 0, (a, b) = \mathbf{R}$ and uniformly for $s \in \mathbf{R}$, $h(s) \leq h < \infty$; $c(s) \leq c < \infty$; $q(s) \leq q < \infty$; $d(s) \geq d > 0$ and moreover

$$\int_{-\infty}^{\infty} h(s)ds < \infty$$

one obtains

Corollary 1 *There exist: $\varepsilon_0 > 0; k_1 < \infty; k_2 > 0$ depending upon h, c, q, d such that for $0 \leq \varepsilon \leq \varepsilon_0$:*

$$\mathcal{P} \leq k_1\int_{-\infty}^{\infty} h(s)exp(-k_2(\frac{d}{\varepsilon})^{1/1+\varepsilon})ds. \tag{1.11}$$

Remarks: i. Without going into the details of the history of Corollary 1 (see [JS],[JP1], [Ma] and references therein) let us point out that the best control to date on the constant k_2 in (1.12) was obtained recently by Martinez [Ma], using a priori microlocal exponential weighted estimates. The problem of a sharp estimate on k_2 is still open.
ii. Families of projections $P_{\varepsilon,k}(s)$ satisfying (1.10) with $\delta(\varepsilon, s)$ of order ε^k; $k = 1, 2, ...$ have been constructed for the first time (to our best knowledge) at the formal level and for $H(s)$ with discrete nondegenerate spectrum by Garrido [Ga] (see also [Sa]),and later on, independently, in the general case by Nenciu [Ne1]. Recently the same procedure has been rediscovered by Berry [Be1] for the case when the underlying Hilbert space is two-dimensional. Let us stress that in [Ga], [Ne1] $P_{\varepsilon,k}(s)$ were constructed via some auxiliary (nonsingular) evolution equations, so that it was not clear whether they are "local" quantities i.e. depend only upon $H(s)$ and a finite number of its derivatives at the point s. That this is the case was first proved in [Ne2] where a "local" iterative scheme has been given (see [NR] and also [Ne3] for a review; for a related "local" iterative scheme see [JP2]). While the constructions in [Ga],[Ne1-4],[JP2], seem different at the first sight, due to Theorem 1, they are equivalent in the sense that irrespective of the method of construction, if $P_{\varepsilon,k}(s)$ has an asymptotic expansion in ε, then the first k coefficients must coincide with $E_j(s)$.

1.3 Intertwining evolutions

We shall start with some generalities concerning the intertwining of families of projections. Let $Q(s)$ be a norm differentiable family of bounded projections, $s \in (a, b) \subset \mathbf{R}$. The problem is to find families, $V(s, s_0)$, of bounded with bounded inverse operators intertwining $Q(s)$ i. e.

$$Q(s) = V(s, s_0)Q(s_0)V^{-1}(s, s_0) \tag{1.12}$$

We shall call $V(s, s_0)$ an intertwining evolution for $Q(s)$. We shall give two methods of constructing intertwining evolutions. The first one is a (mild) generalisation, [Ne4] of the well known Krein-Kato lemma [Ka], [Kr].

Lemma 2 *Suppose: i. $Q(s)$ is a norm differentiable family of bounded projections in a Hilbert space $\mathcal{H}$.*
ii. $N(s)$ is a strongly continuous family of uniformly bounded operators.
Then if

$$K_N(s) = N(s) + (1 - 2Q(s))(i\frac{d}{ds}Q(s) - [N(s), Q(s)]) \tag{1.13}$$

$A_N(s, s_0)$ given by

$$i\frac{d}{ds}A_N(s, s_0) = K_N(s)A_N(s, s_0); \quad A_N(s_0, s_0) = 1 \tag{1.14}$$

is an intertwining evolution for $Q(s)$.

Remark:

The Krein-Kato lemma corresponds to $N(s) = 0$ and gives the parallel transport for $Q(s)$.

The second method goes back (via Kato [Ka]) to Sz-Nagy.

Lemma 3 *Suppose P, Q are (not necessarily self-adjoint) bounded projections satisfying*

$$\| P - Q \| < 1.$$

Then

$$V = (1 - (P - Q)^2)^{-1/2}(PQ + (1 - P)(1 - Q))$$

has a bounded inverse

$$V^{-1} = (QP + (1 - Q)(1 - P))(1 - (P - Q)^2)^{-1/2}$$

and

$$P = VQV^{-1}.$$

If P, Q are self-adjoint then V is unitary.

Remark:

See [ASS] for related topics.

Applying Lemma 2 to the family $P_\varepsilon(s)$, given by Theorem 2 with $\varepsilon^{-1}H(s)$ as $N(s)$ one obtains:

Theorem 3 *Let $U_\varepsilon^A(s, s_0)$ given by*

$$i\varepsilon \frac{d}{ds} U_\varepsilon^A(s, s_0) = H_A(s; \varepsilon) U_\varepsilon^A(s, s_0); \quad U_\varepsilon^A(s_0, s_0) = 1 \tag{1.15}$$

with

$$H_A(s; \varepsilon) = H(s) - B_\varepsilon(s) \tag{1.16}$$

where

$$B_\varepsilon(s) = -(1 - 2P_\varepsilon(s))(i\varepsilon \frac{d}{ds} P_\varepsilon(s) - [H(s), P_\varepsilon(s)]). \tag{1.17}$$

Then without any error

$$P_\varepsilon(s) = U_\varepsilon^A(s, s_0) P_\varepsilon(s_0) U_\varepsilon^A(s, s_0)^{-1} \tag{1.18}$$

i. e. $U_\varepsilon^A(s, s_0)$ is an intertwining operator for $P_\varepsilon(s)$. If $\Omega_\varepsilon(s, s_0)$ is defined by

$$U_\varepsilon(s, s_0) = U_\varepsilon^A(s, s_0) \Omega_\varepsilon(s, s_0) \tag{1.19}$$

then

$$i\varepsilon \frac{d}{ds} \Omega_\varepsilon(s, s_0) = U_\varepsilon^A(s, s_0)^{-1} B_\varepsilon(s) U_\varepsilon^A(s, s_0) \Omega_\varepsilon(s, s_0) \tag{1.20}$$

Remark:

If $H(s)$ is self-adjoint then $U_\varepsilon(s,s_0)$ is unitary and then

$$\| \Omega_\varepsilon(s,s_0) - 1 \| \leq \varepsilon^{-1} \int_{s_0}^{s} \| B_\varepsilon(u) \| du \leq \varepsilon^{-1} \int_{s_0}^{s} \delta(u;\varepsilon) du. \tag{1.21}$$

Let now $V_\varepsilon(s,s_0)$ be any intertwining evolution for $P_\varepsilon(s)$. If $\Phi_{\varepsilon,V}(s,s_0)$ is defined by

$$U_\varepsilon^A(s,s_0) = V_\varepsilon(s,s_0)\Phi_{\varepsilon,V}(s,s_0) \tag{1.22}$$

then

$$[\Phi_{\varepsilon,V}(s,s_0), P_\varepsilon(s_0)] = 0$$

i.e. $\Phi_{\varepsilon,V}(s,s_0)$ has a block diagonal structure. Moreover one can compute the equation of motion for $\Phi_{\varepsilon,V}(s,s_0)$. In particular:

Corollary 2

$$i\varepsilon \frac{d}{ds} P_\varepsilon(s_0)\Phi_{\varepsilon,V}(s,s_0)P_\varepsilon(s_0) = H_{eff}(s,s_0)P_\varepsilon(s_0)\Phi_{\varepsilon,V}(s,s_0)P_\varepsilon(s_0) \tag{1.23}$$

where

$$H_{eff}(s,s_0) = P_\varepsilon(s_0)\{i\varepsilon(\frac{d}{ds}V_\varepsilon^{-1}(s,s_0)V_\varepsilon(s,s_0)) + V_\varepsilon^{-1}(s,s_0)H_A(s;\varepsilon)V_\varepsilon(s,s_0)\}P_\varepsilon(s_0). \tag{1.24}$$

Theorem 3 and Corollary 2 contain the basic facts of the adiabatic reduction theory. It reduces (up the exponentially small errors) the problem of integrating (1.1) to the problem of integrating (1.23) in $P_\varepsilon(s_0)\mathcal{H}$ (and the corresponding equation in $(1-P_\varepsilon(s_0))\mathcal{H}$. Let us consider in more detail the formula (1.24) for the effective hamiltonian. As in the time independent case, since the intertwining evolution for $P_\varepsilon(s)$ is not unique. One "distinguished" choice is to take as $V_\varepsilon(s,s_0)$, the parallel transport, $A_0(s,s_0)$, for $P_\varepsilon(s)$ [Ne4] and then

$$H_{eff}(s,s_0) = P_\varepsilon(s_0)A_0^{-1}(s,s_0)H_A(s;\varepsilon)A_0(s,s_0)P_\varepsilon(s_0).$$

The unpleasent feature of this choice is that $H_{eff}(s,s_0)$ is not a "local" object: it depends upon the values of $H(s)$ in the whole interval $[s_0,s]$. If

$$sup_{s\in(a,b)} \| P_\varepsilon(s) - P_\varepsilon(s_0) \| < 1 \tag{1.25}$$

then one can use the Sz-Nagy formula (see Lemma 3) for the intertwining evolution. Notice that in this case, $H_A(s;\varepsilon)$ is a local object. It is this choice which has been used in the case of the regular perturbation theory for time dependent hamiltonians [MN1], as well as (in a slightly modified form) in the application in the next section.

Remark:

In the self-adjoint case, it follows that

$$\| U_\varepsilon^A(s,s_0) - U_\varepsilon(s,s_0) \| \leq \varepsilon^{-1} \int_{s_0}^{s} \delta(u;\varepsilon) du.$$

We shall follow Berry [Be2] and call $U_\varepsilon^A(s,s_0)$ "the superadiabatic evolution".

2 Semiclassical S-matrix for one dimensional N-state systems

2.1 Reduction theory

Consider the scattering theory for the hamiltonian (in $L^2(\mathbf{R})^n$)

$$-\frac{\hbar^2}{2M}\frac{d^2}{dx^2}+v(x) \tag{2.1}$$

in the limit $\varepsilon^2 \equiv \hbar^2/2M \to 0$ (either semiclassical or Born-Oppenheimer limit). In (2.1), $v(x)$ is a self-adjoint operator in $\mathbf{C}^n$ satisfying

$$\| \frac{d^k}{dx^k}v(x) \| \leq \frac{a}{(1+\mid x \mid)^{1+\eta}}c^k(k!)^{1+\alpha} \tag{2.2}$$

$$a, c < \infty, \eta > 0; k = 1, 2, ...$$

Let $\mathcal{S}(\varepsilon, E)$ be the scattering matrix at energy E. We shall consider only energies above the barrier:

$$E - v(x) \equiv \omega(x) \geq \Delta > 0 \ \ all \ \ x \in \mathbf{R}. \tag{2.3}$$

In what follows:

$$v_\pm = \lim_{x\to\pm\infty} v(x); \omega_\pm = E - v_\pm.$$

As in the $n = 1$ case the scattering matrix can be extracted from the asymptotic behaviour of the generalised eigenfunctions. If

$$\psi_E(x) \sim \frac{1}{2^{1/2}}\omega_+^{-1/4}(exp(i\varepsilon^{-1}\omega_+^{1/2}x)a_{in} + (-i\varepsilon^{-1}\omega_+^{1/2}x)a_{out}$$

$$\psi_E(x) \sim \frac{1}{2^{1/2}}\omega_-^{-1/4}(exp(i\varepsilon^{-1}\omega_-^{1/2}x)b_{out} + (-i\varepsilon^{-1}\omega_-^{1/2}x)b_{in}$$

as $x \to \infty$ and $x \to -\infty$ respectively (notice that $a_{in}, a_{out}, b_{in}, b_{out} \in \mathbf{C}^n$) then as an operator in $\mathbf{C}^n \bigoplus \mathbf{C}^n$, $\mathcal{S}(\varepsilon, E)$ is given by

$$\mathcal{S}(\varepsilon, E)\begin{pmatrix} a_{in} \\ b_{in} \end{pmatrix} = \begin{pmatrix} S_{++} & S_{+-} \\ S_{-+} & S_{--} \end{pmatrix}\begin{pmatrix} a_{in} \\ b_{in} \end{pmatrix} = \begin{pmatrix} b_{out} \\ a_{out} \end{pmatrix}.$$

The various matrix elements of $\mathcal{S}(\varepsilon, E)$ have the same meaning as in the $n = 1$, for example S_{++} describes the transmission from the left to the right, S_{-+} the reflection to the left, etc. While the literature for $n = 1$ is more than extensive , the only results we are aware of for $n > 1$ are due to Fedorjuk, who considered the case when $v(x)$ has, for all x n nondegenerate eigenvalues $v_j(x)$

$$\inf_{i\neq j; x\in\mathbf{R}} \mid v_i(x) - v_j(x) \mid = d > 0 \tag{2.4}$$

In particular for the analytic case ($\alpha = 0$) he proved that

$$\| S_{-+} \| = \mathcal{O}(e^{-k(\Delta,d)/\varepsilon}). \tag{2.5}$$

The condition (2.4) is very restrictive from the physical point of view; in fact the most interesting situations are those when the eigenvalues of $v(x)$ have crossings or quasicrossings.

Concerning S_{++}, on intuitive grounds, one expects (2.5) to hold true without the nondegeneracy condition.

Concerning S_{++}, the theoretical chemists derived so called " trajectory models" to compute it (approximately). Let us describe the simplest example. If:

$$n = 2$$

and

$$v(x) = \begin{pmatrix} v_1(x) & \delta u(x) \\ \delta \overline{u}(x) & v_2(x) \end{pmatrix}$$

where δ is a positive small parameter, define:

$$P_j(x) = [2M(E - v_j(x)]^{1/2} - \textit{ the classical momenta,}$$

$$P_m(x) = (P_1(x) + P_2(x))/2 - \textit{ the mean momentum}$$

and the "mean trajectory" $x(t)$ from the Newton's law:

$$M\frac{d}{dt}x(t) = P_m(x(t)).$$

Then it is argued that if $\hat{S}$ is the scattering matrix corresponding to the following time dependent effective Schrödinger equation

$$i\hbar\frac{d}{dt}\phi(t) = v(x(t))\phi(t); \;\; \phi \in \mathbf{C}^2$$

one has

$$S_{++} - e^{i\varphi}\hat{S} = \mathcal{O}(\varepsilon, \delta^2, \frac{| P_1 - P_2 |}{P_1 + P_2}) \tag{2.6}$$

where φ is an overall phase factor. Notice that this is exactly a "reduction scheme" result: one computes some matrix elements of $S(\varepsilon, E)$ via some effective evolutions.

Under the conditions (2.2), (2.3), our result [MN2] is contained in the following:

Theorem 4 *There exist* $\varepsilon_0 > 0$, $k_1 < \infty$, k_2 , A, $C < \infty$ *depending upon* Δ, a, c *such that for* $0 \leq \varepsilon \leq \varepsilon_0$

i.

$$\| S_{-+} \| \leq k_1 e^{k_2/\varepsilon^{1/1+\alpha}}$$

ii.

There exist a family of self-adjoint operators in $\mathbf{C}^n$, $H_{\varepsilon,eff}(x)$ *with the properties: a.*

$$\| \frac{d^k}{dx^k} H_{\varepsilon,eff}(x) \| \leq \frac{A}{(1+\mid x \mid)^{1+\eta}} C^k (k!)^{1+\alpha}$$

b. $H_{\varepsilon,eff}(x)$ *has an (uniform in x) asymptotic expansion*

$$H_{\varepsilon,eff}(x) = -\omega^{1/2}(x) + \frac{i\varepsilon}{2}[\frac{d}{dx}\omega^{1/4}(x), \omega^{-1/4}(x)] + \mathcal{O}(\varepsilon^2)$$

c. If

$$\tilde{S} = \lim_{x \to \infty} e^{-\frac{i}{\varepsilon}\omega_+^{1/2} x} \tilde{U}_\varepsilon(x, -x) e^{-\frac{i}{\varepsilon}\omega_-^{1/2} x}$$

$$i\varepsilon \frac{d}{dx} \tilde{U}_\varepsilon(x, x_0) = H_{\varepsilon,eff}(x) \tilde{U}_\varepsilon(x, x_0); \;\; \tilde{U}_\varepsilon(x, x_0) = 1$$

then

$$\| S_{++} - \tilde{S} \| \leq k_1 e^{-k_2/\varepsilon^{1/1+\alpha}}$$

2.2 A refined Landau-Zener-Friedrichs formula

By Theorem 3 the computation of $S(E,\varepsilon)$ was reduced (up to exponentially small errors) to an effective adiabatic problem, for which the theory is much more developed. In particular, in some simple cases one can compute (at least partially) $\tilde{S}$ in the limit $\varepsilon \to 0$. In what follows we shall describe results of that sort in the simplest setting. Take $n = 2$, $\alpha = 0$, and disregard the dependence of $H_{\varepsilon,eff}(x)$ upon ε taking

$$H_{\varepsilon,eff}(x) = \tilde{v}_0(x) + \delta\tilde{u}(x).$$

Let $v_j^0(x)$, $n_j^0(x)$ be the eigenvalues and eigenvectors (chosen to be smooth as functions of x) of $\tilde{v}_0(x)$ respectively. Suppose that the eigenvalues $v_j^0(x)$ have a linear crossing at $x = 0$ and no other crossings. Let us stress that the labelling of $v_j^0(x)$ is the one making $v_j^0(x)$ analytic in a neighbourhood of the origin (via Rellich theorem). Notice that for $\delta \neq 0$ but small,

$$\tilde{v}_0(x) + \delta\tilde{u}(x).$$

has a quasicrossing at 0.

Let $\tilde{S}_{11}(\varepsilon,\delta)$ be the matrix element of $\tilde{S}$ describing the transitions from the state corresponding to $v_1^0(-\infty)$ to the state corresponding to $v_1^0(\infty)$. Then one has [MN2]:

Theorem 5 *For* ε, δ *small enough*

$$\mid \tilde{S}_{11}(\varepsilon,\delta) \mid^2 = exp[-\frac{2\pi}{a\varepsilon} \mid \delta u_{1,2} - i\varepsilon h_{1,2} \mid^2 (1 + \mathcal{O}((\varepsilon + \delta))] \tag{2.7}$$

where

$$h_{1,2} = (n_1^0(x), \frac{d}{dx} n_2^0(x))_{x=0},$$

$$u_{1,2} = (n_1^0(0), \tilde{u}(0) n_2^0(0))$$

and

$$a = \lim_{x \to 0} \frac{v_1^0(x) - v_2^0(x)}{x} > 0.$$

Formula (2.7) contains as particular cases the recent results of Joye and Pfister [J] (see also [H1]) as well as the result of Hagedorn [H2]. We end up by remarking that, while the proofs in [J], [H1], [H2] are completely different and rather technical, our proof is rather "cheap"; in the proper setting the computations and the estimations are very similar to the usual perturbation theory.

Acknowledgements
It is a pleasure to thank the organisers for the invitation and for the financial support.

References

[ASS] Avron,J., Seiler,R., Simon, B.:The index of a pair of projections.To appear in J.Func.Anal.

[Be1] Berry, M.V.: Quantum phase corrections from adiabatic iteration. Proc. R. Soc. Lond. **A414**, 31-46(1987).

[Be2] Berry, M.V.: Histories of adiabatic quantum transitions. Proc. R. Soc. Lond. **A429**, 61-72(1990).

[Ga] Garrido, L.M.: Generalised adiabatic invariance. J. Math. P hys. **5**, 335-362(1964).

[H1] Hagedorn, G.: Proof of the Landau-Zener formula in an adiabatic limit with small eigenvalue gaps. Commun. Math. Phys. **136**, 433-449(1991).

[H2] Hagedorn, G.: Adiabatic expansions near eigenvalue crossing. Ann. Phys. **196**, 278-296(1989).

[JS] Jakšić, J., Segert, J.: Exponential approach to the adiabatic limit and the Landau-Zener formula. Rev.Math.Phys. **4**, 529-574(1992).

[J] Joye, A.: Proof of the Landau-Zener formula. Preprint CPT Marseille, Dec. 1992.

[JP1] Joye, A., Pfister, C-E.: Exponentially small adiabatic invariant for the Schrödinger equation. Commun. Math. Phys. **140**, 15-41(1991).

[JP2] Joye, A., Pfister, C-E.: Full asymptotic expansion of transition probabilities in the adiabatic limit. J. Phys. A: Math. Gen. **24**, 753-766(1991).

[Ka] Kato, T.: Perturbation theory for linear operators. Berlin, Heidelberg, New York: Springer 1976.

[Kr] Krein, S. G.: Linear differential equations in Banach spaces. Translations of Mathematical Monographs. Vol 29: Providence 1971.

[MN1] Martin, Ph-A., Nenciu, G.:Perturbation theory for time dependent hamiltonians: rigorous reduction theory. H.P.A. **65**, 528-559(1992).

[MN2] Martin, Ph-A., Nenciu, G.:Semi-classical inelastic S-matrix for one-dimensional N-states systems. Preprint ESI Wien 1994.

[Ma] Martinez, A.: Precise exponential estimates in adiabatic theory. Preprint Université Paris Nord 1993.

[Ne1] Nenciu, G.: Adiabatic theorem and spectral concentration. Commun. Math. Phys. **82**, 125-135(1981).

[Ne2] Nenciu, G.: Adiabatic theorem and spectral concentration II. Arbitrary order asymptotic invariant subspaces and block diagonalisation. Preprint FT-308-1987, Central Institute of Physics, Bucharest.

[Ne3] Nenciu, G.: Asymptotic invariant subspaces, adiabatic theorems and block diagonalisation. In: Boutet de Monvel et al (eds), Recent developments in quantum mechanics, pp 133-149. Dordrecht: Kluver 1991.

[Ne4] Nenciu, G.: Linear adiabatic theory. Exponential estimates. Commun. Math. Phys. **152**, 479-496(1993).

[Ne5] Nenciu, G.:Exponential estimates in linear adiabatic theory.Report No. 7, 1992/93, Institut Mittag-Leffler.

[NR1] Nenciu, G., Rasche, G.: Adiabatic theorem and Gell-Mann-Low formula. H. P. A. **62**, 372-388(1989).

[Sa] Sancho, S.J.: m'th order adiabatic invariance for quantum systems. Proc.Phys.Soc.Lond. **89**, 1-5(1966).

G. Nenciu, Dept. Theor. Phys., University of Bucharest, PO Box MG 11, Romania

Operator Theory:
Advances and Applications, Vol. 70

The functional structure of the monodromy matrix for Harper's equation*

Vladimir Buslaev and Alexander Fedotov

Contents

1 Introduction

1.1 Harper's equation

In this paper we continue our investigation of Harper's equation:

$$\frac{\psi(x+h)+\psi(x-h)}{2}+\cos x\ \psi(x)=E\psi(x). \tag{1.1}$$

Here h is a fixed positive parameter and $x\in\mathbb{R}$ or $x\in\mathbb{C}$. This equation appeared as a model for Bloch electron in a weak constant magnetic field [Ho]. The structure of the spectrum σ_h of Harper's equation on $\mathbf{L}_2(\mathbb{R})$ appeared to be very rich and Harper's equation attracted the attention of both physicists and mathematicians, see, for example, [C-F-K-S].

Formally, Harper's equation can be rewritten in the form

$$\left(\cos\frac{h}{i}\frac{d}{dx}+\cos x\right)\psi(x)=E\psi(x)$$

and so, the analysis of its spectrum for small h is a typical semi-classical problem. The investigation of this problem was begun by physicists; see the papers of Azbel

*The work was supported by the Russian Foundation of Fundamental Research

[Az] and Wilkinson [Wi]. The main rigorous results were obtained by B.Helffer and J.Sjöstrand; see, for example, [H-S]. Using the methods of the pseudo-differential operator theory these authors obtained very subtle estimates describing the geometrical structure of σ_h. [H-K-S]. The papers of B.Helffer and J.Sjöstrand contain an exhaustive review and a full list of literature devoted to the problem.

The most interesting asymptotic characteristics of the spectrum of Harper's equation are connected with the quantities which are exponentially small with respect to h as $h \to 0$. In the case of ordinary differential equations such exponentially small quantities can be controlled by means of the complex WKB method [Fe], [Si]. In that method one constructs analytic solutions of the equation on the complex plane $x = z \in \mathbb{C}$ and describes their asymptotic behavior as $h \to 0$. In [B-F] we develop the complex WKB method version for Harper's equation. Here we use these results to continue our analysis.

1.2 The monodromy matrix notion

Consider the set of meromorphic solutions of Harper's equation. We denote it by $\mathbb{M}$. It is clear that $\mathbb{M}$ is invariant with respect to the translations $\psi(z) \mapsto \psi(z+2\pi)$. Furthermore, in [B-F] we have shown that the set $\mathbb{M}$ a two-dimensional linear space over the field of meromorphic h-periodic functions. These two properties of $\mathbb{M}$ allow to introduce the notion of a monodromy matrix.

Let $\Psi_\pm(z)$ be two linearly independent meromorphic solutions of Harper's equation. Since $\Psi_\pm(z+2\pi)$ are also its solutions one can write

$$\vec{\Psi}(z+2\pi) \;=\; M(z)\,\vec{\Psi}(z),$$

where

$$\vec{\Psi}(z) = \begin{pmatrix} \Psi_+(z) \\ \Psi_-(z) \end{pmatrix}$$

and $M(z)$ is a 2×2-matrix having meromorphic h-periodic coefficients. We call this matrix *a monodromy matrix.*

1.3 The results of the paper

In [B-F] we have proved that there is a positive constant h_0 such that for $0 < h \leq h_0$ Harper's equation has two linearly independent entire solutions having standard asymptotic behavior on a given canonical domain, i.e. a domain chosen in the complex plane in accord with the special rules. We have also begun the investigation of the monodromy matrix corresponding to these solutions.

The main result of this paper is the following statement.

Theorem Let $0 < \varepsilon < E < 2-\varepsilon$, where ε is a fixed positive constant. For $0 < h \leq h_0(\varepsilon)$ Harper's equation has two linearly independent entire solutions such that the corresponding monodromy matrix has the form

$$\mathbf{M}(z) = \begin{pmatrix} \frac{1}{w} + \frac{1}{\overline{w}} + \overline{w} - 2\cos\left(\frac{2\pi}{h}z\right) & -i\sqrt{\frac{w}{\overline{w}}} + i\sqrt{w\overline{w}}\, e^{-\frac{2\pi i}{h}z} \\ i\sqrt{\frac{w}{\overline{w}}} - i\sqrt{w\overline{w}}\, e^{\frac{2\pi i}{h}z} & w \end{pmatrix}.$$

The square roots in this formula satisfy the relations:

$$\sqrt{\frac{w}{\overline{w}}}\sqrt{w\overline{w}} = w, \quad \sqrt{w\overline{w}} > 0.$$

The coefficient $w = w(E, h)$ is independent of z. This coefficient is a continuous function of h and depends analytically on E on some vicinity of the interval $0 < \varepsilon < E < 2 - \varepsilon$.

Remarks

- The spectrum of Harper's equation lies on the interval $[-2, 2]$ and is symmetric with respect to 0. Thus, for spectral applications it suffices to consider the case $0 \leq E \leq 2$. In this paper we do not investigate small vicinities of the ends of this interval.

- The monodromy matrix coefficients have to be h-periodic. As it is seen from the Theorem, the Fourier series for the coefficients of the matrix $\mathbf{M}$ consist only of few terms!

- In [B-F] we got the asymptotic formula for $\mathbf{M}_{22}$:

  $$\mathbf{M}_{22} = t(E)\, u(E)(1 + O(h)),$$

  where

  $$t(E) = \exp\left(-\frac{1}{h}S(E)\right), \tag{1.2}$$

  $$u(E) = \exp\left(\frac{i}{h}\Phi(E)\right)$$

  and

  $$S(E) = \int_{\substack{-\pi \leq x \leq \pi \\ E - \cos x \geq 1}} arccosh(E - \cos x)\, dx,$$

  $$\Phi(E) = \int_{\substack{-\pi \leq x \leq \pi \\ E - \cos x \leq 1}} arccos\,(E - \cos x)\, dx, \quad arccos : [-1, 1] \to [0, \pi]. \tag{1.3}$$

  Note that the coefficients $t(E)$ and $\Phi(E)$ have a direct semi-classical interpretation. Let $0 < E < 2$. Consider the real isoenergetic curve

  $$\cos p + \cos x = E, \quad x \in \mathbf{R}. \tag{1.4}$$

The main semi-classical objects associated with an isoenergetic curve in quantum mechanics are the phase integral and the tunneling coefficient. In the case of Harper's equation, i.e. for the isoenergetic curve (1.4), the tunneling coefficient and the phase integral can be described by the formulae (1.2) and (1.3), see [Wi].

- Since $S(E) > 0$, the tunneling coefficient $t(E)$ is exponentially small for small h.

1.4 The structure of the paper

In section 2 we investigate the set of meromorphic solutions of the one- dimensional difference Schrödinger equations. There we introduce the notion of a monodromy matrix.

In section 3 we formulate one of the main results of [B-F]: the theorem about existence of the entire solutions having standard behavior on the complex plane; we describe also the main geometrical objects related to this theorem.

In section 4 we investigate in detail behavior of the above solutions for large $Im\, z$. It allows to describe the functional structure of the corresponding monodromy matrix.

In Section 5 we find out the relations between the monodromy matrix coefficients. The space of meromorphic solutions of Harper's equation is invariant with respect to the certain transformations. These invariance properties allow to describe the coefficients of the monodromy matrix in terms of only one coefficient and to prove the above Theorem.

2 Monodromy matrices for the one-dimensional difference Schrödinger equations

We shall discuss here the one-dimensional difference Schrödinger equation

$$\frac{\psi(z+h)+\psi(z-h)}{2}+v(z)\psi(z)=E\psi(z), \quad z\in\mathbb{C}, \tag{2.1}$$

where $v(z)$ is a meromorphic function and h is a fixed real parameter.

2.1 Wronskians

Let $\psi(z)$ and $\phi(z)$ be two solutions of (2.1). We call the expression

$$(\psi,\phi)\equiv\psi(z+h)\phi(z)-\psi(z)\phi(z+h) \tag{2.2}$$

the Wronskian of the solutions ψ *and* ϕ.

Lemma 2.1 The Wronskian of two solutions of (2.1) is an h-periodic function.

The proof follows from the equalities

$$\psi(z)\left(\phi(z+h)+\phi(z-h)\right)=2(E-v(z))\psi(z)\phi(z)=\phi(z)\left(\psi(z+h)+\psi(z-h)\right).$$

□

2.2 The set of meromorphic solutions

In this section we assume that there exist two meromorphic solutions of equation (2.1), $\psi(z)$ and $\phi(z)$, and

$$(\psi,\phi)\not\equiv 0. \tag{2.3}$$

Lemma 2.2 Any meromorphic solution of (2.1) can be represented in the form:

$$g(z)=\frac{(g,\phi)}{(\psi,\phi)}\psi(z)+\frac{(\psi,g)}{(\psi,\phi)}\phi(z).$$

Proof. Let g be a meromorphic solution of equation (2.1). Consider the function

$$\tilde{g}=g-\frac{(g,\phi)}{(\psi,\phi)}\psi-\frac{(\psi,g)}{(\psi,\phi)}\phi.$$

It satisfies (2.1) and

$$\begin{aligned}(\tilde{g},\psi)&=0,\\(\tilde{g},\phi)&=0.\end{aligned} \tag{2.4}$$

Let us represent $\tilde{g}$ in the form

$$\tilde{g}(z)=C(z)\psi(z).$$

Substituting this formula in (2.4) we get that $C(z+h)=C(z)$. Hence

$$0=(\tilde{g},\phi)=C\cdot(\psi,\phi)$$

and thus $\tilde{g}=0$. □

Denote by $\mathbb{M}$ the set of meromorphic solutions of equation (2.1).

Theorem 2.3 $\mathbb{M}$ is a two-dimensional linear space over the field of meromorphic h-periodic functions.

Proof. We should prove that $g(z)$ is a meromorphic solution of (2.1) if and only if it can be represented in the form

$$g(z)=a(z)\psi(z)+b(z)\phi(z), \tag{2.5}$$

where a and b are meromorphic h-periodic functions. Substituting (2.5) into (2.1) we see that $g\in\mathbb{M}$. Therefore the proof of the theorem follows from lemma 2.2. □

The pair of meromorphic solutions ψ and ϕ satisfying condition (2.3) forms the basis of the space of solutions.

2.3 Transition matrices

Let ψ, ϕ and $\tilde{\psi}, \tilde{\phi}$ be two basises. Then we can write

$$\tilde{\psi}(z) = a(z)\psi(z) + b(z)\phi(z),$$

$$\tilde{\phi}(z) = c(z)\psi(z) + d(z)\phi(z).$$

In accordance with Lemma 2.1 the coefficients a, b, c, d can be expressed in terms of the wronskians of the solutions $\psi, \phi, \tilde{\psi}$ and $\tilde{\phi}$. We call the matrix

$$T(z) = \begin{pmatrix} a(z) & b(z) \\ c(z) & d(z) \end{pmatrix}$$

the *transition matrix from the basis* ψ, ϕ *to the basis* $\tilde{\psi}, \tilde{\phi}$.

The transition matrix possesses the following important properties:

$$T(z+h) = T(z), \tag{2.6}$$

$$det\, T(z) = \frac{(\tilde{\psi}, \tilde{\phi})}{(\psi, \phi)}. \tag{2.7}$$

2.4 Monodromy matrices

Let the potential of Schrödinger equation (2.1) be periodic, $v(z+2\pi) = v(z)$. In this case the space $\mathbb{M}$ is invariant with respect to the translations $f(z) \mapsto f(z+2\pi)$.

Let $\psi(z)$, $\phi(z)$ be a basis and

$$(\psi, \phi) = Const \neq 0. \tag{2.8}$$

The pair of solutions $\psi(z+2\pi), \phi(z+2\pi)$ is also a basis. We call the transition matrix from the basis $\psi(z), \phi(z)$ to the basis $\psi(z+2\pi), \phi(z+2\pi)$ *the monodromy matrix* corresponding to the basis $\psi(z)$, $\phi(z)$.

In view of (2.7) and (2.8) the monodromy matrix, $M(z)$, is unimodular:

$$det\, M(z) = 1, \tag{2.9}$$

Due to (2.6) $M(z)$ is h-periodic.

2.5 The trace formula

Equation (2.1) is equivalent to the system:

$$\vec{\psi}(z) = \Lambda(z-h)\vec{\psi}(z-h),$$

where

$$\Lambda(z) = \begin{pmatrix} 2(E-v(z)) & -1 \\ 1 & 0 \end{pmatrix}. \tag{2.10}$$

Theorem 2.4 (The trace formula) Let $h = \frac{2\pi m}{n}$, $m, n \in \mathbb{N}$. Then

$$\begin{aligned} &Tr\ (\Lambda(z+(n-1)h)\dots\Lambda(z+h)\Lambda(z)) = \\ &\quad = Tr\ (M(z+(m-1)2\pi)\dots M(z+2\pi)\,M(z)), \end{aligned} \tag{2.11}$$

where $M(z)$ is a monodromy matrix.

Proof. Let $\psi(z)$ and $\phi(z)$ be the solutions corresponding to the monodromy matrix $M(z)$. Let

$$S(z) = \begin{pmatrix} \psi(z) & \phi(z) \\ \psi(z-h) & \phi(z-h) \end{pmatrix}.$$

It is evident that

$$S(z+h) = \Lambda(z)\,S(z)$$

and

$$S(z+2\pi) = S(z)\,M^t(z),$$

where the symbol ^{t} denotes the transposition. Since $det\,S(z) = (\psi, \phi) \not\equiv 0$ and $nh = 2\pi m$ we can write

$$\begin{aligned} &Tr\ (M(z+(m-1)2\pi)\dots M(z+2\pi)\,M(z)) = \\ &\quad = Tr\ \left(S^t(z+2\pi m)\,S^{t^{-1}}(z)\right) = \\ &\quad = Tr\ \left(S(z+2\pi m)\,S^{-1}(z)\right) = \\ &\quad = Tr\ \left(S(z+nh)\,S^{-1}(z)\right) = \\ &\quad = Tr\ (\Lambda(z+(n-1)h)\dots\Lambda(z+h)\Lambda(z)). \end{aligned}$$

□

3 The analytic solutions of Harper's equation

In this section we formulate one of the results of [B-F]: the existence theorem for the entire solutions having standard asymptotic behavior. We begin with the description of the geometrical constructions related to this Theorem.

3.1 The complex momentum

All the geometric objects we use can be described in terms of the complex momentum $p(z)$,

$$\cos p(z) + \cos z = E. \tag{3.1}$$

The equation (3.1) can be solved explicitly

$$p(z) = \frac{1}{i} \ln(v(z) + \sqrt{v^2(z) - 1}), \tag{3.2}$$

where

$$v(z) = E - \cos z.$$

The finite branching points (square root branching points) of the function $p(z)$ satisfy the equation

$$v(z) = \pm 1.$$

There is also an infinite branching point (logarithm branching point). As it was said in Introduction we assume that

$$0 < E < 2.$$

In this case in the domain $-\pi < Re\, z < \pi$ there are two real branching points $z = a$ and $z = -a$, $a = arccos(E - 1)$, and two complex branching points $z = ib$ and $z = -ib$, $b = arccosh(E+1)$. All other finite branching points can be obtained from these by the translations $z \mapsto z + 2\pi k$, $k \in \mathbb{Z}$.

Let K be a simply connected domain, $K \subset \mathbb{C}$, and let it do not contain any branching point. Denote by $p_o(z)$ some branch of the momentum continuous on this domain. All the other branches continuous on K are connected with p_o by the relations:

$$p_n^{\pm}(z) = \pm p_0(z) + 2\pi n, \quad n \in \mathbf{Z}. \tag{3.3}$$

Consider the behavior of the momentum for large $|Im\, z|$. The following lemma implies from the formula (3.2).

Lemma 3.1 Let $y = Imz$. Then

$$p_o(z) = \pm(z - \pi) + 2\pi k + O(e^{-|y|}) \quad \text{as} |y| \to \infty.$$

The signs $\pm$ and the integer constant k depend on the choice of the branch p_o. The "error" estimate is uniform with respect to $Re\, z$.

3.2 The action and Stokes lines

The action integral $\int^z p\, dz$ is a many-valued function. It has the same branching points as the momentum.

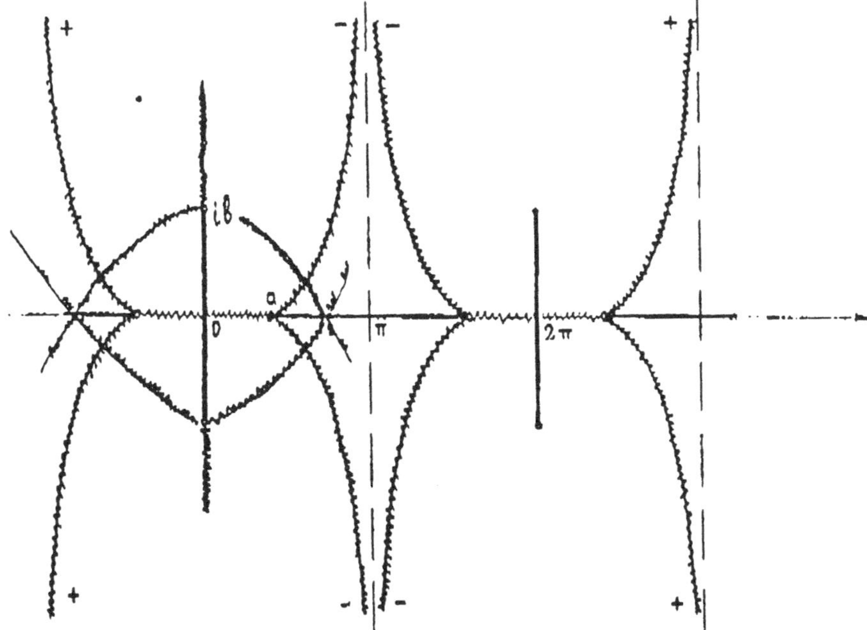

Figure 1:

Consider the curves described by the equation

$$Im \int_{z_*}^{z} (p(z') - p(z_*))\, dz' = 0, \tag{3.4}$$

where z_* is a finite branching point. We call these curves *Stokes lines beginning at* z_*. In view of (3.3) the Stokes line definition does not depend on the choice of the branch $p(z)$ in (3.4).

Three Stokes lines begin at every branching point. The angles between them at this point are equal to $\frac{2\pi}{3}$.

Some of the Stokes lines are finite: they are connecting two finite branching points; some of the Stokes lines are infinite: they are going from finite branching points to the infinity. The asymptotes of the infinite lines can be investigated by means of Lemma 3.1.

Let P be some branching point. The point $P + 2\pi$ is also a branching point. The Stokes lines beginning at it can be obtained from the Stokes lines beginning at P by the translation $z \mapsto z + 2\pi$.

The character of the Stokes lines in the strip $-\pi < Re\, z < \pi$ corresponds to fig. 3.1.

Consider the branching point a in fig.3.1 and the corresponding Stokes lines. Two of them are infinite. They are going to $\pm i\infty$ and have the asymptote $Re\, z = \pi$. One of these Stokes lines is finite. It goes along the real axis. The Stokes lines beginning at the complex branching points are infinite. Consider the branching point ib. One of the corresponding Stokes lines goes along the imaginary axis.

3.3 Canonical domains

The canonical domain notion is the most important and the most complicated geometrical notion we use in our analysis. Its origin lies far beyond the geometry of the Riemann surface of the complex momentum and is connected with the construction of the analytic solutions having standard asymptotic behavior.

Let γ be a smooth curve, $\gamma \subset \mathbb{C}$. We call γ *vertical* if it intersects any of the lines $Im\, z = Const$ at an non-zero angle. We orient vertical lines from $-i\infty$ to $i\infty$.

Consider a vertical curve γ. Fix on γ a continuous branch of the momentum. We call γ *canonical* if along it the value

$$Im \left(\int_{\gamma}^{z} p\, dz \right)$$

monotonically decreases and

$$Im \left(\int_{\gamma}^{z} (p + \pi)\, dz \right)$$

monotonically increases.

Let K be a simply connecting domain containing no finite branching points. Fix on K a continuous branch of the momentum. We call K *canonical* if for any $z \in K$ there exists a canonical curve γ_z such that $z \in \gamma_z \subset K$.

In this paper we are not going to describe the asymptotics of the every meromorphic solution of (1.1) on the whole complex plane. Our aim is just to construct the specific entire solutions for which the monodromy matrix possesses the properties described in our main Theorem. For this purpose it suffices to consider only one canonical domain. In the sequel we always denote this domain by K. The "portrait" of the domain K is shown in fig. 3.2.

The boundaries of K consist of segments of Stokes lines. We fix on K a continuous branch of the momentum by the condition

$$p(z) = z - \pi + o(1), \quad z \to -i\infty, \; z \in K. \tag{3.5}$$

The proof that the domain K is canonical follows from the investigation of the families of the lines $Re \int_{z_0}^{z} p\, dz = Const$ and $Re \int_{z_0}^{z} (p+\pi)\, dz = Const$, *the Antistokes-type lines*, in this domain; see [B-F].

In the sequel we consider only the above canonical domain K. Working with K and its subdomains we always denote by $p(z)$ the branch fixed in (3.5).

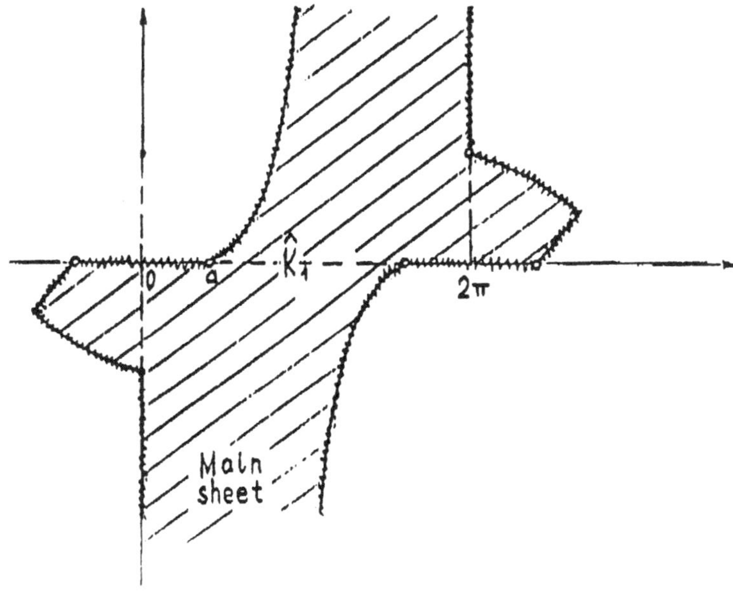

Figure 2:

3.4 Admissible domains

Let ∂K be the boundary of the canonical domain K and let δ be a positive constant. We call the subdomain

$$K_\delta = \{z \in K \mid dist\,(z, \partial K) \geq \delta\}$$

admissible.

In the sequel we always assume that h is sufficiently small for Harper's equation to make sense on K_δ.

3.5 The analytic solutions

In [B-F] we prove a theorem about the existence of analytic solutions of Harper's equation. It can be formulated as follows.

Theorem 3.2 Let ε be a fixed positive constant and let $0 < \varepsilon < E < 2 - \varepsilon$. Let K_δ be an admissible subdomain and let z_o be a point in K_δ. There exists a positive constant $h_o(\delta, \varepsilon)$ such that for $0 < h < h_o$ Harper's equation has two solutions,

$\psi_\pm$, which are analytic on K_δ and have the representations

$$\psi_\pm(z) = \frac{e^{\pm\frac{i}{h}\int_{z_0}^{z} p\,dz}}{\sqrt{2i\sin p\,(z)}}(1+O(h)), \quad z \in K_\delta. \tag{3.6}$$

The "error" estimate is uniform with respect to z.

The solutions described in this theorem continuously depend on h. Moreover, they are analytic functions of E in some vicinity of the interval $0 < \varepsilon < E < 2-\varepsilon$. The size of this vicinity depends on ε and δ. We didn't notice these two properties in [B-F] but they follow directly from the proofs of that paper.

In the sequel we call the solutions $\psi_\pm$ possessing the properties described in the Theorem 3.2 "the solutions having standard behavior".

4 The solutions' estimates and the monodromy matrix

In the sequel $C_1, C_2, \dots$ denote positive constants.

4.1 The solutions' estimates on an admissible domain

Lemma 4.1 Let K_δ be an admissible subdomain of the canonical domain K and let $\psi_\pm$ be analytic solutions having standard behavior on K_δ. Then

$$|\psi_\pm(z)| \le C_1\, e^{\pm\frac{x-\pi}{h}|y| - \frac{|y|}{2}}, \quad z \in K_\delta. \tag{4.1}$$

In these formulae $x = Re\, z$, $y = Im\, z$.

Proof. The proofs of these estimates are quite similar. We prove (4.1) for $|\psi_+(z)|$ in case $y < 0$.

The solution ψ_+ has the asymptotic representation (3.6). The function $\sin p\,(z)$ is equal to zero only at the branching points of the momentum p. Inside the admissible domain $\sin p\,(z) \ge C_2$. Moreover, using Lemma 3.1 we get

$$\left|\frac{1}{\sqrt{\sin p\,(z)}}\right| \le C_3 e^{-\frac{|y|}{2}}, \quad z \in K_\delta. \tag{4.2}$$

Taking into account the choise of the branch $p\,(z)$, see (3.5), we deduce from Lemma 3.1:

$$p\,(z) = z - \pi + O(e^{-|y|}) \quad z \to -i\infty,\ z \in K_\delta,$$

and therefore

$$\left|e^{\frac{i}{h}\int_{z_o}^{z} p\,dz}\right| = \left|e^{\frac{i}{h}\int_{z_o}^{z}(p+\pi-z)\,dz}\; e^{\frac{i}{h}\int_{z_o}^{z}(z-\pi)\,dz}\right| \leq$$
$$\leq C_4\left|e^{\frac{i}{h}\int_{z_o}^{z}(z-\pi)\,dz}\right| \leq C_5 e^{\frac{\pi-x}{h}y}, \quad z\in K_\delta,\ y<0. \tag{4.3}$$

Taking into account (3.6) and estimates (4.2) and (4.3) we prove inequality (4.1) for $|\psi_+(z)|$ in case $y<0$. □

4.2 The basis of the space of meromorphic solutions

Let us calculate the wronskian of the solutions having standard behavior on K_δ.

Proposition 4.2 The wronskian of the solutions ψ_- and ψ_+ is constant and

$$(\psi_+,\psi_-) = 1+O(h) \quad \text{as} h\to 0. \tag{4.4}$$

Proof. Using the wronskian definition and Lemma 4.1 we get

$$|(\psi_+,\psi_-)| \leq C_6, \quad z\in K_\delta.$$

But the wronskian is an h-periodic function and so it is uniformly bounded on $\mathbb{C}$. And then, as an analytic function, it has to be constant. Let us calculate it in some finite point $z\in K_\delta$. Obviously,

$$(\psi_+,\psi_-) = \frac{1}{2i\sqrt{\sin p\,(z)\ \sin p(z+h)}}\cdot$$
$$\cdot\left(e^{\frac{i}{h}\int_z^{z+h} p\,dz + O(h)} - e^{-\frac{i}{h}\int_z^{z+h} p\,dz+O(h)}\right) =$$
$$\frac{e^{i\,p\,(z)}+e^{-i\,p\,(z)}+O(h)}{2i\sqrt{\sin p\,(z)\ \sin p(z+h)}} = 1+O(h).$$

□

This lemma means, in particular, that the solutions $\psi_\pm$ form a basis of the space of the meromorphic solutions of Harper's equation. Without loss of generality we assume in the sequel that

$$(\psi_+,\psi_-)=1.$$

We call this basis *standard.*

4.3 The solutions outside the admissible domain

Analytic properties of solutions outside the admissible domain. In the previous part of the paper we have described solutions analytic on a given admissible domain. These solutions are entire functions of z. Really, for the analytic continuation we can use Harper's equation itself: by means of the formulae

$$\psi(z+h) = -\psi(z-h) + 2(E - \cos z)\psi(z)$$

and

$$\psi(z-h) = -\psi(z+h) + 2(E - \cos z)\psi(z)$$

we can continue the solutions $\psi_\pm$ into the h-vicinity of the admissible domain, then into the $2h$-vicinity and so on. The result of the next subsection allows to get the estimates of $\psi_\pm$ outside the admissible domain.

A Priori estimates for smooth solutions. In this subsection we call a segment of a line $Im\, z = const$ *a horizontal path.* We orient the horizontal path (z', z'') from z' to z''.

We denote by V_δ the δ-vicinity of the finite branching points.

Lemma 4.3 Let $\psi(z)$ be a smooth solution of Harper's equation and let d and δ be some positive constants. Assume that the horizontal path (z', z'') satisfies the conditions

$$(z', z'') \cap V_\delta = \emptyset$$

and

$$|z' - z''| \leq d.$$

Then $\psi(z)$ satisfies the estimate

$$\begin{gathered} |\psi(z)| \leq C(\delta, d)\, e^{\frac{1}{h}\int_{\tilde z}^{z} |Im\, p\, dz|} \sqrt{|\psi(\tilde z)|^2 + |\psi(\tilde z - \sigma h))|^2}, \\ z,\ \tilde z \in (z', z''),\ \tilde z = z - hk,\ k \in \mathbb{Z}, \\ \sigma = +1 \text{ if } Re\, z > Re\, \tilde z \text{ and } \sigma = -1 \text{ if } Re\, z < Re\, \tilde z. \end{gathered} \tag{4.5}$$

The coefficient $C(\delta, d)$ depends only on δ and d.

Proof. Consider the case when $k > 0$. Let

$$\vec\psi(z) = \begin{pmatrix} \psi(z) \\ \psi(z-h) \end{pmatrix}, \quad \Lambda(z) = \begin{pmatrix} 2(E - \cos z) & -1 \\ 1 & 0 \end{pmatrix}.$$

Evidently,

$$\left\|\vec\psi(z)\right\|_{\mathbb{C}^2} \leq \|\Lambda(z-h)\| \left\|\vec\psi(z-h)\right\|_{\mathbb{C}^2}.$$

The eigenvalues of the matrix $\Lambda\,(z)$ are equal to $e^{\pm ip\,(z)}$. Therefore

$$\left\|\vec\psi(z)\right\|_{\mathbb{C}^2} \leq e^{\sum_{l=0}^{k-1} |Im\, p\,(\tilde z + hl)|} \left\|\vec\psi(\tilde z)\right\|_{\mathbb{C}^2}, \quad \tilde z = z - hk.$$

Let $z,\,\tilde{z}\,\in(z',z'')$. We get

$$\left|\sum_{l=0}^{k-1}|Im\,p(\tilde{z}+lh)|-\frac{1}{h}\int_{\tilde{z}}^{z}|Im\,p\,dz|\right|\le\sup_{z\notin V_\delta}|p'(z)|\,|z'-z''|.$$

Outside V_δ the value $|p'(z)|$ is bounded and thus

$$\left\|\vec{\psi}(z)\right\|_{\mathbb{C}^2}\le C(\delta,d)\,e^{\frac{1}{h}\int_{\tilde{z}}^{z}|Im\,p\,dz|}\left\|\vec{\psi}(\tilde{z})\right\|_{\mathbb{C}^2}.$$

In the case when $k<0$ the proof is quite similar. □

The estimates of the solutions outside the admissible domain. We shall use the following statement to investigate the functional structure of the monodromy matrix.

Lemma 4.4 Let z be to the left from the domain K_δ. The solutions $\psi_\pm$ satisfy the estimates:

$$|\psi_+(z)|\le C_7\,e^{\frac{\pi+4\delta-x}{h}y+\frac{y}{2}},\quad y>0, \tag{4.6}$$

$$|\psi_+(z)|\le C_7\,e^{\frac{-\pi+4\delta-x}{h}|y|+\frac{|y|}{2}},\quad y<0, \tag{4.7}$$

$$|\psi_-(z)|\le C_7\,e^{\frac{\pi-x}{h}|y|-\frac{|y|}{2}}, \tag{4.8}$$

where $x=Re\,z$ and $y=Im\,z$.

Proof. To prove the estimates for $|\psi_+(z)|$ we use estimate (4.1) and the a priori estimate of Lemma 4.2:

$$\begin{aligned}|\psi_+(z)|&\le C(\delta,d)\,e^{\frac{1}{h}\int_{\tilde{z}}^{z}|Im\,p\,dz|}\sqrt{|\psi_+(\tilde{z})|^2+|\psi_+(\tilde{z}+h))|^2}\le\\&\le C_8\,e^{\frac{1}{h}\int_{\tilde{z}}^{z}|Im\,p\,dz|}\;e^{\frac{(\tilde{x}+h)-\pi}{h}|y|-\frac{|y|}{2}},\\&\tilde{z}=z+hk\in K_\delta,\;\tilde{x}=Re\,\tilde{z}.\end{aligned} \tag{4.9}$$

To estimate the integral $\int_{\tilde{z}}^{z}|Im\,p\,dz|$ we use the inequality

$$|Im\,p\,(z)|\le C_9+|y|,\quad z=x+iy,$$

which is a direct consequence of Lemma 3.1. It leads to the result

$$\left|\psi_+^1(z)\right|\le C_{10}e^{\frac{\tilde{x}-x}{h}|y|+\frac{\tilde{x}-\pi}{h}|y|+\frac{|y|}{2}}. \tag{4.10}$$

Now let us concentrate on the case $y>0$. Since the solution ψ_+ is entire it suffices to prove (4.6) only for $y\ge C_{11}$. We remind that the boundary of K_δ consists of segments of Stokes lines. The line $Re\,z=\pi$ is one of its asymptotes for $Im\,z\to+i\infty$, see fig. 3.2. Therefore there exists a C_{11} such that we can choose $\tilde{z}\in K_\delta$ satisfying the conditions $y>C_{11}$ and $\pi<\tilde{x}<\pi+2\delta$. It leads from (4.10) to the estimate (4.6).

The estimates (4.7) and (4.8) can be proved in the similar way. □

4.4 The functional structure of the monodromy matrix

In this subsection we begin to investigate the monodromy matrix $\mathbf{M}(z)$ corresponding to the basis $\psi_\pm$.

Proposition 4.5 The monodromy matrix $\mathbf{M}(z)$ has the following functional structure:

$$\mathbf{M}(z) = \begin{pmatrix} a_0 + a_1 e^{\frac{2\pi i}{h} z} + a_{-1} e^{-\frac{2\pi i}{h} z} & b_0 + b_{-1} e^{-\frac{2\pi i}{h} z} \\ c_0 + c_1 e^{\frac{2\pi i}{h} z} & d_0 \end{pmatrix}, \tag{4.11}$$

where a, b, c and d with subscripts denote coefficients independent of z.

Proof. We remind that the coefficients of the monodromy matrix can be expressed in terms of the wronskians of the basis solutions. Evidently,

$$(\psi_\pm, \psi_\pm) = 0.$$

Taking into account these formulae, we obtain from the definiton of the monodromy matrix:

$$\mathbf{M}_{11}(z) = \frac{(\psi_+(z+2\pi), \psi_-(z))}{(\psi_+(z), \psi_-(z))}, \quad \mathbf{M}_{12}(z) = \frac{(\psi_+(z), \psi_+(z+2\pi))}{(\psi_+(z), \psi_-(z))}, \tag{4.12}$$

$$\mathbf{M}_{21}(z) = \frac{(\psi_-(z+2\pi), \psi_-(z))}{(\psi_+(z), \psi_-(z))}, \quad \mathbf{M}_{22}(z) = \frac{(\psi_+(z), \psi_-(z+2\pi))}{(\psi_+(z), \psi_-(z))}. \tag{4.13}$$

Together with $(\psi_+, \psi_-) = Const$, see Lemma 4.2, the formulae (4.12)–(4.13) mean, in particular, that the coefficients of the matrix $\mathbf{M}(z)$ are entire h-periodic functions. To prove the Proposition it suffices to get the estimates of these functions for large $|y|$, $y = Im\, z$. It can be done by means of the inequalities obtained in Lemma 4.1 and Lemma 4.4. In result we get for $z + 2\pi \in K_\delta$:

$$\begin{aligned} |\mathbf{M}_{11}(z)| &\le C_{12}\, e^{\frac{2\pi}{h}|y|}, \\ |\mathbf{M}_{12}(z)| &\le C_{12}\, e^{\frac{2\pi+4\delta+h}{h} y}, \quad y > 0; \\ |\mathbf{M}_{12}(z)| &\le C_{12}\, e^{\frac{4\delta+h}{h}|y|}, \quad y < 0; \\ |\mathbf{M}_{21}(z)| &\le C_{12}\, e^{\frac{-2x}{h}|y|}. \end{aligned}$$

Here C_{12} is a positive constant. These estimates together with the analyticity and the h-periodicity of the monodromy matrix coefficients lead to representation (4.11). Really, consider the coefficient $\mathbf{M}_{11}(z)$. Since it is h-periodic and entire, the above estimate means that

$$\mathbf{M}_{11}(z) = a_0 + a_1 e^{\frac{2\pi i}{h} z} + a_{-1} e^{-\frac{2\pi i}{h} z}.$$

The analogous arguments lead to the representations

$$\mathbf{M}_{12}(z) = b_0 + b_{-1}e^{-\frac{2\pi i}{h}z}$$

and

$$\mathbf{M}_{21}(z) = c_o + c_1\, e^{\frac{2\pi i}{h}z}.$$

The formula

$$\mathbf{M}_{22}(z) = d_o$$

follows from the equality $det\,\mathbf{M} = 1$.

5 The relations between the monodromy matrix coefficients

5.1 The reflection invariance

The space of meromorphic solutions of Harper's equation is invariant with respect to the reflections: $\psi(z) \mapsto \psi(-z)$. It leads to a certain relations between the coefficients of the monodromy matrix $\mathbf{M}$.

Remind that in [B-F] we show that the solutions $\psi_\pm$ described in Theorem 3.2 can be constructed so that

$$\psi_+(z) = \psi_-(2\pi - z)\, e^{-\frac{i}{h}\int_{2\pi-z_o}^{z_o} p\, dz},$$

where z_o is the same parameter as in (3.6). In the sequal we choose $z_o = \pi$. In this case

$$\psi_+(z) = \psi_-(2\pi - z). \tag{5.1}$$

This formula implies the following statement.

Lemma 5.1 The monodromy matrix $\mathbf{M}(z)$ can be represented in the form

$$\mathbf{M}(z) = \begin{pmatrix} a_o + a_1 \cos\left(\frac{2\pi}{h}z\right) & -b_0 - b_1\, e^{-\frac{2\pi i}{h}z} \\ b_0 + b_1\, e^{\frac{2\pi i}{h}z} & d_o \end{pmatrix}, \tag{5.2}$$

where the letters a, b and d denote coefficients independent of z.

Proof. Let

$$\Psi(z) = \begin{pmatrix} \psi_+(z) \\ \psi_-(z) \end{pmatrix},$$

It is evident that

$$\Psi(z + 2\pi) = \mathbf{M}(z)\, \Psi(z)$$

and

$$\Psi(2\pi - z) = \sigma \cdot \Psi(z), \quad \sigma = \begin{pmatrix} 0 & 1 \\ 1 & 0 \end{pmatrix}.$$

It allows to write:

$$\Psi(z+2\pi) = \sigma \cdot \Psi(-z) = \sigma \cdot \mathbf{M}^{-1}(-z) \cdot \Psi(2\pi - z) = \sigma \cdot \mathbf{M}^{-1}(-z) \cdot \sigma \cdot \Psi(z).$$

Therefore

$$\mathbf{M}(z) = \sigma \cdot \mathbf{M}^{-1}(-z) \cdot \sigma.$$

This formula together with Proposition 4.5 and the formula $det\, M(z) = 1$ implies representation (5.2). □

5.2 The trace formula and the coefficient a_1

Proposition 5.2 The coefficient a_1 of the monodromy matrix is equal to -2,

$$a_1 = -2. \tag{5.3}$$

The proof of this proposition is based on the trace formula (2.11) and the following lemma.

Lemma 5.3 Let $h = 2\pi \frac{m}{n}$, where m and n are relatevely prime natural numbers, and let $\mu(z)$ be a matrix of the form

$$\mu(z) = \begin{pmatrix} \alpha_o + \alpha_1 \cos z & \beta_0 + \beta_1 e^{-iz} \\ \gamma_0 + \gamma_1 e^{iz} & \delta_o \end{pmatrix}, \tag{5.4}$$

where the Greek letters denote coefficients independent of z. Then

$$Tr\ \left(\mu(z+(n-1)h)\ldots\mu(z+h)\,\mu(z)\right) = F_0^\mu(h) - 2\left(-\frac{\alpha_1}{2}\right)^n \cos(nz),$$

where F_0^μ is a coefficient independent of z.

Proof. Obviously, $Tr\ (\mu(z+(n-1)h)\ldots\mu(z+h)\,\mu(z))$ is an h-periodic function of z. Therefore

$$\begin{aligned} Tr\ (\mu(z+(n-1)h)\ldots\mu(z+h)\,\mu(z)) &= \\ = \sum_{k=-N}^{k=N} F_k^\mu\, e^{\frac{2\pi i k}{h} z} &= \sum_{k=-N}^{k=N} F_k^\mu\, e^{\frac{nik}{m} z}, \end{aligned} \tag{5.5}$$

where N is a finite integer number and F_k^μ are coefficients independent of z. Taking into account formula (5.4) we see that, in fact,

$$Tr\ (\mu(z+(n-1)h)\ldots\mu(z+h)\,\mu(z)) = F_0^\mu + F_m^\mu\, e^{inz} + F_{-m}^\mu\, e^{-inz}.$$

Therefore

$$Tr\ \left(e^{inz}\mu(z+(n-1)h)\ldots\mu(z+h)\,\mu(z)\right) = F_0^\mu\, e^{inz} + F_m^\mu\, e^{2inz} + F_{-m}^\mu.$$

Both, the left-hand side and the right-hand side in this formula are continuous functions of $w = e^{iz}$. Putting $w = 0$ we get

$$F^{\mu}_{-m} = Tr \left(e^{-ih(n-1)} \begin{pmatrix} \frac{\alpha_1}{2} & \beta_1 \\ 0 & 0 \end{pmatrix} \cdot \ldots \cdot e^{-ih} \begin{pmatrix} \frac{\alpha_1}{2} & \beta_1 \\ 0 & 0 \end{pmatrix} \cdot \begin{pmatrix} \frac{\alpha_1}{2} & \beta_1 \\ 0 & 0 \end{pmatrix} \right) =$$

$$= \left(\frac{\alpha_1}{2}\right)^n e^{-ih\frac{n(n-1)}{2}} = \left(\frac{\alpha_1}{2}\right)^n e^{-i\pi m(n-1)} = -\left(-\frac{\alpha_1}{2}\right)^n.$$

Therefore $F^{\mu}_{-m} = -\left(-\frac{\alpha_1}{2}\right)^n$. In the analogous way one can see that $F^{\mu}_{m} = -\left(-\frac{\alpha_1}{2}\right)^n$. It proves the lemma. □

Proof of Proposition 5.2. Let $h = \frac{2\pi m}{n}$ and $\Lambda(z)$ be the matrix defined by (2.10) with $v(z) = \cos(z)$. It follows from Lemma 5.3 that

$$Tr\ (\Lambda(z + (n-1)h) \ldots \Lambda(z+h)\Lambda(z)) = F^{\Lambda}_0(h) - 2\cos(nz).$$

Analogously, we get for the monodromy matrix $\mathbf{M}(z)$

$$Tr\ (\mathbf{M}(z + (m-1)2\pi) \ldots \mathbf{M}(z+2\pi)\,\mathbf{M}(z)) =$$

$$= F^{\mathbf{M}}_0(\frac{4\pi^2}{h}) - 2\left(-\frac{a_1}{2}\right)^m \cos(m \cdot \frac{2\pi z}{h}) = F^{\mathbf{M}}_0(\frac{4\pi^2}{h}) - 2\left(-\frac{a_1}{2}\right)^m \cos(nz).$$

But it follows from (2.11) that the right-hand sides in these formulae are equal. Therefore for $\frac{h}{2\pi} \in \mathbb{Q}$ we have $a_1 = -2\,1^{\frac{1}{m}}$. At the same time the coefficients of the monodromy matrix $\mathbf{M}$ are the wronskians of the solutions which are continuous functions of h. Hence, the coefficient a_1 is a continuous function of h. And then $a_1 = -2$. This formula is valid also for irrational $\frac{h}{2\pi}$. □

5.3 The invariance $\psi(z) \mapsto \overline{\psi(\overline{z})}$ and the final formula for the monodromy matrix

The space of solutions of Harper's equation is invariant with respect to the transformation $\psi(z) \mapsto \overline{\psi(\overline{z})}$. It leads to additional relations between the coefficients of the monodromy matrix.

Theorem 5.4 The monodromy matrix $\mathbf{M}(z)$ has the form

$$\mathbf{M}(z) = \begin{pmatrix} \frac{1}{w} + \frac{1}{\overline{w}} + \overline{w} - 2\cos\left(\frac{2\pi}{h}z\right) & -i\sqrt{\frac{w}{\overline{w}}} + i\sqrt{w\overline{w}}\, e^{-\frac{2\pi i}{h}z} \\ i\sqrt{\frac{w}{\overline{w}}} - i\sqrt{w\overline{w}}\, e^{\frac{2\pi i}{h}z} & w \end{pmatrix}, \tag{5.6}$$

where the roots' branches are defined by the relations

$$\sqrt{\frac{w}{\overline{w}}}\sqrt{w\overline{w}} = w \tag{5.7}$$

and

$$\sqrt{w\overline{w}} > 0; \tag{5.8}$$

the coefficient $w = w(E, h)$ is independent of z.

Proof. Since $\overline{\psi}_{\pm}(\overline{z})$ are entire solutions of Harper's equation then

$$\begin{pmatrix} \overline{\psi}_{+}(\overline{z}) \\ \overline{\psi}_{-}(\overline{z}) \end{pmatrix} = J \cdot \begin{pmatrix} \psi_{+}(z) \\ \psi_{-}(z) \end{pmatrix}, \tag{5.9}$$

where J is a 2×2-matrix having h-periodic meromorphic coefficients.

Lemma 5.5 The coefficients of the matrix J are constant,

$$J = \begin{pmatrix} \alpha & i\beta \\ i\beta & \alpha \end{pmatrix}, \tag{5.10}$$

$$\alpha, \beta \in \mathbb{R}, \tag{5.11}$$

and

$$\alpha^2 + \beta^2 = 1. \tag{5.12}$$

Proof. Note that

$$(\overline{\psi}_{+}(\overline{z}), \overline{\psi}_{-}(\overline{z})) = (\psi_{+}(z), \psi_{-}(z)) = 1.$$

It means, in particular, that

$$det\, J = 1. \tag{5.13}$$

Obviously,

$$J_{11} = \left(\overline{\psi}_{+}(\overline{z}), \psi_{-}(z)\right), \qquad J_{12} = \left(\psi_{+}(z), \overline{\psi}_{+}(\overline{z})\right),$$
$$J_{21} = \left(\overline{\psi}_{-}(\overline{z}), \psi_{-}(z)\right), \qquad J_{22} = \left(\psi_{+}(z), \overline{\psi}_{-}(\overline{z})\right).$$

The last four formulae mean that the coefficients of the matrix J are entire h-periodic functions. These formulae allow to obtain the estimates of the coefficients J_{jk}: by means (4.1) and the estimates of Lemma 4.4 we get

$$|J_{11}| \le C, \quad y > 0,$$
$$|J_{11}| \le C\, e^{\frac{2\pi - 2x + 4\delta}{h}|y|}, \quad y < 0,$$
$$|J_{12}| \le C\, e^{\frac{h + 4\delta}{h}|y|},$$
$$|J_{21}| \le C\, e^{\frac{2\pi - 2x}{h}|y|},$$

when $\delta < x < \pi - \delta$. In these formulae $z = x + iy$ and δ is a constant from the definition of the admissible domain K_δ. Recall that δ is a fixed positive constant which could be chosen arbitrarily small. It follows from the estimates for the coefficients J_{11}, J_{12} and J_{21} that they, as entire h-periodic functions of z, have to be constant; the formula $J_{22} = Const$ follows from (5.13).

Let us check the relations between the coefficients of the matrix J. Obviously,

$$\overline{J_{11}} = \overline{(\overline{\psi}_+(z), \psi_-(\overline{z}))} = (\psi_+(z), \overline{\psi}_-(\overline{z})) = J_{22}$$

$$\text{and} \quad \overline{J_{12}} = \overline{(\psi_+(\overline{z}), \overline{\psi}_+(z))} = -(\psi_+(z), \overline{\psi}_+(\overline{z})) = -J_{12}.$$

Analogously,

$$\overline{J_{21}} = -J_{21}.$$

Therefore the matrix J has the form

$$J = \begin{pmatrix} \alpha & i\beta \\ i\gamma & \overline{\alpha} \end{pmatrix}.$$

To complete the proof of (5.10)–(5.11) we recall that

$$\begin{pmatrix} \psi_+(2\pi - z) \\ \psi_-(2\pi - z) \end{pmatrix} = \sigma \cdot \begin{pmatrix} \psi_+(z) \\ \psi_-(z) \end{pmatrix}, \quad \sigma = \begin{pmatrix} 0 & 1 \\ 1 & 0 \end{pmatrix}.$$

Since the same relation is valid for $\begin{pmatrix} \overline{\psi}_+(\overline{z}) \\ \overline{\psi}_-(\overline{z}) \end{pmatrix}$ we have$\sigma \cdot J \cdot \sigma = J$.
It proves (5.10)–(5.11). Formula (5.12) follows from (5.13). □

Since

$$\begin{pmatrix} \overline{\psi}_+(\overline{z + 2\pi}) \\ \overline{\psi}_-(\overline{z + 2\pi}) \end{pmatrix} = \overline{\mathbf{M}(\overline{z})} \cdot \begin{pmatrix} \overline{\psi}_+(\overline{z}) \\ \overline{\psi}_-(\overline{z}) \end{pmatrix}$$

We have

$$\overline{\mathbf{M}(\overline{z})} \cdot J = J \cdot \mathbf{M}(z). \tag{5.14}$$

Remind that the monodromy matrix $\mathbf{M}(z)$ has the representation (5.2). Taking into account the formula $a_1 = -2$ and equating the coefficients of the trigonometric polynomials in the left-hand and right-hand sides of (5.14) we come to the relations:

$$\alpha^2\overline{a_0} - 2i\alpha\beta\overline{b_0} + \beta^2\overline{d_0} = a_0; \tag{5.15}$$

$$i\alpha\beta\overline{a_0} + (\beta^2 - \alpha^2)\overline{b_0} - i\alpha\beta\overline{d_0} = -b_0; \tag{5.16}$$

$$\beta^2\overline{a_0} + 2i\alpha\beta\overline{b_0} + \alpha^2\overline{d_0} = d_0; \tag{5.17}$$

$$\begin{aligned} \alpha^2 + i\alpha\beta\overline{b_1} &= 1; \\ \beta^2\,\overline{b_1} - i\alpha\beta &= -b_1; \\ i\alpha\beta + \alpha^2\overline{b_1} &= 0; \\ -\beta^2 + i\alpha\beta\overline{b_1} &= 0. \end{aligned} \tag{5.18}$$

From relations (5.18) we get $\alpha, \beta \neq 0$ and $b_1 = i\frac{\beta}{\alpha}$. The last formula means that

$$b_1 \in i\mathbb{R}. \tag{5.19}$$

Summing relations (5.16) and (5.17) and taking into account (5.12) we get

$$a_0 + d_0 = \overline{a_0} + \overline{d_0}. \tag{5.20}$$

Remind that $det\,\mathbf{M}(z) = 1$. It means that

$$a_0 d_0 + {b_0}^2 + {b_1}^2 = 1 \tag{5.21}$$

and

$$-d_0 + b_0 b_1 = 0. \tag{5.22}$$

Relations (5.19)–(5.22) imply formulae (5.6) and (5.7). We omit the elementary calculations proving this implication. The reation (5.8) follows from the asymptotic formula for $\mathbf{M}_{12}$, see [B-F]. □

References

[Az] Ya. Azbel, *Energy spectrum of a conduction electron in a magnetic field*, Soviet Physics JETP **19** No 3 (1964).

[C-F-K-S] Cycon, R. Froese, W. Kirsch, B.Simon, *Schrödinger operators*, Springer-Verlag, 1987.

[B-F] V. Buslaev, A. Fedotov, *Complex WKB method for Harper's equation*, Reports of Mittag-Leffler Institute **11** (1993).

[Fe] M. Fedoriouk, *Méthodes asymptotiques pour les équations différentielles ordinaires linéares*, MIR, Moscou, 1987.

[Ho] D. Hofstadter, *Energy levels and wave functions of Bloch electrons in rational and irrational magnetic fields*, Phys. Rev. B **14** (1976), 2239–2249.

[H-S] B. Helffer, J. Sjöstrand, *Analyse semi-classique pour l'equation de Harper (avec application a'l etude de l'equation de Schrödinger avec champ magnétique)*, Mémoires de la SMF **34** (1988).

[H-K-S] B. Helffer, P. Kerdelhué, J. Sjöstrand, *Le papillon de Hofstadter revisité*, Mémoires de la SMF **43** (1990).

[Si] Y. Sibuya, *Global theory of a second order linear ordinary differential equation with a polynomial coefficient*, North-Holland, Amsterdam, 1975.

[Wi] M. Wilkinson, *Critical properties of electron eigenstates in incommensurate systems* Proc. Royal Society of London. A **391** (1984), 305–350.

Vladimir Buslaev and Alexander Fedotov, Department of Mathematical Physics, St. Petersburg University, 1, Ulianovskaja st., St. Petersburg, 198904, Russia
buslaev@onti.phys.lgu.spb.su
fedotov@onti.phys.lgu.spb.su

Operator Theory:
Advances and Applications, Vol. 70

Eigenfunction expansion of right definite multiparameter problems

A.Yu.Konstantinov*

Abstract

We extend expansion results of [2,3] on nonuniformly right definite multiparameter problems. We also obtain an abstract approximation criterion for the existence of commuting self-adjoint extensions of a family of symmetric operators.

Let H_j be a separable Hilbert space, A_j be a densely defined symmetric operator in H_j, B_{jk} be a bounded self-adjoint operator in H_j $(j,k = 1,\dots,n)$. Consider in the tensor product $H = \otimes_{j=1}^{n} H_j$ symmetric operators

$$A_j^+ = \mathbf{1} \otimes \dots \otimes \mathbf{1} \otimes A_j \otimes \dots \otimes \mathbf{1}$$

(A_j in the jth place) and similarly defined bounded self-adjoint operators B_{jk}^+. Denote $\mathcal{D} = \text{span }(\{u_1 \otimes \dots \otimes u_n | u_j \in \mathcal{D}(A_j)\})$. Here $\mathcal{D}(A)$ is the domain of A, span (M) is the linear span of a set M. We shall say that $0 \neq \varphi = \varphi(\lambda) \in \mathcal{D}$ is the *eigenvector corresponding to eigenvalue* $\lambda = (\lambda_1, \dots, \lambda_n) \in \mathrm{R}^n$ if

$$A_j^+ \varphi = \sum_{k=1}^{n} \lambda_k B_{jk}^+ \varphi \quad (j = 1,\dots,n) \tag{1}$$

We define

$$\Delta = \det\ [B_{jk}^+]$$

and Δ_{jk} as the corresponding (j,k) cofactor. These operators are bounded self-adjoint on H. We shall assume that the problem (1) is *right definite*, i.e., the operator Δ is positive ($(\Delta u, u)_H > 0$ for all nonzero $u \in H$). Let H_Δ be a completion of H in the inner product $< \cdot,\cdot >= (\Delta\cdot,\cdot)_H$. Consider the rigged Hilbert space

$$H_- \supset H \supset H_+ \supset D,$$

where D is a separable linear topological space, densely and continuously embedded into H_+. We assume that $D \subset \mathcal{D}$ and $\forall j,k\ A_j^+ \in \mathcal{L}(D, H^+), B_{jk}^+ \in \mathcal{L}(H_+)$.

*Research supported in part by the FFR of Ukraine Grant 1/238 "Operator"

Here $\mathcal{L}(K,G)$ is the class of linear continuous operators from K into G, $\mathcal{L}(K) = \mathcal{L}(K,K)$. We suppose that $H_+ \subset \mathcal{D}(\Delta^{-1/2})$ and hence one can extend Δ^{-1} to a bounded operator from H_+ into H_Δ. We retain notation Δ^{-1} for this extension. Define on D operators

$$X_k = \Delta^{-1} \sum_{j=1}^{n} \Delta_{jk} A_j^+ \quad (k = 1, \dots, n).$$

It is easy to see that X_k is a symmetric operator in H_Δ. Now we state the main result of this paper.

Theorem 1. *$(X_k)_{k=1}^n$ admits an extension to a family of commuting self-adjoint operators in a Hilbert space $\mathcal{H}$ containing H_Δ as a closed subspace, and $\forall u \in D$*

$$\sum_{k=1}^{n} B_{jk}^+ X_k u = A_j^+ u \quad (k = 1, \dots, n).$$

We remark that if Δ is strongly positive (the problem (1) is *uniformly right definite*), then X_k are essentially self-adjoint in H_Δ and their closures commute [7].

The proof of theorem 1 is based on the approximation of the problem (1) by finite-dimensional multiparameter problems and the following abstract proposition.

Theorem 2. *Let $(T_k)_{k=1}^n$ be a family of symmetric operators in a separable Hilbert space H. Suppose that $D = \cap_{k=1}^n \mathcal{D}(T_k)$ is a dense set in H and suppose that there exists a sequence of families of commuting self-adjoint operators $(T_{k,s})_{k=1}^n$ such that*

$$D \subset \cup_{l=1}^{\infty} \cap_{s=l}^{\infty} \cap_{k=1}^{n} \mathcal{D}(T_{k,s})$$

and $\forall u \in D$ $T_{k,s}u$ converges to $T_k u$. Then $(T_k)_{k=1}^n$ admits an extension to a family of commuting self-adjoint operators in a Hilbert space $\mathcal{H}$ containing H as a closed subspace.

The proof of theorem 2 is based on the following lemma.

Lemma 1. *Let E_m be the joint resolution of the identity of $(T_{k,m})_{k=1}^n$. Then there exists a subsequnce $E_{m(k)}$ and a nonnegative $\mathcal{L}(H)$-valued Borel measure F on R^n such that $\forall f \in C_0(\mathrm{R}^n)$ $\forall u, v \in H$*

$$\int_{\mathrm{R}^n} f(\lambda) d(E_{m(k)}(\lambda)\, u, v)_H \to \int_{\mathrm{R}^n} f(\lambda) d(F(\lambda)\, u, v)_H, \ k \to \infty.$$

Next, one can show that F is a generalized resolution of the identity of $(T_k)_{k=1}^n$ (see [1]) and it is enough to prove theorem 2.

Now it is easy to obtain the abstract expansion result for the problem (1); cf. [2,3]. Define a generalized eigenvector for the problem (1) as a vector $0 \neq \varphi = \varphi(\lambda) \in H_-$ such that $\forall u \in D$

$$(\varphi, A_j^+ \ u)_H = \sum_{k=1}^{n} \lambda_k (\varphi, B_{jk}^+ \ u)_H \quad (j = 1, \ldots, n). \tag{2}$$

Let O be the imbedding operator of H_+ into H.

Theorem 3. *Let $\Delta^{-1/2}O$ be the Hilbert-Schmidt operator from H_+ into H, $B_{jk}^+ \in \mathcal{L}(D)$. Then there exists a finite Borel measure ρ in R^n such that for ρ-almost all $\lambda \in \mathrm{R}^n$ we can construct a family $(\varphi_\alpha(\lambda))_{\alpha=1}^{N(\lambda)}$ of generalized eigenvectors for problem (1), and a corresponding Fourier transform:*

$$H_+ \ni u \to \tilde{u}(\lambda) = (< u, \varphi_\alpha(\lambda) >)_{\alpha=1}^{N(\lambda)} \in l_2(N(\lambda)),$$

such that the Parseval equality holds:

$$< u, v >= \int_{\mathrm{R}^n} (\tilde{u}(\lambda), \tilde{v}(\lambda))_{l_2(N(\lambda))} d\rho(\lambda) \quad (u, v \in H_+).$$

Here $l_2(\infty) = l_2, \ l_2(N) = \mathrm{C}^N \ (N < \infty)$.

We note that $H_\Delta \subset H_-$ and consider the new rigged Hilbert space

$$H_- \supset H_\Delta \supset G_+ \supset D,$$

where G_+ is conjugate to H_- with respect to $< \cdot, \cdot >$. G_+ can be constructed as a completion of H_+ in the inner product $(\Delta \cdot, \Delta \cdot)_{H_+}$. A generalized joint eigenvector of the family $(X_k)_{k=1}^n$ is defined as a vector $0 \neq \varphi = \varphi(\lambda) \in H_-$ such that $\forall u \in D$

$$< \varphi, X_k \ u >= \lambda_k < \varphi, u > \quad (k = 1, \ldots, n).$$

It is easy to see that a joint generalized eigenvector $\varphi(\lambda)$ of $(X_k)_{k=1}^m$ satisfies (2). Next, $X_k \in \mathcal{L}(D, H_+)$ and the imbedding operator $G_+ \subset H_\Delta$ is of the Hilbert-Schmidt class. Now theorem 3 follows from the general theory of generalized eigenvector expansions [1].

Finally, note that theorem 3 can be immediately applied to multiparameter problems for partial differential operators (see [2,3]). For ordinary differential operators of even order such expansions have been obtained in [5,6]. Expansion results for nonuniformly right definite problems with discrete spectrum can be found in [4,8].

References

[1] Yu.M.Berezansky, "Expansions in Eigenfunctions of Self-Adjoint Operators", Translations of Mathematical Monographs, Vol.17, Amer. Math. Soc., Providence, RI, 1968.

[2] Yu.M.Berezansky and A.Yu.Konstantinov, On eigenvector expansion of multiparameter spectral problems, Funk.Anal. i Ego Pril. **26**, No. 1 (1992), 81-83.

[3] Yu.M.Berezansky and A.Yu.Konstantinov, Expansion in eigenvectors of multiparameter spectral problems, Ukr.Mat.J. **44**, No. 7 (1992), 901-913.

[4] P.Binding, Nonuniform right definiteness, J. Math. Anal. Appl. **102** (1984), 233-243.

[5] P.J.Browne, A singular multi-parameter eigenvalue problem in second order ordinary differential equations, J. Differential Equations **12** (1972), 81-94.

[6] H.A.Isaev, On singular multiparameter differential operators, Mat. Sbornik, **131**, No.1 (1986), 52-72.

[7] H.Volkmer, On multiparameter theory, J. Math. Anal. Appl. **86** (1982), 44-53.

[8] H.Volkmer, "Multiparameter Eigenvalue Problems and Expansion Theorems", Springer-Verlag, Berlin–New-York, 1988.

A.Yu.Konstantinov, Department of Mathematics, University of Kiev,
Kiev, Ukraine 252017

Operator Theory:
Advances and Applications, Vol. 70

Singularly Perturbed Operators

V.Koshmanenko

Let $A = A^* \geq 1$ be a self-adjoint unbounded operator in a separable complex Hilbert space $\mathcal{H}$.

Definition 1 *A self-adjoint operator $\tilde{A} \neq A$ in $\mathcal{H}$ is called singularly perturbed with respect to A if the linear set*

$$\mathcal{D} := \{f \in \mathcal{D}(A) \cap \mathcal{D}(\tilde{A}) | Af = \tilde{A}f\} \tag{1}$$

is dense in $\mathcal{H}$, where $\mathcal{D}(\cdot)$ denotes an operator domain.

If in addition the range $\mathcal{R}(\tilde{A}) = \mathcal{H}$ then we write $\tilde{A} \in \mathcal{A}_S(A)$. Thus a self-adjoint operator $\tilde{A}$ in $\mathcal{H}$ belongs to the class $\mathcal{A}_S(A)$ if it is boundedly invertible and the pair $\tilde{A}, A$ has a common symmetric part, i.e.

$$\dot{A} := A|\mathcal{D} = \tilde{A}|\mathcal{D} \tag{2}$$

is a densely defined symmetric and closed operator in $\mathcal{H}$. We write $\tilde{A} \in \mathcal{A}_S^+(A)$ if $\tilde{A}$ is positive.

Define the A-scale of the Hilbert spaces

$$\cdots \mathcal{H}_{-\alpha} \supset \cdots \mathcal{H}_o \equiv \mathcal{H} \supset \cdots \mathcal{H}_\alpha \supset \cdots \tag{3}$$

where $\mathcal{H}_\alpha = \mathcal{D}(A^{\alpha/2})$ with the norm $\|\cdot\|_\alpha = \|A^{\alpha/2}\cdot\|$ and $\mathcal{H}_{-\alpha}$ is the completion of $\mathcal{H}$ in the norm $\|\cdot\|_{-\alpha} = \|A^{-\alpha/2}\|$. The duality between $\mathcal{H}_\alpha, \mathcal{H}_{-\alpha}$ is denoted by $< \cdot, \cdot >$.

Definition 2 *A linear closed operator $T : \mathcal{H}_\alpha \longrightarrow \mathcal{H}_{-\alpha}$ is called a singular perturbation of A if the linear set*

$$F_{o,T} := Ker\, T \tag{4}$$

is dense in $\mathcal{H}$.

If in addition T is symmetric , i.e.

$$< T\varphi, \psi > = < \varphi, T\psi >, \quad \varphi, \psi \in \mathcal{D}(T) \tag{5}$$

and the range $\mathcal{R}(T)$ is a closed subspace in $\mathcal{H}_{-\alpha}$ then we write $T \in \mathcal{T}_S(A)$. In such case (see [14])

$$\mathcal{R}(T) \cap \mathcal{H} = \{0\}.$$

Thus an operator $T : \mathcal{H}_\alpha \longrightarrow \mathcal{H}_{-\alpha}$ belongs to the class $\mathcal{T}_S(A)$ iff it has the following representation :

$$T = D_{-,+}V, \qquad D_{-,+} = D_{-,o}D_{o,+} : \mathcal{H}_\alpha \longrightarrow \mathcal{H}_{-\alpha} \tag{6}$$
$$D_{o,+} = A^{\alpha/2} : \mathcal{H}_\alpha \longrightarrow \mathcal{H}_0, \ \ D_{-,o} = (A^{\alpha/2})^{cl} : \mathcal{H}_0 \longrightarrow \mathcal{H}_{-\alpha}$$

(cl denotes the closure), where V is a self-adjoint operator in $\mathcal{H}_\alpha$ such that the subspace $KerV = KerT = F_{o,T}$ is dense in $\mathcal{H}$ and the range $\mathcal{R}(V)$ coincides with $\mathcal{H}_\alpha \ominus F_{o,T} = F$.

We write $T \in \mathcal{T}_{S_{\beta,r_\alpha}} \equiv \mathcal{T}_{\beta,\alpha}, O \leq \beta < \alpha$ if $T \in \mathcal{T}_S(A)$ and $F_{o,T}$ is dense in $\mathcal{H}_\beta$,i.e. T is regular in $\mathcal{H}_\alpha$ and singular in $\mathcal{H}_\beta$.

For the bilinear form $\gamma_T(\varphi,\psi) =< T\varphi,\psi >$ which is generated by T in the A-scale we also write $\gamma_T \in \mathcal{T}_{\beta,\alpha}$. It is clear that in $\mathcal{H}_\alpha$ the form γ_T is closed and V is its associated operator.

Theorem 1 *Under the condition, that $\mathcal{D}$ is dense in $\mathcal{H}_1$, there is a one to one correspondence between $A \in \mathcal{A}_S(A)$ and $T \in \mathcal{T}_{1,2}(A)$. Namely for each $\tilde{A} \in \mathcal{A}_S(A)$ the difference*

$$\tilde{A}^{-1} - A^{-1} = \tilde{B}^{-1} \tag{7}$$

defines an operator $T_B \in \mathcal{T}_S(A)$,

$$T_B := A^{cl}(O \oplus B)A \tag{8}$$

where B acts in $\mathbf{N}_o = \mathbf{M}_o^\perp, \mathbf{M}_o = A\mathcal{D} = \tilde{A}\mathcal{D}$, it is self-adgoint and boundedly invertible operator. Conversely each $T \in \mathcal{T}_{1,2}(A)$ defines in $\mathbf{N}_o = AF$ an operator

$$B_T = A(V|F)A^{-1} \tag{9}$$

which gives $\tilde{A}$ by (7) if we put

$$\tilde{B}^{-1} = \begin{cases} B_T^{-1} & in \ \mathbf{N}_o \\ 0 & on \ \mathbf{M}_o \end{cases} \tag{10}$$

In the above case the Friedrichs extention of $\dot{A}$ coincides with A. Under the assumption that operators $A, \tilde{A}, T$ are positive the above result has a formulation in terms of the bilinear forms.

Put $\mathcal{H}_2 \equiv \mathcal{H}_+, \ \ \mathcal{H}_1 = Q(\gamma_A)$ in the norm $\|\varphi\|_1, \varphi \in Q(\gamma_A)$ where γ_A is the bilinear form generated by A.

Definition 3 *A densely defined positive bilinear form γ in $\mathcal{H}$ with domain $Q(\gamma) \subset \mathcal{H}_+$ is called singular and belongs to class $\mathcal{T}^+_{S_1,r_2}(A)$ if it is closed in $\mathcal{H}_+$ and $Ker\ \gamma$ is dense in $\mathcal{H}_1$. We write $\gamma \in \mathcal{T}^+_{S_o,r_1}(A)$ if γ is closed in $\mathcal{H}_1$ and $Ker\ \gamma$ is dense in $\mathcal{H}$.*

Theorem 2 *Under the condition, that $\mathcal{D}$ is dense in $\mathcal{H}$ there is a one to one correspondence between $\tilde{A} \in \mathcal{A}_S^+(A)$ and $\gamma \in \mathcal{T}_{S_1,r_2}^+(A)$ which is fixed by the direct sum*

$$\gamma_{\tilde{A}} = \gamma_A \dot{+} \gamma_B \tag{11}$$

where γ_B is defined as the restriction on $\mathbf{N}_o = \mathbf{M}_o^\perp, \mathbf{M}_o \equiv \mathcal{R}(\dot{A})$ of the bilinear form $\gamma(A^{-1}\varphi, A^{-1}\psi)$.

Note that in case $\gamma \in \mathcal{T}_{S_o,r_1}^+(A)$ the usual form-sum

$$\tilde{\gamma} = \gamma_A + \gamma \tag{12}$$

defines the closed bilinear form in $\mathcal{H}$.

In (11) the direct sum is correctly defined because $\mathbf{N}_0 \cap Q(\gamma_A) = \{o\}$. The equality (11) follows from Theorem 1 if the operator $T = D_{-,+}V$ where V is the operator associated with γ in $\mathcal{H}_+$.

The domain $\mathcal{D}(\tilde{A})$ has the following description

$$\mathcal{D}(\tilde{A}) = \{g \in \mathcal{H} | g = f + B^{-1}P_{\mathbf{N}_o}Af, f \in \mathcal{D}(A)\} \tag{13}$$

$$\tilde{A}g = Af$$

where $P_{\mathbf{N}_o}$ is the orthogonal projector on $\mathbf{N}_o$ and the operator B is connected with A, $\tilde{A}$ and T by (7)–(10).

The resolvent $\tilde{R}_z$ of $\tilde{A}$ is expressed by the M.Krein formula

$$\begin{aligned} \tilde{R}_z &= R_z + \tilde{B}_z^{-1} \qquad (14) \\ \tilde{B}_z^{-1} &= \begin{cases} B_z^{-1} & on\ \mathbf{N}_z, \quad \mathbf{N}_z = \text{Ker } (A^* - z) \\ 0 & on\ \mathbf{M}_z, \quad \mathbf{M}_z = \mathbf{N}_z^\perp \end{cases} \\ B_z^{-1} &= U_{z,o}(B_T - zG_{z,o})^{-1}P_{\mathbf{N}_z} \qquad (15) \end{aligned}$$

where $U_{z,o} = AR_z$, $G_{z,o} = P_{\mathbf{N}_z}(1 - zA^{-1})$ and $P_{\mathbf{N}_z}$ is the orthogonal projection on $\mathbf{N}_z$.

Thus we have two abstract procedures for construction of the singularly perturbed operator $\tilde{A}$ with respect to A by the given singular operator $T : \mathcal{H}_+ \longrightarrow \mathcal{H}_- = \mathcal{H}_{-2}$ if T^{-1} on $\mathcal{R}(T)$ is bounded or by the singular bilinear form γ if it is positive. Note that both T, γ play roles of the abstract boundary conditions for construction of the self-adjoint extension of $\dot{A}, \mathcal{D}(\dot{A}) = \mathcal{D}$ where $\mathcal{D}$ coincides with Ker T = Ker γ. It is for this reason why we call the presented procedures a the self-adjoint extension method in singular perturbation theory.

The various applications of the above presented general approach can be found in the publications [1 - 18].

Now let $\tau \in \mathcal{T}_{2,\alpha}, \alpha > 2$. Then the set Ker τ is dense in $\mathcal{H}_2$ and the restriction $A|$Ker τ is an essentially self-adjoint operator. It means that τ is a trivial perturbatiion for A. What can one do in such case?

We propose to introduce the auxiliary intermediate perturbation γ which is singular and non-trivial for A ($\gamma \in \mathcal{T}_{0,2}(A)$) and such that τ is a non-trivial perturbation for γ. Then it is possible to give the non-trivial sense to the formal expression $A+\gamma+\tau$ and define $\tilde{A} = A_{\gamma+\tau}$.

Namely, let $F_0 = \text{Ker } \gamma \subset \text{Ker } \tau$. Consider in $F = \mathcal{H}_2 \ominus F_0$ the operator $A_1 = V + 1$ where V is the operator associated with γ in F. Assume that $\tau \in \mathcal{T}_{0,2}(A_1)$ in the A_1-scale of Hilbert spaces. Then we can construct $\tilde{A}$ in two steps. In the first step we define the operator A_τ from A_1 and τ using Theorem 1 or 2. In the second step we define $\tilde{A} = A_{\gamma+\tau}$ from A and A_τ using again Theorem 1 or 2. Note that the bilinear form γ_τ generated by A_τ in $\mathcal{H}_2$ belongs to $\mathcal{T}_{0,2}(A)$. Thus we have

Theorem 3 *Let $A = A^* \geq 1$ in $\mathcal{H}$ and $T \in \mathcal{T}_{0,2}(A)$. Suppose moreover that the positive bilinear form $\tau \in \mathcal{T}_{0,2}(A_1)$, where $A_1 = V + 1$ in $F = \mathcal{H}_2 \ominus Ker\, \gamma_T$ $(\gamma_T = \gamma)$. Then the singular perturbed operator $\tilde{A} = A_{\gamma+\tau}$ exists and is defined by the self-adojoint extension method.*

We now consider an application. Let $\mathcal{H} = L^2(R^n, dx), n > 1, A = -\Delta + 1, \gamma = \gamma_{\mu_1}, \tau = \gamma_{\mu_2}, \gamma_{\mu_i}(\varphi,\psi) = \int \varphi(x)\psi(x)d\mu_i(x), i = 1,2, \varphi,\psi \in C_0(R^n)$ where μ_i is a positive Radon measure, $supp\mu_i = N_i$, with Lebesgue measure $|N_i| = 0$. Assume $\gamma_{\mu_1} \in \mathcal{T}_{0,2}(A)$, i.e. $cap_2(N_1) > 0$ and $\gamma_{\mu_1} \in \mathcal{T}_{2,\alpha}(A), \alpha > 2$, i.e. $0 = cap_2(N_2) < cap_\alpha(N_2)$ where *cap* denotes the capacity [6]. For example, take $n = 4, N_1$ a circle, μ_1 the Lebesque measure on N_1 and μ_2 is corresponds to a sum of δ_{y_k} -functions for points $y_k \in N_1$. Then the above presented approach gives sense to the formal expression $-\Delta + \mu_1 + \mu_2$. Moreover the same method gives sense to the formal expression $-\Delta + \gamma_{-\Delta|N_1} + \sum_{k \in N_1} \delta_{y_k}$ where the second term corresponds to the restriction of the Laplace operator on the circle N_1.

References

[1] S.Albeverio, F.Gesztesy, R.Høegh-Krohn, H.Holden, *Solvable models in quantum mechanics.* Springer, Berllin (1988)

[2] S.Albeverio, J.R.Fenstad, R.Høegh-Krohn, W.Karwowski, T.Lindstrøm, SFB 273 - Preprint Nr. 96 (1991)

[3] S.Albeverio, J.Brasche, M.Röckner, Lect. Notes Phys., 345 (1989)

[4] S.Albeverio, R.Høegh-Krohn, L.Streit, J. Math. Phys., 18, 907 - 917 (1977)

[5] S.Albeverio, W.Karwowski, V.Koshmanenko, SFB 273 - Preprint Nr.176 (1992)

[6] J.Brasche, P.Exner, Yu.A.Kuperin, P. Šeba, SFB 237 - Preprint Nr. 132 (1991)

[7] S.E.Cheremshantsev, LOMI - Preprint, Leningrad (1989)

[8] L.Dabrowski, J.Shabani, J. Math. Phys., 29, 2241 - 2244 (1988)

[9] P.Exner,P. Šeba, P.Štǒviček, J. Phys.,A 21, 4009 -4019 (1988)

[10] P.Exner, P. Šeba, Lect, Notes Phys. 324 (1989)

[11] W.Karwowski, V.Koshmanenko, Ukr. Math. J., 42, N.9, 1199 - 1204 (1990)

[12] A.N.Kochubej, J. Funct. Annal. Appl., 16, 78 - 79 (1982)

[13] V.Koshmanenko, *Singular bilinear forms in perturbations theory of self-adjoint operators.* Naukova Dumka, Kiev, (1993)

[14] V.Koshmanenko, ITP UWr - Preprint, No. 835 (1993)

[15] H.Neidhardt, V.A.Zagrebnov, Commun. Math. Phys., 149, 573 - 583 (1992)

[16] S.Ôta, Zeitschrift für Anal. und ihre Anwend.,7,15 - 17 (1987)

[17] B.S.Pavlov, Teor. Mat. Fiz., 59, 345 - 354 (1984)

[18] A.Teta, Publ. RIMS Kyoto Univ., 26, 803 - 817 (1990)

V.Koshmanenko, Institute of Mathematics, Kyïv, Ukraine

Operator Theory:
Advances and Applications, Vol. 70

Some Mathematical Problems of p-adic Quantum Theory

A.N. Kochubei

Abstract

We study pseudo-differential equations over the field of p-adic numbers with properties similar to those of classical Laplace and Schrödinger equations.

1. Recent activity in $p-$ adic models of quantum mechanics and quantum field theory has given a considerable impetus to the development of p-adic mathematics. We shall concentrate on pseudo-differential operators acting on complex-valued functions defined over $Q_p^n = Q_p \times \cdots \times Q_p$ (n times) where $p \neq 2$ is a prime number, Q_p is a field of p-adic numbers, $n \geq 1$. Our aim is to identify and study $p-$adic counterparts of elliptic and evolution partial differential equations. The role of such equations in p-adic physics may happen to be similar to the role of the Laplace and Schrödinger equations in the conventional quantum theory.

This contribution is a brief review of some recent results. The detailed treatment will be published elsewhere.

2. Let us consider a self-adjoint pseudo-differential operator

$$A = F^{-1} M_{a,\alpha} F \tag{1}$$

in the Hilbert space $L_2(Q_p^n)$ constructed with the use of the Haar measure on the additive group of Q_p .Here $M_{a,\alpha}$ is an operator of multiplication by $|a(\xi_1, \dots, \xi_n)|_p^\alpha$, $a(\xi_1, \dots, \xi_n)$ is a quadratic form with coefficients from Q_p , $\alpha > 0$,

$$(Fu)(\xi_1, \dots, \xi_n) = \int_{Q_p^n} \chi(x_1\xi_1 + \cdots + x_n\xi_n) u(x_1, \dots, x_n) dx_1 \dots dx_n \ ,$$

χ is a canonical additive character on Q_p.

The study of the operator (1) is stimulated by the prospect (outlined in [1]) of constructing a p-adic version of the Euclidean quantum field theory. It is natural to take the Green function $G(x-y, \lambda)$, the integral kernel of $(A+\lambda I)^{-1}, \lambda > 0$, as a kernel of the free covariance operator. Of course such a function $G(z, \lambda)$ should be non-negative; it is expected also to be continuous for $z \neq 0$.

The simplest example ($n = 2, \alpha = 1, a(\xi_1, \xi_2) = \xi_1^2 + \xi_2^2$) was considered in [2]. It was shown that the properties listed above take place only if $p \equiv 3(\mod 4)$ or,equivalently,if $\xi_1^2 + \xi_2^2 \neq 0$ for $|\xi_1|_p + |\xi_2|_p \neq 0$.

Theorem 1. *If*

$$a(\xi_1, \dots, \xi_n) \neq 0 \qquad for \qquad |\xi_1|_p + \dots + |\xi_n|_p \neq 0 \tag{2}$$

then $G(z, \lambda)$ *is continuous for* $z \neq 0$, *non-negative, locally integrable, and* $G(z, \lambda) \to 0$ *when* $\max_j |z_j|_p \to \infty$.

Note that quadratic forms satisfying the condition (2) exist only for $n \leq 4$, and for the "physical" dimension $n = 4$ such a form is unique (see [3]) up to a linear isomorphism.

The proof of the theorem is based on a complete classification of quadratic forms satisfying (2) and a procedure of reduction to one-dimensional pseudo-differential operators over local fields (extensions of Q_p; in the case $n = 4$ it is convenient to use a non-commutative quaternion algebra over Q_p).

The asymptotics of $G(z, \lambda)$ for $\max_j |z_j|_p \to \infty$ and estimates of its behaviour near the point $z = 0$ have also been found.

Another interesting example of a pseudo-differential operator over Q_p^n corresponds to the symbol

$$a(\xi_1, \dots, \xi_n) = \left[\max_j |\xi_j|_p\right]^\alpha, \ \alpha > 0.$$

All the assertions of Theorem 1 remain valid in this case for any $n \geq 1$.

3. The formalism of p-adic quantum mechanics proposed in [1,4] is based on the direct construction of the unitary propagator $U(t)$ in the Hilbert space $L_2(Q_p)$ related to the classical dynamics via the representation of the Heizenberg-Weyl group. Both time and spatial variables are assumed to be p-adic while values of the wave function are complex.

The standard Hamiltonian approach fails in the p-adic situation since Q_p is totally disconnected which excludes any possibility to define infinitessimal generators of space or time translations. Nevertheless there is a natural desire to find some equation for the wave function,at least in the free particle case. Such an equation was found in [5]. However,it turns out that the solution of the Cauchy problem for the equation from [5] is not unique even in the Bruhat test-function space $\mathcal{D}$ of locally constant functions with compact supports (see [6]) which is in fact too narrow to include the wave function whose support with respect to t is not compact.

Let us consider the Cauchy problem

$$(Bu)(x, t) = f(x, t), \qquad x \in Q_p^n, t \in Q_p, \tag{3}$$

$$u(x, 0) = u_0(x), \tag{4}$$

where $n \geq 1$, $f \in \mathcal{D}(Q_p^{n+1})$, $u_0 \in \mathcal{D}(Q_p^n)$. The operator B is defined initially on $\mathcal{D}(Q_p^{n+1})$ as

$$(Bu)(x,t) = F^{-1}_{(\xi,\tau)\to(x,t)} \left[|\tau - (b_1\xi_1^2 + \cdots + b_n\xi_n^2)|_p^\alpha F_{(y,\theta)\to(\xi,\tau)} u \right],$$

$\alpha > 0, 0 \neq b_j \in Q_p \quad (j = 1, \ldots, n)$.

The operator B can be expressed as a hyper-singular integral operator

$$\begin{aligned}(Bu)(x,t) = \frac{1-p^\alpha}{1-p^{-\alpha-1}} F^{-1}_{y\to x} \int_{Q_p} |\tau|_p^{-\alpha-1} [\chi(-\tau(b_1y_1^2 + \cdots \\ \cdots + b_ny_n^2))v(y,t-\tau) - v(y,t)]d\tau\end{aligned} \tag{5}$$

where $v(y,t) = F_{x\to y}u(x,t)$, $u \in \mathcal{D}(Q_p^{n+1})$.

The expression (5) makes sense for a wider functional class $\mathcal{B}_\alpha$ consisting of those functions $u(x,t)$, $x \in Q_p^n$, $t \in Q_p$ belonging to $L_2(Q_p^n)$ for each t which satisfy the following conditions. There exist $N, l \in \mathbf{Z}$ and a positive function $c(t)$ (depending on u) such that

$$v(y,t) = 0 \qquad \text{for all } t \in Q_p \text{ if } \max_j |y|_p > p^N;$$

$$|v(y,t)| \leq c(t) \, , \; y \in Q_p^n, t \in Q_p; \quad \int_{|t|_p>1} c(t)|t|_p^{-\alpha-1}dt < \infty;$$

$$v(y,t+t') = v(y,t) \text{ for all } y \in Q_p^n \, , \; t \in Q_p \, , \text{if } |t'|_p \leq p^l.$$

Note that the free wave function from [1,4] belongs to $\mathcal{B}_\alpha$ and satisfies (3) with $n = 1$ and some $b_1 \in Q_p$ where the operator B is understood in the sense of (5).

The following uniqueness result is based on the properties [6] of the fractional differentiation operator on Q_p.

Theorem 2. If $u \in \mathcal{B}_\alpha$, $Bu = 0$, $u(x,0) \equiv 0$ then $u(x,t) \equiv 0$.

The explicit formulas for the solution of the Cauchy problem (3),(4) are too long to be presented here. The solution possesses some additional properties making it possible to eliminate the Fourier transform from the action of the operator B which is thus represented as a hyper-singular integral operator with respect to all the variables.

Acknowledgment. This research was supported in part by the Fundamental Research Fund of the Ukrainian State Committee of Science and Technology,under Grant No. 1/238.

References

[1] Vladimirov, V.S.; Volovich I.V.: *Comm. Math. Phys.* **123** (1989), 659-676.

[2] Bikulov, A.Kh.: *Theor.Math.Phys.* **87** (1991),600-610.

[3] Serre, J.-P.: *Course d'arithmetique*, Presses Universitaires de France, Paris, 1970.

[4] Ruelle, Ph.; Thiran, E.; Verstegen, D.; Weyers, J.: *J. Math. Phys.* 30 (1989), 2854-2874.

[5] Vladimirov, V.S.; Volovich, I.V.: *Lett. Math. Phys.* 18 (1989), 43-53.

[6] Vladimirov, V.S.: *Russian Math. Surveys* 43, No.5 (1988), 19-64.

A.N. Kochubei, Institute of Mathematics, Ukrainian Academy of Sciences,
vul. Tereshchenkivska 3, Kiev, 252601 Ukraine

Titles previously published in the series

OPERATOR THEORY: ADVANCES AND APPLICATIONS
BIRKHÄUSER VERLAG

33. **I. Gohberg** (Ed.): Topics in Interplation Theory of Rational Matrix-Valued Functions, 1988, (3-7643-2233-0)
34. **I. Gohberg** (Ed.): Orthogonal Matrix-Valued Polynomials and Applications, 1988, (3-7643-2242-X)
35. **I. Gohberg, J.W. Helton, L. Rodman** (Eds.): Contributions to Operator Theory and its Applications, 1988, (3-7643-2221-7)
36. **G.R. Belitskii, Yu.I. Lyubich:** Matrix Norms and their Applications, 1988, (3-7643-2220-9)
37. **K. Schmüdgen:** Unbounded Operator Algebras and Representation Theory, 1990, (3-7643-2321-3)
38. **L. Rodman:** An Introduction to Operator Polynomials, 1989, (3-7643-2324-8)
39. **M. Martin, M. Putinar:** Lectures on Hyponormal Operators, 1989, (3-7643-2329-9)
40. **H. Dym, S. Goldberg, P. Lancaster, M.A. Kaashoek** (Eds.): The Gohberg Anniversary Collection, Volume I, 1989, (3-7643-2307-8)
41. **H. Dym, S. Goldberg, P. Lancaster, M.A. Kaashoek** (Eds.): The Gohberg Anniversary Collection, Volume II, 1989, (3-7643-2308-6)
42. **N.K. Nikolskii** (Ed.): Toeplitz Operators and Spectral Function Theory, 1989, (3-7643-2344-2)
43. **H. Helson, B. Sz.-Nagy, F.-H. Vasilescu, Gr. Arsene** (Eds.): Linear Operators in Function Spaces, 1990, (3-7643-2343-4)
44. **C. Foias, A. Frazho:** The Commutant Lifting Approach to Interpolation Problems, 1990, (3-7643-2461-9)
45. **J.A. Ball, I. Gohberg, L. Rodman:** Interpolation of Rational Matrix Functions, 1990, (3-7643-2476-7)
46. **P. Exner, H. Neidhardt** (Eds.): Order, Disorder and Chaos in Quantum Systems, 1990, (3-7643-2492-9)
47. **I. Gohberg** (Ed.): Extension and Interpolation of Linear Operators and Matrix Functions, 1990, (3-7643-2530-5)
48. **L. de Branges, I. Gohberg, J. Rovnyak** (Eds.): Topics in Operator Theory. Ernst D. Hellinger Memorial Volume, 1990, (3-7643-2532-1)
49. **I. Gohberg, S. Goldberg, M.A. Kaashoek:** Classes of Linear Operators, Volume I, 1990, (3-7643-2531-3)
50. **H. Bart, I. Gohberg, M.A. Kaashoek** (Eds.): Topics in Matrix and Operator Theory, 1991, (3-7643-2570-4)

51. **W. Greenberg, J. Polewczak** (Eds.): Modern Mathematical Methods in Transport Theory, 1991, (3-7643-2571-2)
52. **S. Prössdorf, B. Silbermann:** Numerical Analysis for Integral and Related Operator Equations, 1991, (3-7643-2620-4)
53. **I. Gohberg, N. Krupnik:** One-Dimensional Linear Singular Integral Equations, Volume I, Introduction, 1992, (3-7643-2584-4)
54. **I. Gohberg, N. Krupnik:** One-Dimensional Linear Singular Integral Equations, Volume II, General Theory and Applications, 1992, (3-7643-2796-0)
55. **R.R. Akhmerov, M.I. Kamenskii, A.S. Potapov, A.E. Rodkina, B.N. Sadovskii:** Measures of Noncompactness and Condensing Operators, 1992, (3-7643-2716-2)
56. **I. Gohberg** (Ed.): Time-Variant Systems and Interpolation, 1992, (3-7643-2738-3)
57. **M. Demuth, B. Gramsch, B.W. Schulze** (Eds.): Operator Calculus and Spectral Theory, 1992, (3-7643-2792-8)
58. **I. Gohberg** (Ed.): Continuous and Discrete Fourier Transforms, Extension Problems and Wiener-Hopf Equations, 1992, (3-7643-2809-6)
59. **T. Ando, I. Gohberg** (Eds.): Operator Theory and Complex Analysis, 1992, (3-7643-2824-X)
60. **P.A. Kuchment:** Floquet Theory for Partial Differential Equations, 1993, (3-7643-2901-7)
61. **A. Gheondea, D. Timotin, F.-H. Vasilescu** (Eds.): Operator Extensions, Interpolation of Functions and Related Topics, 1993, (3-7643-2902-5)
62. **T. Furuta, I. Gohberg, T. Nakazi** (Eds.): Contributions to Operator Theory and its Applications. The Tsuyoshi Ando Anniversary Volume, 1993, (3-7643-2928-9)
63. **I. Gohberg, S. Goldberg, M.A. Kaashoek:** Classes of Linear Operators, Volume 2, 1993, (3-7643-2944-0)
64. **I. Gohberg** (Ed.): New Aspects in Interpolation and Completion Theories, 1993, (3-7643-2948-3)
65. **M.M. Djrbashian:** Harmonic Analysis and Boundary Value Problems in the Complex Domain, 1993, (3-7643-2855-X)
66. **V. Khatskevich, D. Shoiykhet:** Differentiable Operators and Nonlinear Equations, 1993, (3-7643-2929-7)
67. **N.V. Govorov** †: Riemann's Boundary Problem with Infinite Index, 1994, (3-7643-2999-8)
68. **A. Halanay, V. Ionescu:** Time-Varying Discrete Linear Systems Input-Output Operators. Riccati Equations. Disturbance Attenuation, 1994, (3-7643-5012-1)
69. **A. Ashyralyev, P.E. Sobolevskii:** Well-Posedness of Parabolic Difference Equations, 1994, (3-7643-5024-5)
70. **M. Demuth, P. Exner, G. Neidhardt, V. Zagrebnov** (Eds): Mathematical Results in Quantum Mechanics. International Conference in Blossin (Germany), May 17-21, 1993, 1994, (3-7643-5025-3)